52 Springer Series in Solid-State Sciences
Edited by Peter Fulde

Springer Series in Solid-State Sciences

Editors: M. Cardona P. Fulde H.-J. Queisser

Volumes 1 - 39 are listed on the back inside cover

Superconductivity in Magnetic and Exotic Materials

Proceedings of the
Sixth Taniguchi International Symposium,
Kashikojima, Japan, November 14–18, 1983

Editors: T. Matsubara and A. Kotani

With 106 Figures

Springer-Verlag
Berlin Heidelberg New York Tokyo 1984

Professor Takeo Matsubara
Department of Physics, Kyoto University,
Faculty of Science, Kyoto 606, Japan

Professor Akio Kotani
Department of Physics, Osake University,
Faculty of Science, Toyonaka 560, Japan

Series Editors:
Professor Dr. Manuel Cardona
Professor Dr. Peter Fulde
Professor Dr. Hans-Joachim Queisser
Max-Planck-Institut für Festkörperforschung, Heisenbergstrasse 1
D-7000 Stuttgart 80, Fed. Rep. of Germany

ISBN-13:978-3-642-82261-2 e-ISBN-13:978-3-642-82259-9
DOI: 10.1007/978-3-642-82259-9

Library of Congress Cataloging in Publication Data. Taniguchi International Symposium (6th: 1983: Kashikojima, Japan). Superconductivity in magnetic and exotic materials (Springer series in solid-state sciences; 52). Includes index. 1. Superconductors–Magnetic properties–Congresses. 2. Ferromagnetic materials–Congresses. I. Matsubara, T. (Takeo), 1921-. II. Kotani, A. (Akio), 1941-. III. Title. IV. Series. QC612.S8T36 1984 537.6'23 84-10649

Softcover reprint of the hardcover 1st edition 1984

2153/3130-543210

Preface

The sixth Taniguchi Symposium on the Theory of Condensed Matter was held between 14-18 November 1983 at Kashikojima, Japan. During the Symposium, about twenty participants lived together and discussed the magnetic superconductors and related problems in an active and friendly atmosphere. This volume contains the papers presented at this Symposium.

A strong impetus for organizing a Symposium of this subject is afforded by recent intense interest and accumulated information on magnetic and other novel superconductors newly discovered, and indeed the Symposium has produced many excellent contributions to this very exciting field of condensed matter theory, as reported in this volume.

In order to give the readers a general outline of the subject, a brief sketch of the problem is made in the Introduction. Then the remainder of this volume is divided into four Parts and an Appendix. Part I is devoted to discussions on several aspects of ferromagnetic superconductors including superconductivity in heavy fermion systems. Part II treats problems on antiferromagnetic superconductors. In Part III three papers on organic superconductors are presented. Part IV includes discussions on the exotic superconductors. The Appendix is concerned with the new research project towards high T_c superconductors in Japan.

The last but not least remark is to mention the activity of the Taniguchi Foundation whose support makes this Symposium possible. For many years Mr. Toyosaburo Taniguchi, who is the former president of Toyobo Co. Ltd., has supported various academic activities in the natural and human sciences. To remind us afresh of his high philosophy, his speech delivered at the reception in Kyoto after the Symposium is given in the Foreword.

On behalf of all the participants, we would like to express our hearty thanks to Mr. Taniguchi and the Taniguchi Foundation.

March 1984 *Takeo Matsubara · Akio Kotani*

Foreword

Speech by Toyosaburo Taniguchi

Dr. Kubo, Chairman, Distinguished Guests, Ladies and Gentlemen:

I am very happy, pleased and honored to be here this evening with so many distinguished friends, guests, and scholars from different parts of the world.

The Taniguchi Foundation wishes to extend a warm and sincere welcome to the many participants of the Sixth International Symposium on the Theory of Condensed Matter; this was inaugurated five years ago through the dedicated efforts of Dr. Ryogo Kubo, who is gracing us today with his presence.

Certainly mere words will not suffice to express the deep sense of gratitude we owe Dr. Kubo, from whose ideas this Division, one of the 15 now in existence, was created back in 1978.

It is also greatly appreciated that the continued success of this Division at ever higher levels contributes to enhancing the valuation of our Foundation.

We are deeply indebted to Dr. Kubo and his associates for their unsung and unstinting efforts which have been so crucial in rendering this particular symposium possible.

Although the Symposium ended, we are convinced that what made our symposium a success in the final analysis is neither the Foundation nor its founder, but the outstanding qualities and fine personalities of the participants, who have been chosen and brought together by Dr. Kubo and his committee.

In this sense, your words of praise, if any, should be directed to them and to those who have directly or indirectly lent their support to this program.

I am willing to accept your thanks if your words of praise are given me for the dishes prepared by this top level hotel in Kyoto, for I have chosen them personally.

Whenever I meet with the younger people participating in our symposia, I am deeply impressed by the unselfish and sincere dedication and sense of mission they display in pursuing the vocation they are in.

I sincerely hope and trust that even after ascending the climb to academic fame within, say, the next 30 years, they sill perpetuate their ties, join forces, and help to forge closer bonds of friendship and cooperation in a manner that will make a major contribution to the well-being of mankind and to permanent world peace.

We would like to add that the number of the participants is being preferably limited to about 20 persons, the less the better, because we wish to arrange for a close-knit community set-up.

Many participants have told us over the years that this arrangement has been most successful, stimulating, productive, and truly worthwhile. We beg your best understanding of these basic philosophies that guide us.

And, consequently, it is our hope that these symposia be continued, as long as the fund permits, say, about ten years, unless an unexpected situation arises.

In conclusion, we sincerely trust that all the participants, both Japanese and foreign, will return home bearing heartwarming memories of the Symposium and of the pleasant times we have enjoyed together.

Thank you.

Participants in the Taniguchi International Symposium, November 14–18, 1983 (at Kashikojima, Japan)

Seated *(left to right)*

K. Yamaji A. Ron Mrs. Ron K. Levin Mrs. Fenton Mrs. Umezawa E. Matsushita A. Kotani S. Nakajima T. M. Rice

Standing *(left to right)*

K. Machida H. Nakanishi S. K. Sinha E. W. Fenton H. Umezawa T. Matsubara M. Tachiki T. Koyama Y. Okabe Y. Suzumura K. Okada

Contents

[1]Topics on "Antiferromagnetic Superconductors" are also included.

[2]Topics on "Heavy Fermion Superconductors" are also included.

Introduction

A. Kotani

Department of Physics, Faculty of Science, Osaka University
Toyonaka 560, Japan

T. Matsubara

Department of Physics, Kyoto University, Kyoto 606, Japan

The study on the interplay of superconductivity and magnetism was started more than two and a half decades ago. A strong impact on this field was given about seven years ago by the discovery of magnetic superconductors, i.e. the rare earth ternary superconductors where superconductivity coexists with long range magnetic order. Since then, both the theoretical and experimental investigations for these materials have developed remarkably. More recently, several new exotic superconductors, such as organic superconductors, heavy fermion superconductors etc., were discovered. Now, the study of "magnetic and other exotic superconductors", whose theoretical aspects are the theme of the present proceedings, becomes one of the most exciting subjects in solid state physics.

The purpose of this chapter is to provide the reader with a brief introduction, including the survey on the historical development and the present status, to the topics given in these proceedings. The reader can also obtain information on the related topics from the references [1 - 3].

1. Interplay of Superconductivity and Magnetism (Historical Survey)

The first theoretical study on the interplay of superconductivity and ferromagnetism was published by GINZBURG [4] in 1957. He showed, in his paper entitled "Ferromagnetic Superconductors", that superconductivity is almost impossible to coexist with ferromagnetic order. When ferromagnetic metals become superconducting, the persistent current is induced by the ferromagnetic moments to screen them through the electromagnetic interaction (i.e. through the Meissner effect), and the energy increase due to this screening current makes the coexistence state unstable. BALTENSPERGER and STRÄSSLER [5] first pointed out the possibility of the coexistence of superconductivity and antiferromagnetic order. They discussed that although the time reversal symmetry is broken by antiferromagnetic order, the superconductivity is possible to persist in the form of slightly modified pairing states (time reversal combined with a lattice translation).

The experimental attempts on the coexistence problem were initiated by MATTHIAS et al. [6] in 1959. They studied the mutual interaction between superconductivity and magnetism by introducing magnetic impurities into usual superconductors. However, with this method, the coexistence of superconductivity and long range magnetic order could not be obtained because of the following reasons:

(i) Superconductivity was destroyed, in most cases, by small amounts ($\sim$1 at. %) of paramagnetic impurities.
(ii) Even when superconductivity persisted, the long range magnetic or-

der could not be obtained due to the inhomogeneity in the spatial distribution of the magnetic impurities.

The effect (i) originates from that the Cooper pairs are broken through the exchange interaction between the conduction electron spins and the impurity spins (i.e. the pairbreaking effect), as interpreted successfully by ABRIKOSOV and GOR'KOV [7]. In some cases where the exchange interaction is particularly weak, the "coexistence" was apparently accomplished, but the magnetic order was turned out to be of short range [8], as mentioned in (ii).

In 1977, the obstacles (i) and (ii) were overcome, in a dramatic way, by MATTHIAS'group at LaJolla [9] and by FISCHER's group at Geneva [10], independently. They discovered the coexistence of superconductivity and long range magnetic order in the ternary rare earth (RE) compounds $RERh_4B_4$ and $REMo_6S_8$ (the Chevrel compounds), where RE ions are arranged on a regular sublattice. In these materials, the 4f electrons in RE ions are responsible for magnetic order, while the 4d electrons in Rh or Mo are responsible for superconductivity. The strength of the exchange interaction between the conduction electron spins and the RE magnetic moments (denoted by the d-f interaction, hereafter) is very weak (of the order 0.001 $\sim$ 0.01 eV-atom [11, 12]) due to the special crystalline structure where the distances between RE and Rh (or Mo) ions are large. Therefore, the obstacle (i) is removed even though the RE concentration is so large as to form the regular lattice. Further, there is no problem of (ii) because of the regular array of RE ions.

These materials are denoted by magnetic superconductors, and classified to ferromagnetic superconductors and antiferromagnetic superconductors.

2. Magnetic Superconductors

2.1 Ferromagnetic Superconductors

The materials $ErRh_4B_4$ and $HoMo_6S_8$ are ferromagnetic superconductors, which exhibit re-entrant superconductivity; they become superconducting below the upper critical temperature T_{c1} and return to the normal state at the lower critical temperature T_{c2} due to the onset of the ferromagnetic order. The values of T_{c1} and T_{c2}, respectively, are 8.7 K and 0.9 K for $ErRh_4B_4$, and 1.2 K and 0.6 K for $HoMo_6S_8$ [9, 10]. In a narrow temperature range just above T_{c2}, the magnetic order with a long period λ_s ($\lambda_s \sim 100$ Å for $ErRh_4B_4$ and 200 Å for $HoMo_6S_8$) occurs, coexisting with superconductivity, as revealed by neutron diffraction measurements [13, 14]. This magnetic order was first considered to be a spin-spiral order. However, according to more recent neutron diffraction measurements with a single crystal $ErRh_4B_4$ sample [15], this was found to be a spin-sinusoidal order, and furthermore, a strong ferromagnetic scattering was also observed in the same temperature range.

In order to interpret these unusual behaviors (especially the coexistence of superconductivity and the long period magnetic order), two mechanisms were considered theoretically: the Anderson-Suhl mechanism [16] and the mechanism due to the electromagnetic interaction [17, 18]. According to the former, a long period "cryptoferromagnetic order" was expected to occur as a result that the RKKY interaction (the second order effect of the d-f interaction) is modified in the superconducting state. For $ErRh_4B_4$ and $HoMo_6S_8$, however, the following points came to be recognized:

(1) The Anderson-Suhl mechanism is suppressed by the strong spin-orbit scattering of conduction electrons.
(2) The electromagnetic interaction is more important than the d-f interaction.

Due to the electromagnetic interaction, the persistent current screens only the long range component (longer than the London penetration depth $\lambda_L \simeq 1000$ Å) of the interaction between RE magnetic moments. Therefore, the coexistence of superconductivity and ferromagnetic order is strongly suppressed as predicted by GINZBURG [4], but the coexistence of superconductivity and the magnetic order with the long period $\lambda_s \sim (\lambda_L a_m)^{\frac{1}{2}}$ (a_m being the magnetic interaction length) can be stabilized, in a narrow temperature range above T_{c2} as a compromise where both of the superconducting and magnetic condensation energies are partly gained. The period $(\lambda_L a_m)^{\frac{1}{2}}$ is consistent with the experimental value. This mechanism was first pointed out by BLOUNT and VARMA [17] and MATSUMOTO et al.[18] independently. This magnetic order with λ_s was shown to be either of the spin-spiral order or the spin-sinusoidal order, depending on the strength of the magnetic anisotropy [19].

The significance of the screening effect due to the persistent current depends on whether or not the magnetic dipole interaction is the dominant interaction between the RE magnetic moments, because the interaction which is screened by the persistent current is the magnetic dipole interaction. In order to check this point, we estimate the Curie temperature T_m of the fictitious normal state by taking account of only the magnetic dipole interaction (disregarding the RKKY interaction etc.) and compare it with T_{c2}. The expression of T_m is given by

$$T_m = 4\pi n p^2 \mu_B{}^2 / 3k_B ,$$

where n is the density of RE ions and p is the effective Bohr magneton number. Since p is very large for Er and Ho (p= 9.59 for Er and p= 10.6 for Ho with free ion values), we obtain considerably high T_m (higher than T_{c2}): $T_m \sim 2.3$ K for $ErRh_4B_4$ and 1.1 K for $HoMo_6S_8$. Even when we take account of the reduction of p due to the crystalline field effect, we can conclude that the magnetic dipole interaction is dominant in these materials.

Reflecting the dominant role of the electromagnetic interaction, the theories (denoted by the theory of category I) in which only the electromagnetic interaction is taken into account could explain many unusual properties of ferromagnetic superconductors qualitatively or semi-quantitatively. In addition to the reasonable interpretations for the spin-sinusoidal order and the re-entrant superconductivity, the theory of category I predicted that the ultrasonic attenuation depends characteristically on the relative direction between an external magnetic field and the ultrasonic propagation [20]. This prediction was confirmed by experiments [21], which gave a strong support on the significance of the electromagnetic interaction. Experiments for the DC Josephson effect in $ErRh_4B_4$ [22] has also been interpreted by the theory of category I [23]. Furthermore, many predictions have been made for the unusual properties of the surface and film of ferromagnetic superconductors [24, 25], including their dynamic properties [25].

More recently, it has been recognized that the theory (denoted by the theory of category II) where both the electromagnetic interaction and the weak d-f interaction are incorporated is necessary to understand fully the experimental magnetization curves of a single crystal $ErRh_4B_4$ [26]. As shown by UMEZAWA and his coworkers [11] and by TACHIKI and his coworkers [27], the weak d-f interaction is still important through the pairbreaking effect coming from both the spin fluctuation and the conduction electron spin polarization. It is a remaining problem to apply the theory of category II to calculations of many other properties of these materials.

The theoretical development in the categories I and II seems to be very successful, but there still remains an open problem on the origin of the ferromagnetic signal superposed on the spin-sinusoidal signal in neutron diffraction measurements [15] just above T_{c2}. A simple interpretation is that the ferromagnetic signal comes from normal ferromagnetic domains which are inhomogeneously mixed with the superconducting spin-sinusoidal domains. However, the mechanism to stabilize such a mixture of two phases has not been established theoretically. On the other hand, by assuming that the system is in a single phase where superconductivity, spin-sinusoidal order and ferromagnetic order coexist, several mechanisms have been proposed. They are based on the following interactions:

(a) the electromagnetic interaction,
(b) the d-f interaction,
(c) both of them.

For (a), the self-induced laminar phase [28] and the self-induced vortex phase [29] have been proposed. For (b), the Larkin-Ovchinnikov state with spatially dependent superconducting order parameter [30], and for (c) the Anderson-Suhl mechanism combined with the electromagnetic effect [31] have been proposed. In order to settle this problem, further investigations are needed both theoretically and experimentally.

2.2 Antiferromagnetic Superconductors

The materials $REMo_6S_8$ with RE= Gd, Tb, Dy and Er, and $RERh_4B_4$ with RE= Nd, Sm and Tm are antiferromagnetic (AF) superconductors [32, 33]. They become superconducting below T_c, and then exhibit the AF order at T_N. The AF order coexists with superconductivity. For instance, the values of T_c and T_N, respectively, are about 1.4 K and 0.8 K for $GdMo_6S_8$, 1.8 K and 1.0 K for $TbMo_6S_8$, and 2.0 K and 0.4 K for $DyMo_6S_8$. The magnetic structures of these materials were also determined by neutron diffraction measurements [34].

When we consider the electromagnetic interaction, the AF order is expected to coexist with superconductivity. Since the period of the AF order, λ_{AF} (of the order 10 Å), is much smaller than λ_L (of the order 1000 Å), the AF interaction between RE magnetic moments is not screened by the persistent current in contrast with the ferromagnetic case. More generally, it seems that the AF order and superconductivity hardly affect each other, because the magnetic moments of the AF order vanish when they are averaged over the scale of the superconducting coherence length ξ(of the order 100 Å). However, according to the experimental results [32, 33], the upper critical field H_{c2} is considerably influenced by the AF order. The H_{c2} of $REMo_6S_8$ (RE= Gd, Tb, Dy) is suppressed below T_N, whereas that of $SmRh_4B_4$ is enhanced.

To interpret the behavior of H_{c2}, various effects coming from the interplay of superconductivity and AF magnetism have been proposed as follows:

(i) The pairbreaking effect due to spin fluctuations [35][36].
(ii) A decrease of the superconducting-electron number due to the formation of an AF energy gap [37].
(iii) The magnon-mediated interaction between paired quasiparticles [36].
(iv) The decrease of the phonon-mediated attractive interaction between the paired quasiparticles [38][39].
(v) The molecular field as well as the electromagnetic interaction coming from the uniform magnetization induced by the external field [40].
(vi) A pairbreaking effect due to nonmagnetic impurities in the AF state [41-43].

In order to decide which of the above effects is dominating in a particular substance one must perform calculations which are as realistic as possible with respect to the electronic band structure, the phonon spectrum etc. Also they should contain as few adjustable parameters as possible. For $TbMo_6S_8$ and $DyMo_6S_8$ it was shown by ZWICKNAGL and FULDE [38] that (iv) is the dominating effect, by solving the strong coupling version of the EILENBERGER equations. For $TbMo_6S_8$ the theory is able to explain quantitatively the changes in $H_{c2}(T)$ below T_N without any adjustable parameter. In $DyMo_6S_8$ the effect (v) contributes also noticeably since the size of the critical field becomes of the same order of magnitude as the molecular field. Near T_N or for substances with very low T_N the other effects such as (i) and (iii) are probably also of importance. For $SmRh_4B_4$ a satisfactory theory has not yet been presented. The effect (iv) is easy to understand. When quasiparticles with equal single-particle energies are paired (BALTENSPERGER and STRÄSSLER [5]) they sit preferentially on different sublattices as the AF-molecular field increases. Therefore their attraction through phonons decreases below T_N provided that the interaction is sufficiently local. RON and coworker [39] have discussed a simplified version of [38]. It has the advantage of simplicity but the prize for it is the introduction of an adjustable parameter.

The effect (i) was first studied by MAEKAWA and TACHIKI [35], and then studied more extensively, together with (iii), by RAMAKRISHNAN and VARMA [36]. The importance of (i) and (iii) was pointed out in the case of nesting Fermi surfaces. The importance of (v) was proposed by SAKAI et al. [40], and that of (vi) first by MOROZOV [41]. LEVIN et al. [42] have paired electrons in time-resersed states which are not eigenstates of the single-particle Hamiltonian in the presence of an AF-molecular field. This implies the introduction of pairing between electrons with single-particle energies which are different by an amount of the AF gap.

3. Other Exotic Superconductors

3.1 Organic Superconductors

The interplay of superconductivity and magnetism has also been studied in some organic materials. In 1980, JÉROME et al. [44] discovered superconductivity in the organic system $(TMTSF)_2PF_6$ under pressure. At ambient pressure, the ground state of $(TMTSF)_2PF_6$ is the spin density wave (SDW) state, whose transition temperature T_{SDW} is about 12 K. By appling the pressure, T_{SDW} decreases, and at the critical pressure $P_c \sim$ 8Kbar, the ground state changes to the superconducting state with the critical temperature $T_c \sim 1$ K. The materials $(TMTSF)_2X$ with X= AsF_6, SbF_6 and TaF_6 also behave similarly, and their phase diagrams in the P-T plane are almost reproducible from that of $(TMTSF)_2PF_6$ only by shifting the origin of the P-axis appropriately. On the other hand, $(TMTSF)_2ClO_4$ was found to be superconducting at ambient pressure below $T_c \simeq 1.2$ K [45].

The conduction band of these organic superconductors is highly anisotropic, and the energy band widths along the three crystalline axes are estimated to be of the order of 1eV, 0.1eV and less than 0.01eV, from transport and plasma frequency measurements. The conduction electrons in this band are responsible for both the SDW and superconductivity, in contrast with the situation in the magnetic superconductors.

There are several fundamental problems on the properties of organic superconductors:

(i) Does the SDW coesist with superconductivity?
(ii) What is the structure of the SDW?
(iii) Does the superconducting state originate from the singlet pairing or the triplet pairing?
(iv) Is the fluctuation of one-dimensional superconductivity important above T_c?

The first theoretical attempt for the coexistence problem (i) was done by MACHIDA and MATSUBARA [46]. They pointed out the possibility of the coexistence, by using a model conduction band consisting of two separate parts: the bands with and without nesting Fermi surfaces. On the other hand, YAMAJI [47] showed the difficulty of the coexistence, by using a two-dimensional tight binding energy band model. Further, he predicted, on the problem (ii), the successive occurrence of three types of SDW states under pressure, and on the problem (iii) he rather denied the possibility of the triplet superconductivity. FENTON [48] pointed out that if the SDW state and the singlet superconducting state coexist, the order parameter of triplet superconductivity should necessarily be induced. On the problem (iv), JÉROME [49] interpreted a large energy gap observed by tunneling measurements as the superconducting pseudo-gap due to the large fluctuation of quasi-one-dimensional superconductivity. On the other hand, MACHIDA and MATSUBARA [46] interpreted this as the energy gap originating from the SDW order. Some of the theories mentioned here are included in the present proceedings.

3.2 Heavy Fermion Superconductors

In $CeCu_2Si_2$, the Ce ions are nearly trivalent, and form an ordered sublattice. The electronic specific heat, $C \simeq \gamma T$, of this material is enormously large, as represented by $\gamma \simeq 1$ Joule/K^{2}mol [50]. Therefore, the effective mass of conduction electrons is several hundred times as large as the free electron mass, and this system is called heavy fermion system. In 1979, STEGLICH et al.[50] discovered that $CeCu_2Si_2$ exhibits superconductivity below $T_c \simeq 0.5$ K.

The heavy fermions are considered to originate from the motion of the 4f electrons of Ce. Although the 4f electrons are almost localized at each Ce site, they can be weakly itinerant through the hybridization with the extended conduction electron states (s and d states), which results in a very narrow energy band (heavy fermion band) with extremely strong electron correlation. This narrow energy band can also be considered to consist of the Kondo resonance states. When we consider only one Ce ion in this system, the conduction electrons are scattered resonantly through the hybridization with the 4f states of this Ce ion, and form a virtual bound state (Kondo resonance state) near the Fermi level. When we consider the periodic array of Ce ions, the virtual bound states around Ce ions will overlap each other to form the heavy fermion band. However, it is very difficult to describe theoretically this strongly correlated electron system, and no quantitatively reliable theory has been established as yet.

The heavy fermions are responsible for superconductivity as confirmed by the remarkable decrease of the electronic specific heat below T_c. Without the heavy fermions, the superconducting state would not occur in this system, as understood from the fact that $LaCu_2Si_2$ does not exhibit superconductivity. The Debye temperature of $CeCu_2Ci_2$ is $\Theta_D \simeq 200K$, and $k_B\Theta_D$ is much larger than the heavy fermion band width. The characteristic temperature of the spin fluctuation T* (Kondo temperature) is estimated to be about 10 K from quasi-elastic neutron scattering measurements, as well as from the temperature dependence of susceptibility. Therefore, we obtain the relation $T_c/T^* \sim T^*/\Theta_D \sim 0.05$. It is a theoretical problem to explain these unusual facts in the heavy fermion superconductor. Further, it is also an interesting problem to clarify whether the superconducting state originates from the singlet pairing or triplet pairing. Recently, several theoretical attempts have been started by TACHIKI [51], SINHA [52] and RAZAFIMANDIMBY et al. [53].

The superconducting state of heavy fermions similar to that in $CeCu_2Si_2$ has also been discovered recently in UBe_{13} [54] and U_6Fe [55]. These materials are regarded as the second and third examples of a whole class of heavy fermion superconductors. A rapid development of the study in this new field is expected.

3.3 $BaPb_{1-x}Bi_xO_3$ and Some Metal Hydrides

In contrast with the heavy fermion superconductors, $BaPb_{1-x}Bi_xO_3$ and some metal hydrides should be called "light fermion superconductors", because the effective mass of their conduction electrons is exceptionally small compared with that of usual superconductors.

$BaPb_{1-x}Bi_xO_3$ system is an insulator for x= 1 and a normal semimetal for x= 0. However, for $0.05 < x < 0.3$ the superconducting state with rather high T_c occurs, where T_c has the maximum value of about 12 K for $x \simeq 0.25$ [56, 57]. Taking account of the samll electronic specific heat coefficient γ ($\simeq$0.4 mJ/K^2mol for x= 0.25), we may attribute the high T_c of this material to an extremely strong electron-phonon coupling.

The characteristic behavior of $BaPb_{1-x}Bi_xO_3$ with varying x was theoretically discussed, in the strong coupling limit, by RICE and SNEDDON [58]. According to them, the insulating behavior for x= 1 originates from the local charge disproportionation of

$$2Bi^{4+} \to Bi^{3+} + Bi^{5+}$$

due to the strong electron-phonon coupling. With decreasing x, however, the introduction of Pb^{4+} prevents the charge disproportionation trend, and at a critical value of x, the system becomes metallic and exhibits superconductivity. It is very interesting that the occurrence of superconductivity with decreasing x is described as a cross-over from the real space pairing (charge disproportionation) to the momentum space pairing (Cooper pair) of conduction electrons.

However, a puzzling question has recently been posed by MATTHEISS and HAMANN [59], who have made a priori energy band calculation for $BaPb_{1-x}Bi_xO_3$. According to their results, the width of the conduction band is too large for the strong coupling theory by Rice and Sneddon to be applicable. The band calculation cannot explain neither the high T_c nor the semiconducting behavior for $x > 0.3$. In these proceedings, the present status of the theory, including this dilemma, is reviewed by RICE [60].

The interest in superconductivity in metal-hydrogen systems was awaken when superconductivity was discovered in Th_4H_{15} by SATTERTHWAITE and TOEPKE [61] and in PdH by SKOSKIEWICZ [62] in 1970 and 1972 respectively. In view of the fact that gaseous impurities generally lower the superconducting transition temperature T_c of the transition metals, it is rather surprising that these two superconducting systems were found. A review on the influence of hydrogen on the superconductivity of various metals and metal alloys was given by STRITZKER and WÜHL [63]. At present there are only two cases, Th-H and Pd-H systems, in which hydrogen gives constructive contributions to superconductivity. This peculiar behavior is believed to come from an additional electron-phonon coupling characteristic to these systems.

Pd-H system exhibits the interplay between magnetism and superconductivity in the following manner: Pd shows strong paramagnetism due to significant spin fluctuations which are, however, not enough strong to produce long range magnetic order but enough strong to suppress superconductivity. If the spin fluctuations are suppressed by doping hydrogen, the coupling of the conduction electrons to the newly produced optic phonons gives rise to superconductivity. T_c in PdH_x system depends on x and increases with increasing x for $x > 0.8$ up to a high value comparable to A-15 superconductors.

This x-dependence of T_c in Pd-H system is quite contrasted to the case of Th_4H_{15} where superconductivity is found only for stoichiometric composition. The further increase in T_c in Pd-H system is obtained by alloying with noble metals [64-66]. Another interesting feature in the superconductivity of Pd-H system is the so-called inverse isotope effect which is revealed in that T_c is increased when H is replaced by D [67]. This can be explained by an anharmonic potential different for H and D.

The first theoretical argument on the origin of superconductivity in Pd-H system was given by GANGULY in 1973 [67], and since then many theories follow. A brief review on the theories of superconductivity in metal hydrides is found in [63] and a more recent review on the theory of superconductivity in Pd-H system is given by ECONOMOU [69]. PAPACONSTANTOPOULOS et al.[70] made apriori calculations on electronic band structure and electron-phonon coupling constant for Pd-H system, and similar calculation was extended by GUPTA [71] to cover a wider class of metal hydrides so as to enable us to understand, to some extent, the special situation of Pd-H system among other metal hydrides. In these band theoretical approaches, the inverse isotope effect was treated phenomenologically by adopting the neutron scattering data which exhibit isotope effect in optic phonon spectra such that

$$m_H\omega_H^2 \simeq 1.2\, m_D\omega_D^2 \ .$$

A quantitative theoretical account of this anharmonic result was given by MATSUSHITA and MATSUBARA [72]. In the present proceedings a short review on the experimental facts on Pd-H system and simple theory of superconductivity in this system are given by MATSUSHITA and MATSUBARA [73].

4. Concluding Remarks

We have given a brief introduction to the topics in the field of "magnetic and other exotic superconductors". This field is a rapidly developing one, where remarkable progress has been achieved very recently. In the following chapters, the current theoretical topics in this exciting field are reviewed in detail as the proceedings of the 6th Taniguchi Symposium. Needless to say, all the subjects on exotic superconductors cannot be covered due to the limited number of participants. For instance, the interplay of superconductivity and Anderson localization, the superconducting state in some layered compounds, the possibility of new pairing mechanism, etc. are not included.

Papers in the present proceedings are written from purely physical viewpoint. However, it is to be mentioned that this field has a close contact with the technological application. Some of the Chevrel compounds (although nonmagnetic) are well known as high T_c and high H_{c2} superconductors. The discovery of magnetic superconductors was a "by-product" in the course of the research towards high T_c and high H_{c2} superconductors. The heavy fermion superconductors are also "high H_{c2} superconductors" in the sense that $-dH_{c2}/dT$ at T_c is surprisingly large (although T_c is low). The organic superconductors are also expected to be a candidate for high T_c superconductors of the future. A brief description on the new research project towards high T_c superconductors in Japan is given by NAKAJIMA [74] in Appendix of the present proceedings.

1. G. K. Shenoy, B. D. Dunlap and F. Y. Fradin (eds.): Ternary Superconductors (North-Holland, Amsterdam, 1981).
2. M. B. Maple and Ø. Fischer (eds.): Superconductivity in Ternary Compounds II, Topics in Current Physics, Vol.34 (Springer, Berling, Heidelberg, New York, 1982).
3. Proc. 4th Conf. on Superconductivity in d- and f-Band Metals, Karlsruhe (1982).
4. V. L. Ginzburg: Sov. Phys. JETP 4, 153 (1957).
5. W. Baltensperger and S. Strässler: Phys. Kondens. Materie 1, 20 (1963).
6. B. T. Matthias, H. Suhl and E. Corenzwit: Phys. Rev. Lett. 1, 92 (1959).
7. A. A. Abrikosov and L. P. Gor'kov: Sov. Phys. JETP 12, 1243 (1961).
8. J. W. Lynn, D. E. Moncton, L. Passell and W. Thomlinson: Phys. Rev. B21, 70 (1980).
9. W. A. Fertig, D. C. Johnston, L. E. Delong, R. W. McCallum, M. B. Maple, B. T. Matthias: Phys. Rev. Lett. 38, 987 (1977).
10. M. Ishikawa and Ø. Fischer: Solid State Commun. 23, 37 (1977).
11. J. P. Whitehead, H. Matsumoto, R. Teshima and H. Umezawa: preprint; H. Umezawa, H. Matsumoto and J. P. Whitehead: p.*29*
12. M. B. Maple: Ternary Superconductors, ed. by G. K. Shenoy, B. D. Dunlap and F. Y. Fradin (North-Holland, Amsterdam 1981) p.131.
13. D. E. Moncton, D. B. McWhan, P. H. Schmidt, G. Shirane, W. Thomlinson, M. B. Maple, H. B. MacKay, L. D. Woolf, Z. Fisk and D. C. Johnston: Phys. Rev. Lett. 45, 2060 (1980).
14. J. W. Lynn, G. Shirane, W. Thomlinson, R. N. Shelton, D. E. Moncton: Phys. Rev. B24, 3817 (1981).
15. S. K. Sinha, G. W. Crabtree, D. G. Hinks and H. A. Mook: Phys. Rev. Lett. 48, 950 (1982); S. K. Sinha, H. A. Mook, O. A. Pringle and D. G. Hinks: p.*14*

Page numbers in italic refers the present volume.

16. P. W. Anderson and H. Suhl: Phys. Rev. 116, 898 (1959).
17. E. I. Blount and C. M. Varma: Phys. Rev. Lett. 42, 1079 (1979).
18. H. Matsumoto, H. Umezawa and M. Tachiki: Solid State Commun. 31, 157 (1979); M. Tachiki, A. Kotani, H. Matsumoto and H. Umezawa: 31, 927 (1979).
19. H. S. Greenside, E. I. Blount and C. M. Varma: Phys. Rev. Lett. 46, 49 (1981).
20. M. Tachiki, T. Koyama, H. Matsumoto and H. Umezawa: Solid State Commun. 34, 269 (1980).
21. S. C. Schneider, M. Levy, R. Chen, M. Tachiki, D. C. Johnston and B. T. Matthias: Solid State Commun. 40, 61 (1981).
22. C. P. Umbach and A. M. Goldman: Phys. Rev. Lett. 48, 1433 (1982).
23. T. Koyama and M. Tachiki: p.75
24. A. Kotani, M. Tachiki, H. Matsumoto, H. Umezawa and S. Takahashi: Phys. Rev. B23, 5960 (1981).
25. A. Kotani and K. Okada: p.64
26. G. W. Crabtree, F. Behroozi, S. A. Campbell and D. G. Hinks: Phys. Rev. Lett. 49, 1342 (1982).
27. O. Sakai, M. Suzuki, S. Maekawa, M. Tachiki, G. W. Crabtree and F. Behroozi: J. Phys. Soc. Japan 52, 1341 (1983); M. Tachiki: p.53
28. M. Tachiki: Physica 109 & 110B, 1699 (1982).
29. C. G. Kuper, M. Revzen and A. Ron: Phys. Rev. Lett. 44, 1545 (1980).
30. K. Machida: J. Phys. Soc. Japan 51, 3462 (1982); 52, 2181 (1983); H. Nakanishi and K. Machida: p.85
31. M. Tachiki: J. Magn. Magn. Mater. 31-34, 484 (1983).
32. M. Ishikawa and Ø. Fischer: Solid State Commun. 24, 747 (1977).
33. H. C. Hamaker, L. D. Woolf, H. B. MacKay, Z. Fisk and M. B. Maple: Solid State Commun. 32, 289 (1979).
34. see, for instance, W. Thomlinson, G. Shirane, D. E. Moncton, M. Ishikawa and Ø. Fischer: J. Appl. Phys. 50, 1981 (1979).
35. S. Maekawa and M. Tachiki: Phys. Rev. B18, 4688 (1978).
36. T. V. Ramakrishnan and C. M. Varma: Phys. Rev. B24, 137 (1981).
37. K. Machida, K. Nokura and T. Matsubara: Phys. Rev. B22, 2307 (1980).
38. G. Zwicknagl and P. Fulde: Z. Physik 43, 23 (1981).
39. J. Ashkenazi, C. G. Kuper and A. Ron: Phys. Rev. B28, 418 (1983); p.123
40. O. Sakai, M. Tachiki, T. Koyama, H. Matsumoto and H. Umezawa: Phys. Rev. B24, 3830 (1981). See also Y. Suzumura and K. Ishino: Prog. Theor. Phys. 70, 654 (1983); Y. Suzumura: p.129
41. A. I. Morozov, Fiz. Tverd. Tela 22, 3372 (1980) [English transl.: Soviet Phys. Solid State 22, 1974 (1980)]
42. M. J. Nass, K. Levin and G. S. Grest: Phys. Rev. B25, 4541 (1982); K. Levin, M. J. Nass, C. Ro and G. S. Grest: p.114
43. Y. Okabe and A. D. S. Nagi: Phys. Rev. B28, 6290 (1983); Y. Okabe: p.140
44. D. Jérome, A. Mazaud, M. Ribault and K. Bechgaard: J. Physique Lett. 41, L95 (1980).
45. K. Bechgaard, K. Carneiro, F. B. Rasmussen, M. Olsen and C. S. Jacobsen: Phys. Rev. Lett. 46, 852 (1981).
46. K. Machida; J. Phys. Soc. Japan 50, 2195 (1981); K. Machida and T. Matsubara: J. Phys. Soc. Japan 50, 3231 (1981).
47. K. Yamaji: J. Phys. Soc. Japan 51, 2787 (1982); 52, 1361 (1983); p.149
48. E. W. Fenton and G. C. Psaltakis: Solid State Commun. 45, 5 (1982); E. W. Fenton: p.167
49. D. Jérome: J. Magn. Magn. Mater. 31-34, 20 (1983).
50. F. Steglich, J. Aarts, C. D. Bredl, W. Lieke, D. Meschede, W. Franz and H. Schäfer: Phys. Rev. Lett. 43, 1892 (1979).

51. M. Tachiki; p.*53*
52. S. K. Sinha : presented in this symposium (not included in the proceedings).
53. H. Razafimandimby, P. Fulde and J. Keller: Z. Physik B54, 111 (1984)
54. H. R. Ott, H. Rudigier, Z. Fisk and J. L. Smith: Phys. Rev. Lett. 50, 1598 (1983).
55. L. E. Delong, J. G. Huber, K. N. Yang and M. B. Maple: Phys. Rev. Lett. 51, 312 (1983).
56. A. W. Sleight, J. L. Gillson and P. E. Bierstedt: Solid State Commun. 17, 27 (1975).
57. T. D. Thanh, A. Koma and S. Tanaka: Appl. Phys. 22, 205 (1980).
58. T. M. Rice and L. Sneddon: Phys. Rev. Lett. 47, 689 (1981).
59. L. F. Mattheiss and D. R. Hamann: Phys. Rev. B26, 2686 (1982); Phys. Rev. B28, 4227 (1983).
60. T. M. Rice: p.*178*
61. C. B. Satterthwaite and I. L. Toepke: Phys. Rev. Lett. 25, 741 (1970).
62. T. Skoskiewicz: Phys. Status Solidi (a) 11, K123 (1972).
63. B. Stritzker and H. Wühl: Hydrogen in Metals II ed. by G. Alefeld and J. Völkl, Topics in Applied Physics, Vol.29 (Springer, Berlin, Heidelberg, New York, 1978) p.243.
64. J. E. Schirber: Phys. Lett. 45A, 141 (1973).
65. T. Skoskiewicz, A. W. Szafrański and B. Baranowski: Phys. Status Solidi (b) 59, K135 (1973).
66. B. Stritzker: Z. Physik 268, 261 (1974).
67. R. J. Miller and C. B. Satterthwaite: Phys. Rev. Lett. 34, 144 (1975).
68. B. N. Ganguly: Z. Physik 265, 433 (1973).
69. E. N. Economou: Metal Hydrides ed. by G. Bambakidis (Plenum, New York, 1981) p.1.
70. D. A. Papaconstantopoulos, B. M. Klein, E. N. Economou and L. L. Boyer: Phys. Rev. B17, 141 (1978).
71. M. Gupta: Metal Hydrides ed. by G. Bambakidis (Plenum, New York, 1981) p.255.
72. E. Matsushita and T. Matsubara: Prog. Theor. Phys. 63, 357 (1980).
73. E. Matsushita and T. Matsubara: p.*189*
74. S. Nakajima: p.*206*

Part I

Ferromagnetic Superconductors

Neutron Scattering Studies of Magnetic Superconductors

S.K. Sinha
Exxon Research and Engineering Company, Annandale, NJ 08801, USA
H.A. Mook and O.A. Pringle
Oak Ridge National Laboratories, Oak Ridge, TN 37830, USA
D.G. Hinks
Argonne National Laboratory, Argonne, IL 60439, USA

1. Introduction

After nearly twenty years of theoretical and experimental attempts to understand the mutual effects of superconductivity and magnetism on each other, the existence of long-range order of both kinds in the same material was finally established recently with the advent of the families of ternary compounds which have come to be known as "magnetic superconductors" [1,2]. The general use of this term does not necessarily imply a simultaneous coexistence (e.g., at a given temperature) of long-range magnetic order and superconductivity, although the very existence of superconductivity in crystals which contain ordered lattices of magnetic ions is itself initially very surprising and is believed to be due to the relative weakness of the exchange spin-flip (and therefore pair-breaking) interaction between the closed shell (necessarily f-like) of electrons on the magnetic ions and the conduction-band electrons. Nevertheless, in many of these compounds, magnetic long-range order (LRO) can simultaneously coexist with superconductivity over certain temperature ranges. This is particularly true of the case in which the magnetic LRO is of the antiferromagnetic type. The case of ferromagnetic LRO leads to the phenomenon of "re-entrance" and will be discussed in detail later. The best studied families of these compounds are the ternary borides with the chemical formula $ReRh_4B_4$ (where RE is a rare earth ion), and the Chevrel phase compounds with the formula $REMo_6S_8$ or $REMo_6Se_8$. There exist by now several review articles on these extremely interesting classes of materials, which the reader may consult for more detailed experimental information [3,4]. In this article, we shall concern ourselves with neutron scattering studies of these compounds. Reviews of the recent theoretical developments of the subject may be found in many other papers of this symposium.

2. Antiferromagnetic Superconductors

While neutrons unfortunately cannot couple directly to the superconducting order parameter, they are an ideal probe for studying the magnetic order parameter at a microscopic level. We first consider neutron diffraction experiments carried out on the antiferromagnetic superconductors. These are summarized in Table I. For all of these materials, below T_N, superconductivity and antiferromagnetic LRO coexist all the way down to the lowest temperatures. While there are undoubtedly some dramatic effects of the magnetic order parameter on properties such as the critical field Hc_2 [11,12], the superconducting and magnetic order parameters do not interfere too strongly with each other, presumably because their associated wave vectors are not localized in the same region of the space. HAMAKER et al. [10] have studied, by powder diffraction, the compound $Ho(Rh_{0.3}Ir_{0.7})_4B_4$, where a continuous

Table I Magnetic structures of antiferromagnetic superconductors as determined by neutron diffraction

Compound	T_c	T_N	$\lvert\vec{q}\rvert$	Direction of $\vec{q}$	Moment Direction	Max. Ordered Moment	Reference
$GdMo_6Se_8$	5.6K	0.75K	(1/2,0,0) $2\pi/a$	[100]	--	--	7
$DyMo_6S_8$	2.05K	0.4K	(1/2,0,0) $2\pi/a$	[100]	[111]	8.77μB	5,6
$TbMo_6S_8$	2.05K	1.0K	(1/2,0,0) $2\pi/a$	[100]	[111]	8.28μB	6
$NdRh_4B_4$	5.4K	1.5K	0.135Å^{-1}	[100]	[001]		
		1.0K	0.139Å^{-1}	[110]	[001]	3.4μB	8
$TmRh_4Bh_4$	9.8K	0.7K	0.238Å^{-1}	[101]	[010]	6.6μB	9

transition occurs to an antiferromagnetic state at 2.7K from a normal paramagnetic state. Superconductivity then sets in at a lower temperature of T_c = 1.34K and coexists with the magnetic LRO for lower temperatures. The onset of superconductivity has no discernible effect on the magnetic ordering which is estimated to be long range. Another way of understanding this is that the antiferromagnetic interaction has a wave vector which is much shorter than either the London penetration depth or the coherence length of the superconductor and hence is not strongly affected by either the supercurrent screening effects or the singlet pairing effects.

We note that in at least some cases (e.g., $NdRh_4B_4$ [8]) the antiferromagnetic LRO can itself change from one kind to another at a lower temperature. Further, in some compounds (e.g., $TmRh_4B_4$ [9]) there is also a competing tendency towards ferromagnetic order which is presumably suppressed by the superconductivity, but which can re-establish itself when the latter is quenched by an applied field and exist metastably even at very low fields subsequently ("forced ferromagnetism"). This has been discussed theoretically also by FREEMAN and JARLBORG [13]. All the neutron diffraction from these compounds has been carried out on powder samples, single crystals not yet being available.

To summarize the situation with regard to the antiferromagnetic compounds, we can say that neutron diffraction has clearly established the coexistence of antiferromagnetic LRO and superconductivity. It has also yielded detailed information regarding the magnetic order parameter, which has been useful for theorists interested in calculating quantities such as the temperature dependence of Hc_2, etc. We have not yet arrived at a detailed microscopic theory of the magnetic structures themselves which may turn out in a subtle way to be affected by superconductivity, but not in any direct manner.

3. Ferromagnetic Superconductors

We now turn to studies of the so-called re-entrant ferromagnetic superconductors. These are represented by the ternary boride compound $ErRh_4B_4$ and the Chevrel-phase compound, $HoMo_6S_8$. From the theoretical point of view, the problem of the coexistence of ferromagnetic LRO and superconductivity is a more subtle one, since the competing interactions both occur at small wave vectors. The wave vector dependent exchange interaction J(q) is proportional (in the case where this exhange is conduction-electron mediated, as in the RKKY interaction) to the generalized susceptibility function $\chi(q)$ of the conduction electrons. For a material which are likely to order ferromagnetically, $\chi(q)$ must have a maximum at $q = 0$. As discussed by ANDERSON and SUHL [14], in the superconducting phase and in the absence of appreciable spin-orbit coupling, however, $\chi(q)$ must vanish at $q = 0$, reflecting the singlet pairing of the conduction electrons. The net result is a broad maximum at a finite wave vector (typically of order 0.2 Å^{-1} or greater) in the superconducting phase, which would predict some kind of sinusoidal or modulated magnetic structure. Perhaps more important than the above effect, however, is the electromagnetic competition between magnetism and superconductivity, as was realized and discussed recently by several authors [15-19]. The Meissner effect causes the crystal to expel all magnetic flux inside it, which is difficult to do if the spins are lined up ferromagnetically, producing a B field arising from this dipole field. From the electromagnetic point of view, there are two ways in which the tendency to ferromagnetism may reach a compromise with superconductivity. One is for the spins to produce a long-wavelength spiral or other modulated structure so that the B field averages to zero over the length scale of the London penetration depth while still allowing neighboring spins to stay approximately parallel to each other. The other is for the superconductor to allow flux from aligned spins to penetrate in the form of a vortex lattice. The earliest calculations, based on minimizing a Ginzburg-Landau form of the free energy expressed in terms of both the superconducting and magnetic order parameters and all electromagnetic interactions between them, yielded predictions of both of these possible states as intermediate states occurring in a temperature region between the superconducting paramagnetic and ferromagnetic normal states, depending on the choice of parameters used in the free energy expression. The wavelength of all of these periodic structures came out of these calculations as being of the order of $(\lambda_L \lambda_M)^{1/2}$, where λ_L is the London penetration depth and λ_M a characteristic length for the normal magnetic interaction. The wavelength comes out to be ~ 100 Å. A slightly different approach (but with very similar results) was taken at the same time by TACHIKI, UMEZAWA and their collaborators [16,18] using the Boson theory of superconductivity. In their picture, the predominant interaction between spins $J(q)$ arises from the dipole-dipole magnetic interaction between the moments, which is then screened by the supercurrents for $q<1/\lambda_L$, producing a maximum at finite q (similar to the picture in the Anderson-Suhl theory but with q_{max} being much smaller in magnitude).

4. Experiments on Polycrystalline Samples

The first measurements of superconducting and magnetic properties of $ErRh_4B_4$ were all carried out on polycrystalline samples. The superconducting transition temperature Tc_1 is at 8.67K. In these samples susceptibility and resistivity measurements [20] showed hysteretic behavior at Tc_2, indicating a first-order transition at about 0.93K from a superconducting to a normal ferromagnetic state. Thermal expansion measurements [21] showed a cusp at a slightly higher temperature (0.98K) which is typical of a magnetic ordering transition, lending support to a small coexistence region. The first heat capacity measurements [22] showed a sharp spike at 0.93K, with a broad base extending up to higher temperatures. Early neutron scattering experiments [23] revealed that the low temperature phase had an ordered basal-plane ferromagnetic moment of 5.6 μ_B/Er, about 30% smaller than values determined by Mossbauer [24] measurements (8.3 μ_B/Er) and saturation magnetization measurements [25] on single crystals (8.5 μ_B/Er).

The neutron measurements showed that the magnetic order parameter appeared to go smoothly to zero at about 1.2K, but not with the rapidity associated with a conventional second-order transition. Later neutron scattering results [26] showed a peak in the small angle scattering with a $|\vec{q}_s| = 0.06Å^{-1}$ occurring near Tc_2 for temperatures between 0.65K and 1.1K. This peak was taken as evidence of the intermediate spirally ordered magnetic phase predicted theoretically and discussed in the last section. The spontaneous vortex lattice state could also not be completely ruled out. The small angle peak showed strong temperature hysteresis, and so did the ferromagnetic intensity, but in the opposite sense, i.e., the small angle peak was larger on cooling, while the ferromagnetic intensity was larger on warming. This hysteresis afforded the first strong evidence that coexisting regions of superconductivity and ferromagnetism might occur in $ErRh_4B_4$.

Since the neutron diffraction intensity from a magnetic structure is proportional to the square of the magnetization component perpendicular to the neutron scattering wave vector $\vec{Q}$, observation of a small angle satellite necessarily implies that the modulated structure in the intermediate phase must have a non-zero transverse component. However, the lack of further detailed information about the magnetic structure of this phase, the fact that the polycrystalline measurements showed some sample dependence, and the possible smearing out of the transitions due to sample inhomogeneities pointed to the need for single-crystal measurements, which will be described in the next section. The latter measurements have shown that the behavior in the "intermediate phase" is in fact more subtle and complex than the above pictures of a simple pure spiral or vortex intermediate phase separating the superconducting paramagnetic and normal ferromagnetic phases.

Neutron scattering measurements have also been carried out [27,28] on $HoMo_6S_8$, which in many respects shows similar behavior. For this material Tc_1 is 1.82K, and on cooling below 2K, small angle neutron measurements show ferromagnetic spin correlations (corresponding to a correlation length of less than 20Å) which strongly increase with decreasing temperature. At 0.74K, a small angle peak appears centered at $|\vec{Q}| = 0.03Å^{-1}$ and increases and narrows rapidly with decreasing temperature. At 0.73K its width is resolution limited so that the coherence of the modulated structure is already >200Å. On cooling further, the system makes a first-order transition to a normal ferromagnetic structure which exhibits considerable hysteresis in both the order parameters as in $ErRh_4B_4$. At $Tc_2 = 0.67K$, all the material is ferromagnetic-normal.

Neutron scattering studies of pseudoternary alloys of the rhodium borides have also been carried out. The compound $HoRh_4B_4$ is an ordinary ferromagnet which shows no superconductivity. Neutron diffraction studies [29,30] show that the moment is oriented along the c axis in this case, and the magnetic order parameter near T_m behaves according to mean-field theory (i.e., $\beta = 0.5$). On alloying Er into the system, a superconducting phase intervenes between the paramagnetic and ferromagnetic regions and in $(Er_{1-x}Ho_x)Rh_4B_4$ for x = 0.6 [30,32] and 0.75 [31] the initial increase in the magnetization with decreasing temperature is suppressed. For x = 0.84, LYNN et al. [31] apparently observe a ferromagnetic transition above Tc_2 implying a small coexistence region. No magnetic modulated (MM) phase is observed in these compounds. These data are consistent with inhomogeneities which initially keep part of the sample superconducting while the rest orders ferromagnetically, although they might imply genuine microscopic coexistence, not yet accounted for theoretically. For x ≤ 0.3 there are two transitions where basal plane and then c-axis ordering occurs [32] while at x ≈ 0.3 there is presumably an interesting multicritical point where there is a competition between superconductivity, the tendency of the Er^{3+} ions to order in the basal plane and that of the Ho^{3+} ions to order along the c axis. The reader is referred to more comprehensive reviews [4] for further discussion and references.

5. Experiments on Single Crystal Samples

The single crystals used in these experiments were grown at Argonne National Laboratory from an Er-Rh-B melt. No annealing was done after the sample solidified. Sample perfection was indicated by a 10' mosaic spread determined by neutron diffraction. RF penetration depth and AC susceptibility measurements showed a single superconducting transition at Tc_1 = 8.67K, coinciding with the resistive transition. The sample was found to be a bicrystal and was spark cut along its grain boundary to give two single crystals. The larger crystal weighed ~300 mg and was used for the neutron diffraction work described here and the smaller crystal was used for resistivity and magnetic measurements.

The resistive behavior near Tc_2 is shown in detail in Figure 1. A hysteresis in temperature of about 50 mK is observed, indicating a first-order transition at Tc_2. The temperatures at which the transitions occurred were not perfectly reproducible, varying by as much as 20 mK from run to run. However, the magnitude of the resistivity change at the transition and the hysteresis between warming and cooling were reproducible, even though the absolute temperatures were not. Such a non-reproducibility is consistent with a first-order transition, since the exact temperature at which nucleation occurs may depend on sample history.

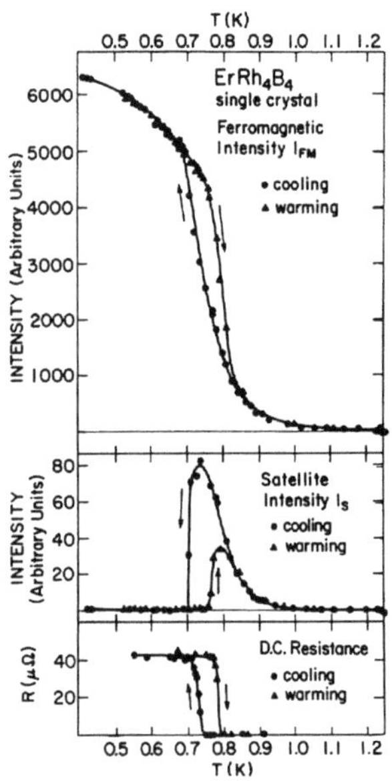

Fig. 1 Behavior of the intensities of the ferromagnetic and the satellite (MM) reflections and of the D.C. resistance as a function of temperature for the $ErRh_4B_4$ single crystal

There have been many detailed magnetization measurements carried out on the single crystal by CRABTREE and his collaborators [25], which will not be discussed here in detail. Some of these results will be discussed in accompanying theoretical papers at this symposium.

For details of the earlier neutron diffraction experiments on the single crystal, the reader may consult previous publications [33-35]. We shall summarize the results here.

The measurements were carried out at the HFIR Reactor at Oak Ridge National Laboratory. Diffraction Measurements were carried out in the a*-c* plane of the reciprocal lattice, with the <010> a axis being perpendicular to this scattering plane. For the field-dependent measurements described below, the field was applied along the latter a axis. From the point of view of a quantitative analysis of the results, the principal difficulties in the single-crystal measurements are due to the neutron absorption in the Rh nuclei, (the Boron used in the sample being isotopically depleted in the heavily absorbing B^{10} nuclei) and due to secondary extinction effects in the very strong Bragg peaks (both magnetic and nuclear) from the sample. The latter preclude a highly accurate determination of the ordered ferromagnetic moments at low temperatures.

Magnetic satellites were found in the a*-c* plane around the Bragg reflections at the positions $\vec{Q} \equiv \vec{\tau} + \vec{q}_s$, where $\vec{q}_s = \pm 0.042\ \vec{a}^* \pm 0.055\ \vec{c}^*$. Because of the ratio of c/a for $ErRh_4B_4$, these correspond to a nearly perfect square of points in reciprocal space. Two much weaker satellites were also seen at $\pm\ 0.055\ \vec{c}^*$ but these are attributed to the equivalent other set of four satellites $\pm\ 0.042\ \vec{b}^* \pm 0.055\ \vec{c}^*$ ($\vec{b}^*$ being the (vertical) reciprocal lattice vector [010] equivalent to the $\vec{a}^*$ vector in the scattering plane) picked up by the relaxed vertical resolution of the diffractometer. The magnitude of the modulation wavevector corresponds to a modulation wavelength of 91.8 ± 2.7 Å, consistent with the peak seen in the small-angle scattering from the polycrystalline samples [26]. Ferro-magnetic intensity was also observed growing at the nuclear Bragg positions below $T \lesssim 1.2K$. The temperature dependence of the ferromagnetic and satellite intensities, together with the resistivity, is shown for both cooling and warming curves in Fig. 1. The equilibration time (with changes in temperature) for the magnetic intensities was found to be of the order of a few minutes or less, and the results are completely reproducible, provided the temperature is changed in only one direction during the measurements.

From Fig. 1, certain conclusions may be drawn. There is a complete correlation of the modulated magnetic moment with bulk superconductivity in the sample, in the sense that they disappear and reappear together (with hysteresis) at Tc_2. The rapid rise in the ferromagnetic component and the normal behavior of the order parameter below Tc_2, accompanied by the disappearance of the satellite and of bulk superconductivity in the sample, confirms a "re-entrance" to a normal ferromagnetic state in a first-order transition. Nevertheless, there is clearly a temperature region where ferromagnetic order <u>co-exists</u> with the magnetically modulated (MM) phase. At 0.73K, where the satellite is nearly at its maximum intensity, high-resolution diffraction scans (with a diffractometer resolution of 0.005 $Å^{-1}$ (FWHM)) showed no broadening of either the main ferromagnetic reflection or the satellite peak, indicating coherence of both structures on length scales of the order of $\gtrsim$ 1000 Å in real space. At higher temperatures, ($T \gtrsim 0.88$ K), both the ferromagnetic intensity and the satellite intensity show a temperature dependence more characteristic of critical scattering or short-range order with an upward curvature. The ferromagnetic intensity is essentially saturated below T = 0.4K.

A search was also made for harmonics of the basic satellite peak, i.e., at $2\vec{q}_s$, $3\vec{q}_s$ and also at other points in the square grid suggested by the satellite positions, at $(3\vec{\zeta} + \vec{\eta})$, $(\vec{\zeta} + 3\vec{\eta})$, etc. (where $\vec{\zeta} = 0.042\ \vec{a}^*$, $\vec{\eta} = 0.055\ \vec{c}^*$). None were found within a statistical uncertainty of 2% of the principal satellite peak.

From the observed intensities, magnetic structure factors were obtained for the satellite peaks and the ferromagnetic intensities (although, as discussed above, extinction effects made the latter subject to large uncertainties at low temperatures). The ferromagnetic intensity per unit cell from the Er^{+3} ions in $ErRh_4B_4$ for neutron scattering wave vector $\vec{Q}$ (equal to a reciprocal wave vector) is proportional to

$$I = (\gamma e^2/2mc^2)^2\ \lfloor f(Q)\rfloor^2\ |F(Q)|^2 \tag{1}$$

where f(Q) is the magnetic form factor for Er^{+3} (determined from other neutron measurements [36]), and $F(\vec{Q})$ is the magnetic structure factor, given by

$$|F(\vec{Q})|^2 = 4\mu^2\ \lfloor 1-(\hat{Q}\cdot\hat{\mu})^2\rfloor \tag{2}$$

where μ is the ferromagnetic moment, and $\hat{\mu}$ and $\hat{Q}$ are unit vectors along $\vec{\mu}$ and $\vec{Q}$. In Eq. (1), $(\gamma e^2/2mc^2)$ is the magnetic scattering length per Bohr magneton and is numerically equal to 0.2695×10^{-12} cm. The first measurements on the single crystal were made with a neutron wavelength of 2.35 Å, and indicated that practically all the ferromagnetic moment was oriented along the vertical $\vec{a}$ axis [010] (normal to the scattering plane), as indicated by the equality of the magnetic structure factor at the (100), (002), and (101) Bragg reflections, but the low-temperature moment yielded only 4.8 μ_B per Er^{+3} ion, considerably lower than the 5.6 μ_B obtained from the powder diffraction measurements of MONCTON et al. [23]. Later measurements at wavelengths of 0.8 Å (and thus subject to much less severe extinction effects) indicate that there is in fact a distribution of ferromagnetic domains along both easy ($\vec{a}$) axes, in the scattering plane as well as normal to it (but with the majority of the domains still aligned along the normal to the plane), with a saturated ferromagnetic moment of ~ 5.3 μ_B, indicating still some extinction at these wavelengths. (The polycrystalline diffraction data are not subject to extinction effects owing to the random orientation of the grains). The extinction effects indicate that the lower temperature part of the ferromagnetic intensity shown in Fig. 1 is probably lower (by about 30%) than the true curve determined by the square of the real ferromagnetic order parameter.

For the MM structure, the weakness of the satellite peaks renders extinction effects inoperative as a source of error. The direction of propagation of the modulation (which is at roughly 45° to the c axis) makes a spiral structure unlikely, since the magnetic anisotropy locking the moments in the basal (001) plane is very large [25]. A non-right angle spiral with the moments in the basal plane is also ruled out since the magnetic structure factor of satellites around (200), (002), and (101) are found to be equal within experimental error; whereas a basal-plane spiral would require the (002) satellite structure factor (where $\vec{Q}$ is everywhere normal to the moments) to be twice as large as the (200) satellite structure factor (where the $[1-(Q.\mu)^2]$ factor gives an average of one-half (see Eq. (2)). The alternative possibility of a plane polarized sinusoidal MM structure with the ordered moment everywhere along the easy $\vec{a}^*$ axis which is normal to the direction of propagation of the modulation is consistent with the data. The plane polarized form of the MM structure draws attention to the role of single-ion magnetic anisotropy in determining the structure for magnetic LRO in the superconductivity phase. The role of magnetic anisotropy and the possibility of a linear sinusoidal wave structure was first discussed by GREENSIDE, BLOUNT, and VARMA [37] and has been incorporated now into most of the theoretical treatments of the problem [41-44]. The reason for the rather strange direction of $\vec{q}_s$ is not yet understood and may depend on Fermi-surface effects which determine (together with the superconducting interactions) the effective $J(\vec{q})$.

The existence of magnetic LRO of both $\vec{q} = 0$ (ferromagnetic) and $\vec{q} = \vec{q}_s$ (the MM structure) in the intermediate temperature region 0.7K < T < 0.88K (which also implies a coexistence between a ferromagnetic component and superconductivity) is an important but puzzling result. There are two possibilities: either both the ferromagnetic and sinusoidal components arise from the same microscopic regions of the sample, or the sample divides spontaneously into normal ferromagnetic domains and superconducting domains containing the MM structure. One example of the first possibility is the regular vortex lattice [17], where by symmetry both components of $\vec{q}_s$ perpendicular to the moment directions (e.g., an [010] axis) would coexist with a ferromagnetic component throughout a particular domain. We tend not to favor such an interpretation for the following reasons: (a) there seems to be no dependence of $\vec{q}_s$ on the spontaneous magnetization or an applied field as would be expected by the flux-quantization condition for a vortex lattice, (b) critical field measurements by BEHROOZI et al. [25] indicate that below 1K $ErRh_4B_4$ behaves like a type I rather than a type II superconductor, (c) the various satellite intensities around a given Bragg point in the a*-c* plane are not always equal, depending on the history of warming and cooling the sample, which would imply that there are at least domains corresponding to the different (equivalent) directions of propagation of the modulation, and (d) the strong hysteresis making the ratio of modulated to ferromagnetic intensities history-dependent also argues against at least a conventional vortex lattice phase. If we assume that the sample is heterogeneously divided between volumes of ferromagnetic and sinusoidal order, and further assume that the latter is equally distributed among domains with different but equivalent $\vec{q}_s$ vectors (an assumption not always justified as discussed above), we arrive at the following ratio of the squared structure factors

$$|F_{MM}|^2/|F_{FM}|^2 = \frac{1}{16}\left(\frac{\mu_{MM}}{\mu_{FM}}\right)^2 \frac{V_{MM}}{V_{FM}} \qquad (3)$$

where μ_{MM}, V_{MM}, and μ_{FM}, V_{FM} are the amplitude of the ordered moment and the total crystal volume associated with the MM and ferromagnetic structures, respectively. Thus, in order to determine the two ratios individually, we must assume that an externally variable parameter (such as a magnetic field) converts some of the crystal volume from V_{MM} to V_{FM} without affecting the ordered moments in the respective regions. Field-dependent measurements of the intensity ratio yielded, at 0.72K, values of $(V_{MM}/V_{FM}) \simeq 0.2$ and $\mu_{MM}/\mu_{FM} \simeq 1.4$. We now briefly discuss the field-

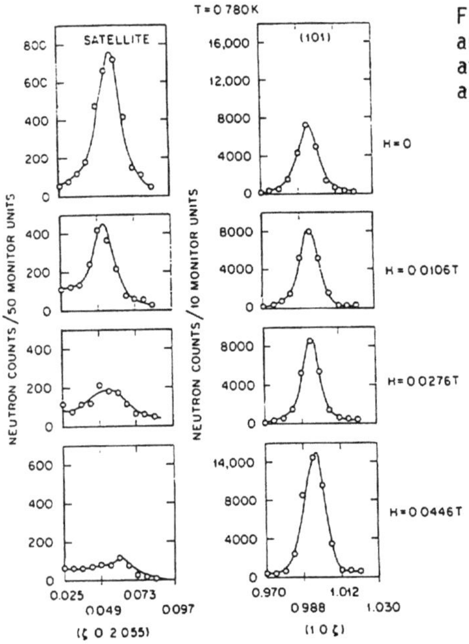

Fig. 2 Scans through the satellite and ferromagnetic (101) reflections at T = 0.78K for various values of applied field H (in Tesla)

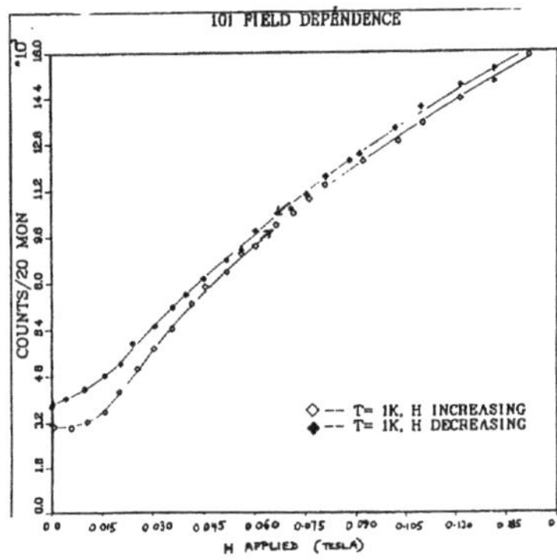

Fig. 3 Ferromagnetic intensity at T = 1K as a function of applied field

dependent measurements which are still preliminary. Fig. 2 shows the effect at T = 0.78K of a magnetic field applied along the [010] axis on the intensities of a magnetic satellite from the (002) Bragg point and of the (101) Bragg peak. One notices that the satellite intensity decreases very rapidly with increasing field, and is essentially gone by H_{app} = 0.045 Tesla. (It should be remembered that owing to the irregular shape of the crystal used, the demagnetization field cannot be calculated very accurately. Hence, the applied field values quoted here are not to be taken as the actual internal field). The position of the satellite peak does not shift, except to slightly larger wave vectors just before it vanishes. The ferromagnetic intensity, on the other hand, is initially very little affected by the field, and then slowly increases. Fig. 3 shows the behavior of the (101) intensity at 1K as a function of increasing and decreasing field. It may be seen that the ferromagnetic intensity is not affected up to a critical field (corresponding to Hc_1) but on applying and then decreasing the field to zero, a finite ferromagnetic intensity remains. This may be due to small ferromagnetic domains residing in regions of trapped flux, or the "forced ferromagnetism" referred to earlier. Fig. 4 shows the corresponding curve at T = 2K. One may see effects at both Hc_1 and Hc_2. At Hc_2 one sees a jump in the ferromagnetic intensity with increasing field consistent with the findings of BEHROOZI et al. [25] of a discontinuity in the magnetization at Hc_2 in the lower temperature part of the H-T phase diagram. This jump in the ferromagnetic intensity decreases with increasing temperature, also consistent with the results of Behroozi et al. and indicating the presence of a tricritical point, where the transition at Hc_2 changes from first to second order. At applied fields greater than 0.07 Tesla, the temperature hysteresis of the ferromagnetic intensity has disappeared, again indicating the finite range of the first-order phase boundary in H-T space (see Fig. 5).

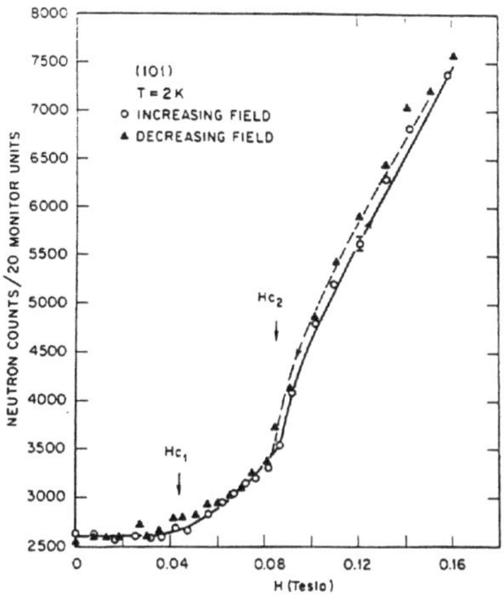

Fig. 4 Ferromagnetic intensity at T = 2K as a function of applied field

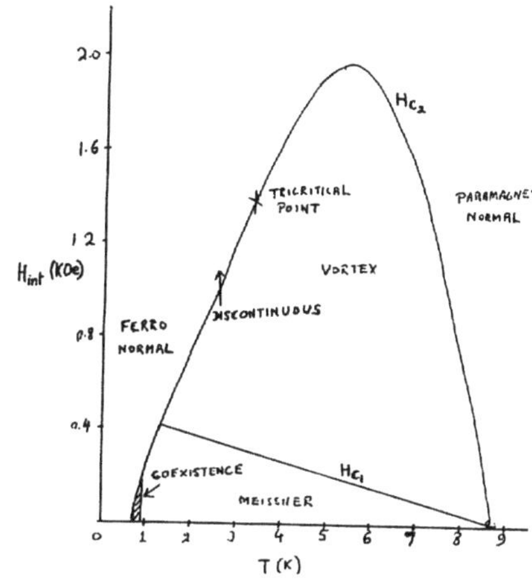

Fig. 5 Rough sketch of phase diagram for $ErRh_4B_4$ as a function of internal field and temperature

Finally, we discuss briefly more detailed studies of the higher temperature region of the "co-existence" phase (between 0.88K and 1.1K) at Riso National Laboratory, Denmark, recently [38]. These measurements, which are still preliminary, show that in this temperature regime both the ferromagnetic intensity and the satellite peak show broadening in wave vector space (see Fig. 6) corresponding to short-range fluctuations. As the temperature is lowered, these correlations develop longer and longer ranges, until true LRO sets in. Thus, short-range order seems to be the reason for the upward curvature of the intensities shown in Fig. 1 in this temperature regime. In this regime, there is no

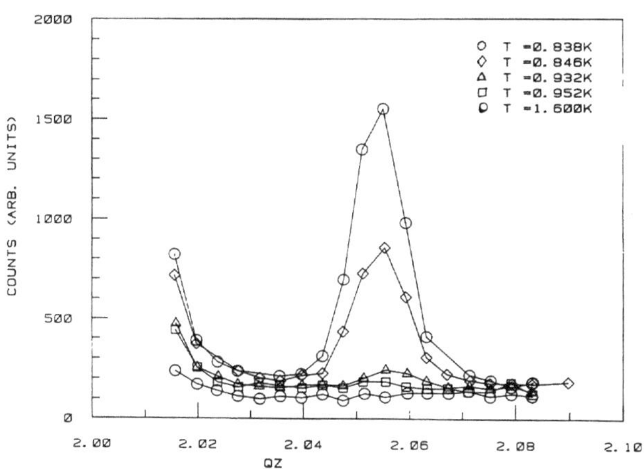

Fig. 6 Scans through satellite positions and in wings of ferromagnetic peak as a function of temperature showing broadening effects at high temperature

temperature hysteresis of the intensities, and the ferromagnetic and satellite intensities seem to grow proportionately to each other. This is consistent with critical fluctuations in the superconducting paramagnetic phase, but the occurrence of ferromagnetic as well as sinusoidal fluctuations seems strange and unanticipated by theory. In this regime, the peak of the scattering at the satellite position also seems shifted to slightly smaller wave vectors.

6. Discussion

For the ferromagnetic superconductors $ErRh_4B_4$, it thus appears that there are (in zero applied field) four regimes as the temperature is lowered below Tc_1. First, there exists a superconducting paramagnetic phase. At lower temperatures, short-range correlations develop, corresponding to both ferromagnetic and sinusoidal order. (Long-wavelength ferromagnetic fluctuations will not in principle be prevented by superconductivity provided their wavelength is not greater than $\lambda_\downarrow$ which is presumably very large on the scale of the resolution used in these experiments.) At about 0.88K, LRO of both the ferromagnetic and modulated components sets in, although the exact coincidence of the transition for both these components is still not established. Finally, there is a lower first-order transition, with temperature hysteresis, to a normal ferromagnetic phase.

Although the existence and wave vector of the MM structure is consistent with the theoretical ideas based on the electromagnetic coupling with the superconducting order parameter, the actual coexistence of both the ferromagnetic and MM components (and bulk superconductivity) is as yet not very well understood. The transverse plane polarized nature of the MM structure may be understood in terms of the electromagnetic coupling between the moments. For the reasons given in the last section we do not tend to favor the picture of the coexistence region as that of a conventional vortex lattice phase. HU [39] has proposed that this phase is in fact due to an antiferromagnetic vortex lattice, with a very long-wavelength periodicity (and therefore the $q = 0$ component presumably corresponds to intensity at a finite but very small wave vector unresolved from the Bragg point). In this model the MM structure is simply due to oscillatory magnetism between the vortex cores. There is as yet no direct proof of the correctness of this model, and some other predictions of this model (relating to the existence of harmonics and satellites at $(3\vec{\zeta} + \vec{\eta})$ and $(3\vec{\eta} + \vec{\zeta})$) are contradicted by experiment. The lack of observable harmonics also seems to rule out the antiphase domain structure predicted by BULAEVSKII et al. [40] based on calculations including both the electromagnetic and exchange interactions. According to this model, one would expect higher order harmonics of the main satellite (m = odd integer) with intensity proportional to $(1/m^2)$. BULAEVSKII et al.[40] explain this discrepancy as due to the presence of irregular domain walls.

More recently some other theoretical models have been proposed to account for the coexistence phase. Several of these are due to TACHIKI and collaborators. In one [41], a laminar vortex structure is proposed (which implies a ferromagnetic domain with a single linear sinusoidal modulation superimposed on it, and numerical calculations based on this model do in fact predict behavior qualitatively similar to that shown in Fig. 2. A more recent model by TACHIKI [42] yields very long-period antiphase domains (with periodicity ~ $5\lambda_L$ to $10\lambda_L$) coexisting with the shorter wavelength MM structure in the superconductivity phase just above Tc_1. Since $\lambda_L \sim 1000$ Å or greater, the former would account for the "ferromagnetic" LRO observed as well as the ferromagnetic fluctuations in the paramagnetic phase. Yet another recent proposal by TACHIKI et al. [43] suggests inhomogeneities (e.g., due to strains) in the crystal as the reason for local variations of Tc_2 and hence for the coexistence of true normal ferromagnetic and superconducting MM phases. This does not seem consistent with the growth of the "ferromagnetic" and sinusoidal fluctuations proportionately in the paramagnetic phase, and the smooth increase but sudden cut-off of the satellite intensity at 0.71K.

MACHIDA and collaborators [44] have proposed a quite different picture for the coexistence phase. According to these authors this phase is a realization of the so-called Fulde-Ferrell state where superconductivity exists in a finite molecular exchange field but with a finite $\vec{q}_o$ pairing to minimize the free energy. The superconductivity order parameter thus has a modulation wave vector $\vec{q}_o$ which, by second-order exchange coupling between the conduction electrons and f shells, induces a sinusoidal MM structure for the local spins $\vec{q}_s = 2\vec{q}_o$. While their original theory predicted a first-order transition from a paramagnetic superconducting state to one with coexisting ferromagnetic and sinusoidal components, and a second-order transition below that to a normal ferromagnetic phase (in contradiction to experiment), more refined versions of this theory which include higher order couplings between the two components in the Ginzburg-Landau free energy expression can approximately reproduce the observed behavior. Their most recent calculations (presented at this symposium) predict several harmonics of the satellites just at the temperature (~ 1.2K) where the magnetic correlations begin to set in. These are probably so weak at this temperature that they would in any case be difficult to observe. However, one prediction (a tendency to a smaller $\vec{q}_s$ when the satellite appears) seems to be borne out experimentally (in the regime where the satellite peak is also broadened, however). This theory completely ignores the electromagnetic interactions and thus the Meissner effect poses a problem here, as yet unresolved. For the ferromagnetic superconductor $HoMo_6S_8$ as we have seen, the situation is probably very similar to that in $ErRh_4B_4$. Again, both ferromagnetic and MM fluctuations occur just above Tc_2 in the superconducting paramagnetic phase, below which a MM phase exhibiting LRO appears and also ferromagnetic order. The actual coexistence region is not quite so well defined as yet, but clearly within it hysteretic effects are observed [28]. A detailed determination of the structure and direction of propagation of the MM phase awaits the availability of single crystals of this material.

A word should be said about domain wall effects in the ferromagnetic regimes. In both the single crystal and polycrystalline samples, diffuse scattering is seen around $\vec{q} = 0$ in the ferromagnetic phase. These are not due to critical scattering because (a) for single-crystal $ErRh_4B_4$ this component could not be fitted with the usual Lorentzian scattering function in reciprocal space. It has been interpreted as due to domain walls in the ferromagnetic phase. Its shape can be fitted with either a Gaussian centered at $\vec{q} = 0$ [35] or with a $(1/q^4)$ low characteristic of small-angle scattering from sharply defined but smoothly curved surfaces [45] (Porod scattering). Finally, the question of why the spontaneous ordered ferromagnetic moment at low temperatures is so much less than the Er^{3+} free-ion moment, or that determined by Mossbauer measurements [24] is still a mystery, and apparently cannot be explained either by crystal-field effects [46].

We thus see that both theoretically and experimentally, although much has been learned about these compounds, many fascinating questions remain to be resolved.

The authors are very grateful to the following for useful discussions concerning many aspects of this work: F. Behroozi, G. W. Crabtree, B. D. Dunlap, F. Y. Fradin, K. Gray, C. Hu, J. W. Lynn, K. Machida, R. M. Moon, R. Pynn, G. R. Shenoy, M. Tachiki, and C. M. Varma, and to J. Kjems for permission to quote unpublished work carried out in collaboration with him.

References

1. B. T. Matthias, E. Corenzwit, J. M. Vandenburg and H. E. Barz, Proc. Nat. Acad. Sci. U.S.A. 74, 1334 (1977).

2. M. Ishikawa and Ø. Fischer, J. de Physique C67, 1379 (1978).

3. G. K. Shenoy, B. D. Dunlap and F. Y. Fradin (Eds.), Proc. Int. Conf. on Ternary Superconductors (North-Holland, Amsterdam, 1981).

4. M. B. Maple and Ø. Fischer (Eds.), Superconductivity in Ternary Compounds, Vols. I and II, Topics in Current Physics (Springer, Berlin, Heidelberg and New York, 1982).

5. D. E. Moncton, G. Shirane, W. Thomlinson, M. Ishikawa and Ø. Fischer, Phys. Rev. Lett. 41, 1133 (1978).

6. W. Thomlinson, G. Shirane, D. E. Moncton, M. Ishikawa and Ø. Fischer, J. Appl. Phys. 50, 1981 (1979).

7. M. B. Maple, L. D. Woolf, C. F. Majkrzak, G. Shirane, W. Thomlinson and D. E. Moncton, Phys. Lett. 77A, 487 (1980).

8. C. F. Majkrzak, D. E. Cox, G. Shirane, H. A. Mook, H. C. Hamaker, H. B. Mackay, Z. Fisk and M. B. Maple, Phys. Rev. B26, 245 (1982).

9. C. M. Majkrzak, S. K. Satija, G. Shirane, H. C. Hamaker, Z. Fisk and M. B. Maple, Phys. Rev. B27, 2889 (1983).

10. H. C. Hamaker, H. C. Ku, M. B. Maple and H. A. Mook, Sol. St. Comm. 43, 455 (1983).

11. M. Ishikawa and Ø. Fischer, Solid State Comm. 24, 747 (1977).

12. M. B. Maple, H. C. Hamaker and L. D. Woolf, Chapter 4 of Ref. 4.

13. A. J. Freeman and T. Jarlborg, J. Appl. Phys. 50, 1976 (1979).

14. P. W. Anderson and H., Suhl, Phys. Rev. 116, 898 (1959).

15. E. I. Blount and C. M. Varma, Phys. Rev. Lett. 42, 1079 (1979).

16. H. Matsumoto, H. Umezawa and M. Tachiki, Solid State Comm. 31, 157 (1979).

17. C. G. Kuper, M. Revson and A. Ron, Phys. Rev. Lett. 44, 1545 (1980).

18. M. Tachiki, H. Matsumoto, T. Koyama and H. Umezawa, Solid State Comm. 34, 19 (1980).

19. R. A. Ferrel, J. K. Bhattacharjee and A. Bagchi, Phys. Rev. Lett. 43, 154 (1979).

20. W. A. Fertig, D. C. Johnston, L. E. Delong, R. W. McCallum, M. B. Maple and B. T. Matthias, Phys. Rev. Lett. 38, 983 (1977).

21. H. R. Ott, W. A. Fertig, D. C. Johnston, M. B. Maple and B. T. Matthias, J. Low Temp. Phys. 33, 159 (1978).

22. L. D. Woolf, D. C. Johnston, H. B. MacKay, R. W. McCallum and M. B. Maple, J. Low Temp. Phys. 35, 651 (1979).

23. D. E. Moncton, D. B. McWhan, J. Eckert, G. Shirane and W. Thomlinson, Phys. Rev. Lett. 39, 1164 (1977).

24. G. K. Shenoy, B. D. Dunlap, F. Y. Fradin, S. K. Sinha, C. W. Kimball, W. Potzel, F. Probst and G. M. Kolvino, Phys. Rev. B21, 3886 (1980).

25. G. W. Crabtree, F. Behroozi, S. A. Campbell and D. G. Hinks, Phys. Rev. Lett. 49, 1342 (1982); F. Bahroozi, G. W. Crabtree, S. A. Campbell and D. G. Hinks, Phys. Rev. B27, 6849 (1983).

26. D. E. Moncton, D. B. McWhan, P. H. Schmidt, G. Shirane, W. Thomlinson, M. B. Maple, H. B. MacKay, L. D. Woolf, Z. Fisk and D. C. Johnston, Phys. Rev. Lett. 45, 2060 (1980).

27. J. W. Lynn, G. Shirane, W. Thomlinson, R. N. Shelton and D. E. Moncton, Phys. Rev. B24, 3817 (1981).

28. J. W. Lynn, J. L. Raggazoni, R. Pynn and J. Joffrin, J. Phys. (Paris) Lett. 42, L45 (1981).

29. G. H. Lander, S. K. Sinha and F. Y. Fradin, J. Appl. Phys. 50, 1990 (1979).

30. H. A. Mook, W. C. Koehler, M. B. Maple, Z. Fisk, D. C. Johnston and L. D. Woolf, Phys. Rev. B25, 372 (1982).

31. J. W. Lynn, J. Less Common Metals 94, 75 (1983).

32. H. A. Mook, O. A. Pringle, S. Karvarazaki, S. K. Sinha, G. W. Crabtree, D. G. Hinks, M. B. Maple, Z. Fisk, D. C. Johnston, L. D. Woolf and H. C. Hamaker, Proc. IV Conf. Superconductivity in d- and f-band Metals, Karlsruhe, p. 201 (1982).

33. S. K. Sinha, G. W. Crabtree, D. G. Hinks and H. A. Mook, Phys. Rev. Lett. 48, 950 (1982).

34. S. K. Sinha, G. W. Crabtree, D. G. Hinks and H. A. Mook, Physica 109 + 110B, 1693 (1982).

35. S. K. Sinha, G. W. Crabtree, D. G. Hinks and H. A. Mook, in Novel Materials and Techniques in Condensed Matter, (G. W. Crabtree and P. Vashishta, Eds.), North-Holland, New York, Amsterdam, Oxford, p. 133 (1982).

36. C. Stassis, H. W. Deckman, B. N. Harmon, J. P. Desclaux and A. J. Freeman, Phys. Rev. B15, 369 (1977).

37. H. S. Greenside, E. I. Blount and C. M. Varma, Phys. Rev. Lett. 46, 49 (1981).

38. S. K. Sinha, J. Kjems, H. A. Mook, O. A. Pringle and D. G. Hinks, to be published.

39. C.-R. Hu and T. E. Ham, Physica 108B, 1041 (1981) and private communication.

40. L. N. Bulaevskii, A. I. Buzdin, D. I. Khomskii and S. V. Panjukov, Solid State Comm. 46, 133 (1983).

41. M. Tachiki, Physica 109 + 110B, 1699 (1982).

42. M. Tachiki, J. Magnetism Mag. Materials, 31-34, 479 (1983).

43. M. Tachiki, B. D. Dunlap, G. W. Crabtree, Phys. Rev. B28, 5342 (1983).

44. K. Machida, J. Phys. Doc. Jap. 51, 3462 (1982); ibid. 52, 2181 (1983); K. Machida and T. Matsubara, J. Magnetism Mag. Materials 31-34, 507(1983); K. Machida, Proceedings this Symposium.

45. J. W. Lynn, private communication.

46. B. D. Dunlap, private communication.

A Theory for Ferromagnetic Superconductors

H. Umezawa, H. Matsumoto, and J.P. Whitehead

Department of Physics, The University of Alberta
Edmonton, Alberta T6G 2J1, Canada

A method for the systematic analysis of magnetic superconductors is presented. The method is formulated in order to include, in a consistent manner, the shielding of the magnetic moments by the persistent currents together with the pair-breaking effect arising from the scattering of the electrons by the fluctuations of the localized spins and the polarization of the superconducting electrons. The shielding effect is caused by the dipole interaction while the pair-breaking effects are induced through the d-f interaction. The effect of the localized spin fluctuation will be taken into account by means of a straightforward scaling law.

The table of contents is the following:

1. Introduction

As is well known, a particularly interesting feature of magnetic superconductors arises from the fact that the magnetism and superconductivity mutually try to suppress one another. The Meissner effect of superconducting current tries to weaken the magnetism of the magnetic moments, while the scattering of the conduction electrons by localized spin fluctuations tends to destroy the superconductivity by breaking the Cooper pairs. Furthermore, the presence of the magnetic moments tends to polarize the conduction-electron spins, resulting in an energy gap, μ, between the two spin-states of the electrons. This also tends to break the Cooper pairs, and therefore, to suppress the superconductivity. These three mechanisms will be referred to as the shielding effect due to the persistent currents, the effect of localized-spin fluctuation and the effect of electron-spin polarization,respectively. Depending on the strength of each effect, magnetic superconductors can exhibit a variety of unusual phenomena [1]. The recent discovery of rare-earth ternary (or pseudoternary) compounds gives rise to a renewed interest in the study of magnetic superconductors [2], and there are currently many different kinds of materials which exhibit a wide variety of unusual properties [3].

In our previous analysis [1] of rare-earth ternary compounds, we mainly considered the magnetic dipole interactions which cause the shielding effect and ignored the d-f interaction which creates two other effects, because the d-f interactions in these compounds are supposed to be weak. Though this analysis has been successful in understanding many unusual behaviors in these compounds, the recent experimental analysis of a single crystal of $ErRh_4B_4$ disclosed new unusual phenomena which cannot be explained by the magnetic dipole interaction only. In this lecture we present a general theory for ferromagnetic superconductors formulated to include in a consistent fashion the three various effects mentioned above. It will provide for a systematic framework in which the various theoretical calculations may be discussed and it will assist in making a qualitative analysis of experimental data in terms of these three effects. At the end of this lecture we apply this formulation to a single crystal of $ErRh_4B_4$. Its magnetic properties are described by using a single set of parameters, most of which are determined by experiments. The results show a reasonable agreement with the experiments.

2. The Hamiltonian

We consider a system with the following Hamiltonian density:

$$\mathcal{H}(x) = \mathcal{H}_{BCS}(x) - 1/2\, \vec{M}\gamma_0(-i\nabla)\vec{M} - I\vec{M}\cdot\psi^{\dagger}\vec{\sigma}\psi - \vec{M}\cdot\vec{B} + \mu_B\psi^{\dagger}\vec{\sigma}\psi\cdot\vec{B} + + 1/8\pi\{|\vec{B}|^2 + |\vec{E}|^2\} . \tag{2.1}$$

Here $\mathcal{H}_{BCS}$ is the BCS Hamiltonian with the minimal electromagnetic interaction which is obtained from the free electron Lagrangian through the replacement $\vec{\nabla} \rightarrow \vec{\nabla} - i(e/\hbar c)\vec{A}$, with $\vec{A}$ being the electromagnetic vector potential:

$$\mathcal{H}_{BCS}(x) = \psi^{\dagger}\varepsilon[-i(\vec{\nabla} - \frac{ie}{\hbar c}\vec{A})]\psi - V\psi_{\uparrow}^{\dagger}\psi_{\downarrow}^{\dagger}\psi_{\downarrow}\psi_{\uparrow} + \delta\varepsilon\psi^{\dagger}\psi \quad + \text{Coulomb interaction.} \tag{2.2}$$

The field ψ is the electron field and $\vec{B}$ the magnetic induction ($\vec{B} = \vec{\nabla}\times\vec{A}$). Throughout this paper use is made of the notation $F(-i\vec{\nabla})$, which means

$$F(-i\vec{\nabla})e^{i\vec{k}\vec{x}} = F(\vec{k})e^{i\vec{k}\vec{x}}. \tag{2.3}$$

The electron spectrum denoted by $\varepsilon(\vec{k})$ will be assumed to be parabolic, therefore

$$\varepsilon[-i(\vec{\nabla} - \frac{ie}{\hbar c}\vec{A})] = -\,1/2m[(\vec{\nabla} - \frac{ie}{\hbar c}\vec{A})^2 + k_F^2] . \tag{2.4}$$

The third term in (2.2) is to compensate for the shift in the chemical potential due to the various interactions.

The field $\vec{M}$ is the magnetic moment density of the localized spins:

$$\vec{M} = \sum_n g_J J \mu_B\, \vec{S}(n)\delta(\vec{x}-\vec{R}_n) , \tag{2.5}$$

where $\{\vec{R}_n\}$ denotes the lattice points of the magnetic ions.

In (2.1) the second term is the exchange interaction among localized spins excluding the interactions induced by the d-f interaction and the dipole interaction. The coupling γ_0 will be referred to as the "bare" exchange coupling. The third term in (2.1) is the d-f interaction (the interaction between the conduction electrons and the localized spins) while the fourth is the dipole interaction. The fifth interaction will be referred to as the paramagnetic $\vec{\sigma}\cdot\vec{B}$ interaction.

3. Susceptibilities

With the Hamiltonian (2.1) we can calculate the order parameter, the electromagnetic field, the net magnetic moment and the free energies of various ordered states together with many observable quantities. As a preparation for these calculations, in this section, we calculate susceptibilities. To do this, we need to calculate the magnetization $<\vec{M}>$ which will be denoted simply by $\vec{M}$. The mean field approximation gives

$$|\vec{M}| = g_J\mu_B J N B_J[\beta g_J \mu_B J|\vec{H}_m|], \quad \vec{M}//\vec{H}_m \tag{3.1}$$

where $\beta = 1/k_BT$, the molecular field $\vec{H}_m$ is given by

$$\vec{H}_m = \vec{B} + \gamma_0(-i\nabla)\vec{M} + I\vec{\sigma}, \qquad (\vec{\sigma} \equiv <\psi^{\dagger}\vec{\sigma}\psi>) . \tag{3.2}$$

In the following we simplify the situation by assuming $I >> \mu_B$ and $\mu_B I \chi_\sigma << 1$. (A consideration without this assumption will be presented elsewhere.) Here χ_σ is the electron spin correlation function and is given in terms of a 3 × 3 matrix. As was shown by SUHL and ANDERSON [4], the low momentum part of χ_σ is suppressed due to the effect of the superconducting energy gap (Suhl-Anderson effect). However, this effect becomes small when there is a considerable amount of the spin-orbit coupling. In such a case, χ_σ for the superconducting and normal states are not much different; $\chi^s_\sigma \simeq \chi^n_\sigma$.

To calculate $\vec{B}$ in (3.2) we make use of the Maxwell equation in the absence of any topological objects such as vortices, which may be written as

$$\{- \nabla^2 - \lambda_L^{-2}c(-i\nabla)\}\vec{B} = - 4\pi\nabla^2(\vec{M} - \mu_B\vec{\sigma}) , \tag{3.3}$$

where $c(\vec{k})$ is the non-local form factor of the Meissner current. We then obtain

$$\vec{H}_m = \tilde{\gamma}\vec{M} \tag{3.4}$$

where

$$\tilde{\gamma}(\vec{k}) \simeq \gamma(\vec{k}) - 4\pi F(\vec{k}) \tag{3.5}$$

with $F(\vec{k}) = c(\vec{k})/[\lambda_L^2\vec{k}^2 + c(\vec{k})]$ and

$$\gamma(\vec{k}) = \gamma_0(\vec{k}) + I^2\chi_\sigma(\vec{k}) + 4\pi[T] \tag{3.6}$$

where [T] is the projection operator of the transverse components:

$$[T]_{ij} = \delta_{ij} - (1/\nabla^2)\nabla_i\nabla_j . \tag{3.7}$$

Note that in the normal state, $\lambda_L = \infty$ and we obtain $\tilde{\gamma} = \gamma$. The function $\gamma(\vec{k})$ has a maximum at $\vec{k} = 0$ and damps rapidly at large $\vec{k}$. In superconducting states the last term in (3.5) effectively suppresses the low momentum part in the effective exchange coupling as shown in Fig. 3.1. This is the origin of the shielding effect due to the persistent current [5]. As was mentioned previously, the Suhl-Anderson effect has a similar effect in that it suppresses χ_σ (and therefore γ according to (3.6)) at low momentum for the superconducting state. In this paper we assume that the spin-orbit coupling is sufficiently large that the Suhl-Anderson effect may be considered negligible; $\chi_\sigma^s \simeq \chi_\sigma^n$. In other words we assume that γ is common to both the superconducting and normal states; $\gamma^s \simeq \gamma^n$. As will be shown later, (3.4) will be modified by the appearance of the vortices.

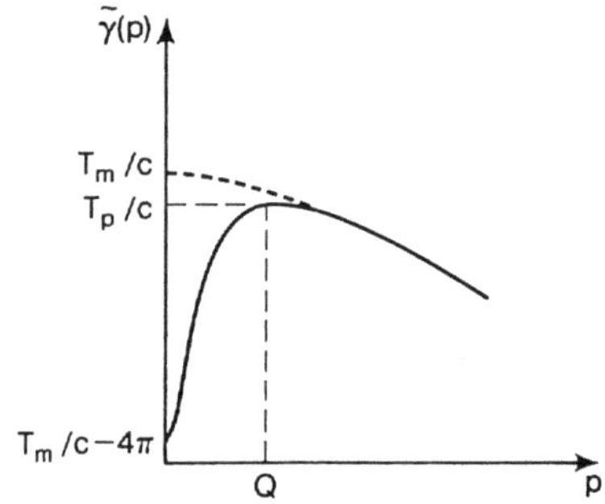

Fig. 3.1. Plot of the effective coupling constant $\tilde{\gamma}(p)$ demonstrating the effect of the screening by the persistent current at low momenta

As is well known, the susceptibility $\tilde{\chi}(\vec{k})$ of the localized spins is related to the effective exchange coupling $\tilde{\gamma}$ through the relation

$$\tilde{\chi}^{-1}(\vec{k}) = \begin{pmatrix} \frac{H_m}{M} & & \\ & \frac{H_m}{M} & \\ & & \frac{T}{C\alpha_J} \end{pmatrix} - \tilde{\gamma}(k) \quad , \tag{3.8}$$

where α_J means $\alpha_J(H_m)$ defined by

$$\alpha_J(H_m) = \frac{3J}{J+1} B_J'(\beta g_J \mu_B J|H_m|) \quad \text{and} \tag{3.9}$$

$$C = g_J^2 \mu_B^2 N\ J(J+1)/3k_B \ . \tag{3.10}$$

In this paper we use the following approximation. First, assuming that $\mu \ll \varepsilon_F$ (μ is the energy gap between two spin states of conduction electron), we use a diagonal χ_σ. Second, assume a certain amount of spin-orbit interaction which leads to $\chi_\sigma^s \simeq \chi_\sigma^n$. This leads to $\gamma^s \simeq \gamma^n$. With these approximations we can parameterize $\gamma(\vec{k})$ as follows;

$$\gamma_{ij}(\vec{k}) = \left(\frac{T_m^i}{C} - \frac{D}{C} \vec{k}^2\right) \delta_{ij} \tag{3.11}$$

where C was given by (3.10). With the above approximations for γ both $\tilde{\gamma}$ and $\tilde{\chi}$ are diagonal. Then (3.8) gives

$$\tilde{\chi}_{33}(\vec{k}) = \frac{C}{T/\alpha_J - C\tilde{\gamma}_{33}(\vec{k})} \tag{3.12}$$

$$\tilde{\chi}_{ii}(\vec{k}) = \frac{1}{H_m/M - \tilde{\gamma}_{ii}(\vec{k})}\ , \quad i = 1,2\ . \tag{3.13}$$

These are the susceptibilities in the superconducting state. The susceptibilities in the normal states are obtained from these by putting $\lambda_L \to \infty$. Note that α_J depends on the magnetic field according to (3.9); α_J decreases with increasing field, due to the saturation of the localised magnetic moments.

4. The Shielding Effect and the Spin Periodic Phase

The second term appearing in $\tilde{\gamma}$ in (3.5) represents the effect of the shielding of the magnetic moments by the superconducting current. As was pointed out in section 3, the screening by the persistent current is most effective at low momenta. Therefore, since $\gamma(k)$ decreases montonically with increasing $|\vec{k}|$, the effective coupling constant $\tilde{\gamma}$ will have a maximum at some finite value of momentum $\vec{k} = \vec{Q} \neq 0$ as shown in Fig. 3.1. This implies that the susceptibility, given by (3.12), will have a maximum at H = 0 and $\vec{k} = \vec{Q}$ and therefore there appears a spin periodic phase in the domain $T \leq T_p$ where T_p is determined by

$$T_p = C\alpha_J\tilde{\gamma}_{33}(Q)\ , \tag{4.1}$$

where

$$Q = (1/\sqrt{\lambda_L})\left|\frac{4\pi C}{D}\right|^{1/4} \tag{4.2}$$

which gives

$$T_p = T_m - 2(4\pi CD/\lambda_L)^{1/2} \tag{4.3}$$

for large κ and $(4\pi C/D)\lambda_L^2 \gg 1$. (Other choices of the parameters are presented in the original paper [6].) More precise values of Q and T_p are usually obtained by means of a numerical computation. As is well known, neutron diffraction experiments indicate the presence of a spin periodic phase in a small temperature domain above T_{C2} (the transition temperature from superconducting state to normal ferromagnetic state) in $ErRh_4B_4$ [7,8] and $HoMo_6S_8$ [9]. The neutron diffraction data in a single crystal of $ErRh_4B_4$, for which the superconducting critical temperature $T_C \simeq 8.7$ K, while $T_p \simeq 1$ K and $T_{C2} \simeq 0.7$ K respectively [8], determined the direction of the wave vector and the magnetic moments, confirming that the spin periodic structure is not a spiral one but is in fact a sinusoidal one. This indicates that the spin component along the hard axis (c axis) is quenched. According to this experiment $\lambda_p = 2\pi/Q$ is about 100Å. It has been proposed that this spin sinusoidal phase is the spin periodic phase caused by the shielding effect of persistent current mentioned above [6,10,11]. The experiment indicates also that a ferromagnetic component coexists together with the spin periodic component.

5. Electron Self-Energy and Energy Gap

Gap Equation. The d-f interaction induces the electron self-energy correction $\Sigma(p)$ which is a loop consisting of two lines; one is the electron Green's function S(p-k) and the other is the spin correlation function calculated in the superconducting state which we denote by $\tilde{\chi}_{ij}(k)$ and which is defined by

$$\langle R[M_i(x)M_j(y)]\rangle = \frac{i}{(2\pi)^4}\int d^4k\ \tilde{\chi}_{ij}(k)\ e^{i\vec{k}(\vec{x}-\vec{y}) - ik_0(t_x-t_y)} . \qquad (5.1)$$

The two vertices in the loop are simply the Pauli spin matrices. Thus the self-energy $\Sigma(p)$ has the form $\sigma_i\tilde{\chi}_{ij}(k)\sigma_j S(p-k)$ integrated over d^4k. The correlation function may be written, using the spectral representation, as

$$\tilde{\chi}_{ij}(k) = \int d\omega\ \rho_{ij}(\omega,\vec{k})\ \frac{1}{k_0 - \omega + i\varepsilon} . \qquad (5.2)$$

At $T \neq 0$, the expression of $1/[k_0 - \omega + i\varepsilon]$ is modified by the temperature effect. The spectral function ρ_{ij} may be approximated by the hydrodynamical form

$$\rho_{ij}(\omega,\vec{k}) = \frac{1}{\pi}\frac{\omega\Gamma}{\omega^2+\Gamma^2}\tilde{\chi}(\vec{k})\delta_{ij} , \qquad (5.3)$$

where $\tilde{\chi}_{ij}(\vec{k})$ is the static susceptibility obtained in the last section and Γ is the decay width which is a function of ω and $\vec{k}$. The temperature is introduced using thermo field dynamics. In the limits $\Gamma \to 0$ and $\Gamma \to \infty$ we obtain the elastic and inelastic limits respectively and the corresponding expression for the self-energy simplifies considerably. In the inelastic limit the self-energy correction is real and leads to a renormalization of the B.C.S. gap. In the elastic limit, on the other hand, the self-energy correction is purely dispersive and the resultant gap equation is formally equivalent to that obtained in the case of magnetic impurities. It has been proposed that since the coherent nature of the spin correlations is dominant, the inelastic limit is appropriate to the situation under consideration [12]. We will therefore restrict our considerations to the inelastic limit. It should be noted, however, that as T approaches T_p the reduction of Γ due to critical slowing down may mean that in the neighborhood of the transition the elastic limit may be more relevant. A more careful calculation may therefore require the use of the elastic limit in the vicinity of T_p and the use of the inelastic limit at higher temperatures, together with a scheme which would allow one to interpolate between the two limits in a straightforward manner.

When we use the free propagator of quasi-electron for S(p-k) in $\Sigma(p)$, we obtain in the inelastic limit

$$\Sigma(\vec{p}) = -\frac{I^2}{(2\pi)^3}\int d^3k \sum_i \tilde{\chi}_{ii}(\vec{p}-\vec{k})\ \frac{\Delta(\vec{k})}{2E(\vec{k})}$$
$$\times \frac{1}{2}\{\tanh\frac{\beta(E(\vec{k})+\mu)}{2} + \tanh\frac{\beta(E(\vec{k})-\mu)}{2}\} , \qquad (5.4)$$

where $E(\vec{k}) = [\varepsilon^2(\vec{k}) + \Delta^2(\vec{k})]^{1/2}$ and $\Sigma(\vec{p})$ is given by $\Sigma(p)$ calculated with $p_0 = E(\vec{p})$. The appearance of μ exhibits the effect of the electron-spin polarization. Adding this self-energy to the usual BCS self-energy, we obtain the gap equation

$$\Delta(\vec{p}) = \frac{1}{(2\pi)^3}\int d^3k\{V - I^2 \sum_i \tilde{\chi}_{ii}(\vec{p}-\vec{k})\}\ \frac{\Delta(\vec{k})}{2E(\vec{k})}$$
$$\times \frac{1}{2}[\tanh\frac{\beta(E(\vec{k})+\mu)}{2} + \tanh\frac{\beta(E(\vec{k})-\mu)}{2}] . \qquad (5.5)$$

6. Effective Coupling and Scaling Law

Effective Coupling. In our calculations we assume that $\Delta(\vec{k})$ is a smooth function of $|\vec{k}|$ and is well approximated by its value on the Fermi surface. We further assume that any anisotropy in $\Delta(\vec{k})$ will be small and that it may be replaced by its angular average, which will be denoted by Δ.

Then introducing the effective coupling constant defined by

$$g(T,H) = \int\frac{d\Omega_k}{4\pi}\int\frac{d\Omega_p}{4\pi}\,[V - I^2 \sum_i \tilde{\chi}_{ii}(\vec{p}-\vec{k})]_{|\vec{p}| = |\vec{k}| = k_F} , \tag{6.1}$$

we approximate the gap equation (5.5) as

$$1 = g(T,H)N(0)\int\frac{d\varepsilon}{2E}\,[\tanh\frac{\beta(E+\mu)}{2} + \tanh\frac{\beta(E-\mu)}{2}] \tag{6.2}$$

with $E = (\varepsilon^2 + \Delta^2)^{1/2}$, and where H denotes the internal magnetic field. Note that in the limit $\mu \to 0$ this has the form of BCS gap equation with V being replaced by $g(T,H)$. The definition in (6.1) differs from the usual definition [12] in the fact that $\tilde{\chi}_{ij}$ in (6.1) is not the susceptibility calculated in the normal state, but that calculated for the superconducting state. The effect of $\tilde{\chi}_{ij}$ in (6.1) represents the effect of the localized-spin fluctuations, while the effect of μ in (6.2) is the effect of the electron-spin polarization. Thus, these two effects are factorized in the gap equation (6.2). This is the merit in the use of the effective coupling approximation.

The appearance of a singularity in $\tilde{\chi}(\vec{k})$ at $\vec{k} = \vec{Q}$ and $T = T_p$ leads to the divergence of $g(T,H)$ and the consequent quenching of the superconductivity. Although this divergence can be weakened if, in the original equation (5.5), we relax the assumption that the momenta $\vec{k}$ and $\vec{p}$ are confined to the Fermi surface, the divergence is not entirely removed. To prevent this catastrophe [13], we assume a certain degree of anisotropy in the momentum dependence of the susceptibility and we replace the expression given in (3.11) by

$$\gamma_{ij}(\vec{k}) = \left(\frac{T_m^i}{C} - \frac{D}{C}\,|\vec{k}|^2\{1 + \frac{3}{2}\alpha(1 - \cos^2\theta)\}\right)\delta_{ij} , \tag{6.3}$$

where θ is the angle between $\vec{k}$ and $\vec{Q}$, and α is the anisotropy parameter. Calculations show that with the anistropy effect Δ remains finite even at $T = T_p$ for a certain domain of parameters.

To calculate the right-hand side of (6.2), we must know the spin-dependent energy shift μ. In our numerical calculation we approximately put

$$\vec{\mu} \simeq I\vec{M} . \tag{6.4}$$

The temperature dependence of $g(T;H)/g(T_c;0)$ is presented for several values of the internal magnetic field H (the reduced parameters $t = T/T_c$ and $h = H/(\phi/\lambda_L^2)$ are used in Fig. 6.1). The figure clearly shows the reduction in $g(T;H)$ as the temperature is reduced, reflecting the increase in the spin fluctuations with decreasing temperature. The reduction of the effective coupling constant by the spin fluctuations disappears as the internal field is increased, consistent with the fact that the higher field will tend to suppress the spin fluctuations.

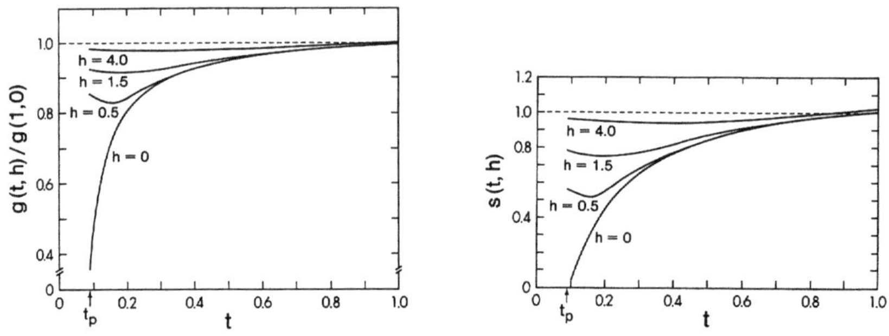

(a) (b)

Fig. 6.1. (a) The temperature dependence of the effective coupling constant $g(t;h)/g_0$ for several values of $h(= H/\phi\lambda_L^{-2})$ (b) The temperature dependence of the scale factor $s(t;h)$ for several values of h

<u>Scaling Law</u>. The equation

$$1 = gN(0)\int\frac{d\varepsilon}{2E}\left[\tanh\frac{\beta(E+\mu)}{2} + \tanh\frac{\beta(E-\mu)}{2}\right] \tag{6.5}$$

has been studied by SARMA [14]. The resultant gap may be expressed in terms of a two-parameter function as

$$\frac{\Delta[T,\mu;g]}{\Delta[0,0;g]} = \mathcal{D}(t_g;\bar{\mu}_g) \ , \tag{6.6}$$

where $\Delta[T;\mu;g]$ is the energy gap and the function $\mathcal{D}(t_g,\bar{\mu}_g)$ is to be determined by

$$\ln\mathcal{D}(t_g,\bar{\mu}_g) = -\ \Phi_1\{\ \frac{\pi}{t_g}\ e^{-\gamma}\mathcal{D}(t_g,\bar{\mu}_g);\ \frac{\pi}{t_g}\ e^{-\gamma}\bar{\mu}\ \}\ , \tag{6.7}$$

γ is Euler's number, $\bar{\mu}_g = \mu/\Delta_g$, Δ_g means $\Delta[0,0;g]$ and Φ_1 is defined by

$$\Phi_1\{x;y\} = \frac{x}{4}\ \mathrm{Re}\int_{-\infty}^{\infty} dz\ \ln(z+\sqrt{z^2-1})\ \cosh^{-2}\frac{1}{2}(xz+y)\ . \tag{6.8}$$

In the above relations, t_g is defined by

$$t_g = T/T_c[g,0]\ , \tag{6.9}$$

where use was made of the symbol $T_c[g,\mu]$ which is the critical temperature determined by (6.5). The quantities Δ_g and $T_c[g,0]$ satisfy the well-known BCS relation;

$$\Delta[0,0;g] = 2\ \omega_D\ \exp\left[-\ \frac{1}{gN(0)}\right] \tag{6.10}$$

$$k_BT_c[g,0] = \frac{e^{\gamma}}{\pi}\ \Delta[0,0;g]\ . \tag{6.11}$$

Note now that the gap equation (6.2) is the same as (6.5) with g being replaced by $g(T,H)$. Thus the energy gap is given by $\Delta[T,\mu;g(T,H)]$. The critical temperature at zero field given by (6.2) is

$$T_c = T_c[g(T_c,0),0]. \tag{6.12}$$

Eq. (6.10) leads to

$$s(t,H) \equiv \frac{\Delta[0,0;g(T,H)]}{\Delta[0,0;g(T_c,0)]} \tag{6.13}$$

$$= \exp\left[\frac{1}{g(T_c,0)N(0)} - \frac{1}{g(T,H)N(0)}\right] \tag{6.14}$$

where $t \equiv T/T_C$. Now, (6.6) leads to

$$\frac{\Delta[T,\mu;g(T,H)]}{\Delta[0,0;g(T_c,0)]} = s(t,H)\,\frac{\Delta[T,\mu;g(T,H)]}{\Delta[0,0;g(T,H)]} = s(t,H)[\mathcal{D}(t_g,\bar{\mu}_g)]_{g\to g(T,H)}. \tag{6.15}$$

The replacement $g\to g(T,H)$ induces the following replacement in (6.7):

$$t_g \to T/T_c[g(T,H),0] = (T/T_c)(T_c[g(T_c,0),0]/T_c[g(T,H),0]) = t/s(t,H) \tag{6.16}$$

$$\Delta_g \to \Delta[0,0;g(T,H)] = s(t,H)\Delta[0,0;g(T_c,0)]\ . \tag{6.17}$$

Therefore, if we define Δ_0 by $\Delta_0 \equiv \Delta[0,0;g(T_C,0)]$ and $\bar{\mu} = \mu/\Delta_0$, (6.15) with (6.7) give the following scaling law:

$$\frac{1}{\Delta_0}\,\Delta[T,\mu;g(T,H)] = s(t,H)\mathcal{D}(t/s(t,H),\bar{\mu}/s(t,H))\ . \tag{6.18}$$

It is remarkable that, according to (6.1) and (6.14), the scale factor $s(t,H)$ does not contain μ, and therefore is not influenced by the electron-spin polarization; the scaling factor exhibits the effect of localized-spin fluctuation. It is for this reason that the spin fluctuation effect will be frequently referred to as the scaling effect. The electron spin polarization effect is expressed by $\mu \simeq I\,M$.

The temperature dependence of the scale factor $s(t;h)$ is presented in Fig. 6.1 for various values of the reduced field h. Since the localized spin fluctuation is enhanced at lower temperature, deviation of $s(t,H)$ from unity tends to increase with decreasing temperature. Since this spin fluctuation decreases with increasing field, $s(t,H)$ approaches unity with increasing H. This behavior can be seen from Fig. 6.1.

The London penetration depth $\lambda_L(T;H)$ may be similarly calculated and the resultant formula is given by

$$\frac{\lambda_L^2(0,0)}{\lambda_L^2(T,H)} = 1 + \int_0^\infty d\varepsilon\left[\frac{\partial}{\partial E}\,f_F(E+\mu) + \frac{\partial}{\partial E}\,f_F(E-\mu)\right]\ , \tag{6.19}$$

where $1/\lambda_L^2(0,0) = (8\pi/3c^2)e^2v_F^2N(0)$. When $\mu = 0$, (6.19) reduces to the well-known formula in the BCS theory. It follows from (6.19) that

$$\frac{\lambda_L(0,0)}{\lambda_L(T,H)} = \mathcal{L}^{-1}(t/s(t,H),\ \bar{\mu}/s(t,H)). \tag{6.20}$$

This is the scaling law for the London penetration depth. Calculation shows that $\mathcal{L}$ is given by

$$\mathcal{L}^{-2}(t,\bar{\mu}) = 1 - \Phi_2\{\frac{\pi e^{-\gamma}}{t}\mathcal{D}(t,\bar{\mu}), \frac{\pi e^{-\gamma}}{t}\bar{\mu}\} \quad \text{with} \tag{6.21}$$

$$\Phi_2(x,y) = \frac{x}{4}\,\mathrm{Re}\int_{-\infty}^{\infty} dz\,\frac{z}{\sqrt{z^2-1}}\cosh^{-2}\frac{1}{2}(xz+y). \tag{6.22}$$

7. The Maxwell Equation and the Vortices

The form factor $c(\vec{k})$ in the Meissner term in the Maxwell equation (3.3) can be obtained from the calculation of the photon self-energy diagram [15, 16]. The results of a numerical calculation of $c(\vec{k})$ have been presented in [17]. In this work it was shown that to a reasonable degree of accuracy the results could be assimilated in the following functional form for $c(\vec{k})$:

$$c(\vec{k}) = \exp[-\xi^2(t)\vec{k}^2/2]. \tag{7.1}$$

Here $\xi(t)$ with $t = T/T_c$ is the coherence length at T. When we replace the BCS coupling V by $g(T,H)$, $\xi(t)$ in (7.1) becomes $\xi(t,h) = \hbar v_F/\pi\Delta(t,h)$, where $\Delta(t,h)$ is the energy gap $\Delta(T,\mu;g(t,H))$ for a magnetic superconductor which can be calculated using the gap equation presented in the last section.

When there exist some topological objects such as vortices, the Maxwell equation in the static cases becomes [15, 16]

$$\vec{\nabla}\times\vec{B} = \frac{1}{\lambda_L^2}c(-i\vec{\nabla})[\vec{A} - (\hbar c/e)\vec{\nabla}f] + 4\pi\vec{\nabla}\times(\vec{M} - \mu_B\vec{\sigma}) . \tag{7.2}$$

Here λ_L means $\lambda_L(T,H)$ and f is the phase of electron field. When we solve this for $\vec{B}$, (3.2) gives the following molecular field:

$$\vec{H}_m = \gamma\vec{M} + 4\pi F(-i\vec{\nabla})(\vec{M} - \mu_B\vec{\sigma}) + \frac{\Phi}{\pi}F(-i\vec{\nabla})\vec{\nabla}\times\vec{\nabla}f . \tag{7.3}$$

When f is a single-value function, $\vec{\nabla}f$ in (7.3) does not contribute to a static magnetic field because $\vec{\nabla}\times\vec{\nabla}f = 0$. However, when there appears a certain topological object such as a vortex, f becomes multivalued, leading to $\vec{\nabla}\times\vec{\nabla}f \neq 0$ and hence f does contribute to the static magnetic field. For example, in the case of a single vortex, we have $f = \theta/2$ where θ is the cylindrical angle around the vortex center which gives $\vec{\nabla}\times\vec{\nabla}f = \pi\vec{e}\delta^2(x)$ and $\nabla^2 f = 0$. Here $\vec{e}$ is the unit vector directed along the vortex line. When we consider a system of vortices lying along the z axis, we have

$$\vec{\nabla}\times\vec{\nabla}f = \pi\vec{e}_z\sum_j\delta^2(\vec{x}-\vec{\xi}_j) , \tag{7.4}$$

where $\vec{x}$ is the two-dimensional position vector and $\vec{\xi}_j$ denotes the position of j-th vortex.

The second term in (7.3) shows the shielding effect due to the persistent current mentioned above. The last term indicates that the presence of topological objects such as vortices or surface topological singularities tend to weaken the shielding effect. This effect tends to induce a kind of ferromagnetic state in the region (with the range $\simeq \lambda_L$) around the topological singularities, defined by $\vec{\nabla}\times\vec{\nabla}f \neq 0$; this is a local ferromagnetic state coexisting with the superconductivity and may arise in the case of a vortex or in the vicinity of a surface.

A System of Vortices. When we have a system of vortices such as a vortex lattice with vortex density n, we have $\vec{M} = \delta\vec{M} + \vec{m}_0$ ($\vec{m}_0$ = space average) and

$$\vec{n}(x)\phi \equiv \frac{\hbar c}{e}\vec{\nabla}\times\vec{\nabla}f = \phi\vec{e}_z\sum_j \delta(\vec{x}-\vec{\xi}_j). \tag{7.5}$$

Note that $\vec{M}$ in (7.2) can be replaced by $\delta\vec{M}$, because $\vec{m}_0$ is independent of $\vec{x}_0$. Then we obtain from (3.1)

$$\delta\vec{M} = \chi(-i\vec{\nabla})\delta\vec{H} = \frac{\chi(-i\vec{\nabla})}{1+4\pi\chi(-i\vec{\nabla})}\delta\vec{B} \tag{7.6}$$

where use was made of $\delta\vec{B} = (1+4\pi\chi)\delta H$. If we neglect the σ contribution to equation (7.2) we obtain

$$B(x) = \frac{[1+4\pi\chi(-i\nabla)]c(-i\nabla)}{-\lambda_L^2\nabla^2 + [1+4\pi\chi(-i\nabla)]c(-i\nabla)}\,n(x)\phi = \sum_j b_s(\vec{x}-\vec{\xi}_j), \tag{7.7}$$

where $\chi(\vec{k})$ is given by (3.12) with $\tilde{\gamma}$ being replaced by γ:

$$\chi(\vec{k}) = C\alpha_J/[T - C\alpha_J\gamma] = \frac{C\alpha_J}{T - T_m + D\vec{k}^2} \quad \text{and} \tag{7.8}$$

$$b_s(\vec{x}) = \frac{\phi}{(2\pi)^2}\int d^2k\, e^{i\vec{k}\vec{x}}\,\frac{[1+4\pi\chi(\vec{k})]c(\vec{k})}{\lambda_L^2k^2 + [1+4\pi\chi(\vec{k})]c(\vec{k})}\;. \tag{7.9}$$

Note that, according to (7.7), $\vec{B}$ has the rather desirable feature that it is a sum of terms, each of which damps exponentially at large distances. Equation (7.7) differs from the one for nonmagentic cases by the replacement

$$\lambda_L \to \lambda_L/[1+4\pi\chi]\,. \tag{7.10}$$

Vortex Core Energy. The electron wave function near the vortex center is influenced by the vortex. Very roughly, one may think of the electrons as being influenced by a potential which may exhibit a spatial dependence rather similar to that of the order parameter $\Delta(\vec{x})$. Since $\Delta(\vec{x})$ vanishes around the vortex center, the overall effect of $\Delta(\vec{x})$ is an increase in the electronic energy accumulated in the vortex core. This is called the core energy. In calculating the mixed state energy, we must take into account this core energy. As for the calculation of the core energy, see refs. [15,16,18 and 19].

8. The Free·Energy

In the calculation of the free energy it is convenient to separate the Hamiltonian (2.1) into electronic, magnetic and electromagnetic terms;

$$\mathcal{H}(\vec{x}) = \mathcal{H}_{el}(\vec{x}) + \mathcal{H}_M(\vec{x}) + \mathcal{H}_{EM}(\vec{x}) \quad \text{where} \tag{8.1}$$

$$\mathcal{H}_{el} = \mathcal{H}_{BCS} - \frac{1}{2}(I\vec{M} - \mu_B\vec{B})\vec{\sigma} \tag{8.2}$$

$$\mathcal{H}_M = -\frac{1}{2}\vec{M}\gamma_0\vec{M} - \frac{1}{2}(I\vec{\sigma} + \vec{B})\vec{M} \tag{8.3}$$

$$\begin{aligned}\mathcal{H}_{EM} &= \frac{1}{8\pi}[\vec{B}^2 + (A_i - \frac{\hbar c}{e}\nabla_i f)\Sigma_{ij}(-i\vec{\nabla})(A_j - \frac{\hbar c}{e}\nabla_j f)] - \frac{1}{2}(\vec{M} - \mu_B\vec{\sigma})\vec{B} \\ &= \frac{1}{8\pi}[\vec{B}^2 + \frac{1}{\lambda_L^2}(\vec{A} - \frac{\hbar c}{e}\vec{\nabla}f)c(-i\vec{\nabla})(\vec{A} - \frac{\hbar c}{e}\vec{\nabla}f) - 4\pi\vec{B}(\vec{M} - \mu_B\vec{\sigma})].\end{aligned} \tag{8.4}$$

Here we are considering a static situation. The space integration of the expectation value of each of these three Hamiltonians gives respective internal energy. These internal energy densities will be denoted by U_{el}, U_M and U_{EM}. Subtracting from these internal energy densities $\beta^{-1}S_{el}$, $\beta^{-1}S_M$ and $\beta^{-1}S_{EM}$ (S means entropy) respectively, we obtain the free energies F_{el}, F_M and F_{EM}. A detailed account showing the calculation of these free energies for the superconducting state (F^s) and for the normal state (F^n) will be published elsewhere. We now introduce the field-dependent critical field $H_c(T,H)$ by means of the condensation energy which is given by

$$\frac{1}{8\pi} H_c^2(T,H) \equiv F_{el}^n - F_{el}^s = 2N(0) \int_0^{\omega_D} d\varepsilon \Big[\varepsilon - E + \frac{\Delta^2}{2E} - \frac{E^2+\varepsilon^2}{2E} \{f_F(E-\mu) +$$

$$+ f_F(E+\mu)\} + 2\,\varepsilon f_F(\varepsilon) + \frac{\mu}{2} \{f_F(E-\mu) - f_F(E+\mu)\}\Big] \; . \tag{8.5}$$

If we now shift the zero point of the electronic energy to the normal electronic free energy, then the free energy of the superconducting state is given by

$$F^s = -\frac{1}{8\pi} H_c^2(T,H) + F_M^s + F_{EM}^s + E_{core} \tag{8.6}$$

$$= -\frac{1}{8\pi} H_c^2(T,H) + \frac{1}{2} \vec{H}_m \cdot \vec{M} - \beta^{-1} N \log Z_J[g_J J \mu_B \beta |\vec{H}_m|] + \frac{1}{8\pi} n\phi\, \vec{e}_z \cdot \vec{H} + E_{core},$$

where E_{core} is the core energy mentioned previously.

The normal state free energy is now given by

$$F^n = \frac{1}{2} \vec{H}_m \cdot \vec{M} - \beta^{-1} N \log Z_J[g_J J \mu_B \beta |\vec{H}_m|] + \frac{1}{8\pi} \vec{B} \cdot \vec{H} \; . \tag{8.7}$$

Condensation Energy and the Scaling Law. The condensation energy in the Meissner state is given by $(1/8\pi)H_c^2(T,0)$ which can be measured by means of the magnetization curve.

The condensation energy (with any H) is given by (8.5), and can be calculated by the following scaling law:

$$\frac{H_c(T,H)}{H_c(0,0)} = s(t,H)\mathcal{H}[t/s(t,H),\bar{\mu}/s(t,H)] \quad \text{with} \tag{8.8}$$

$$\mathcal{H}^2(t,\bar{\mu}) \equiv \mathcal{D}(t,\bar{\mu})^2 - \frac{2\pi^2}{3}\Big[\frac{t}{\pi e^{-\gamma}} - \frac{3}{\pi^2}\mathcal{D}^2(t,\bar{\mu})\Phi_3\{\frac{\pi e^{-\gamma}}{t}\mathcal{D}(t,\bar{\mu}), \frac{\pi e^{-\gamma}}{t}\bar{\mu}\}$$

$$- \frac{3\bar{\mu}}{\pi^2}\mathcal{D}(t,\bar{\mu})\Phi_4\{\frac{\pi e^{-\gamma}}{t}\mathcal{D}(t,\bar{\mu}), \frac{\pi e^{-\gamma}}{t}\bar{\mu}\}\Big], \tag{8.9}$$

$$\Phi_3(x,y) = \frac{x}{4} \,\mathrm{Re} \int_{-\infty}^{\infty} dz\; z\sqrt{z^2-1} \cosh^{-2} \frac{1}{2}(xz+y) \tag{8.10}$$

$$\Phi_4(x,y) = \frac{x}{4} \,\mathrm{Re} \int_{-\infty}^{\infty} dz\; \sqrt{z^2-1} \cosh^{-2} \frac{1}{2}(xz+y) \; . \tag{8.11}$$

The scaling factor $s(t,H)$ is that given in Eq. (6.14).

9. The Critical Fields and the Magnetization Curves

Type I Phase. The critical field for the type I phase is not $H_c(T,0)$, because we must compare the Gibbs free energy of the normal state and the condensation energy. Therefore, the critical field for the type I phase is denoted by H_c^* which is given by

$$\frac{1}{8\pi} H_c^{*2} = \frac{1}{8\pi} H_c^2(T,0) + G^n , \tag{9.1}$$

where G^n is Gibbs free energy of the normal state given by $G^n = F^n - \frac{1}{4\pi} BH$ with F^n in (8.7), $H_m = H_c^* + \gamma M$ and $B = H_c^* + 4\pi M$.

Type II Phase: The Magnetization Curve and Critical Fields. To calculate the magnetization curve for the type II phase, we first presume a triangular lattice of vortices. Then the free energy F^s in (8.6) becomes a function of the vortex density. Since $F^s(n)$ reduces to $H_c^2/8\pi$ at $n = 0$, we write the free energy as

$$F^s = \frac{n}{8\pi} g(n) - \frac{H_c^2}{8\pi} . \tag{9.2}$$

The quantity $g(n)$ may be regarded as the effective energy of each vortex. The Gibbs free energy is $G_s(n) = F^s(n) - (1/4\pi)BH$. The magnetic field H is then determined as a function of n from the condition that $\frac{\partial}{\partial n} G_s(n) = 0$ which gives $H(n) = (1/2)[1 + n(\partial/\partial n)]g(n)$. The magnetization is given by $4\pi M(n) = n\phi - H(n)$. The Gibbs free energy of the normal state is given by $G_n = F^n - (1/4\pi)B\,H$, where F^n was given in (8.7) with $\vec{H}_m = \vec{H} + \gamma\vec{M}$.

We determine H_{c2}^o by the following two conditions:

$$G_s(n_c) = G_n(H_{c2}^o) \tag{9.3}$$

$$\left.\frac{\partial G_s}{\partial H}\right|_{H_{c2}^o} = \left.\frac{\partial G_n}{\partial H}\right|_{H_{c2}^o} \tag{9.4}$$

with n_c being defined by

$$H(n_c) = H_{c2}^o . \tag{9.5}$$

The magnetization curve is obtained from $H(n)$ and $M(n)$.

The condition (9.4) means that two free energies, G_s and G_n, coincide tangentially at $H = H_{c2}^o$. In the case where $G_s \neq G_n$ in the domain $n < n_c$, H_{c2}^o may be identified as the upper critical field H_{c2}. However, when $G_s = G_n$ occurs in the domain $n < n_c$, the magnetization curve for H in the vicinity of H_{c2} behaves in the manner illustrated in Fig. 9.1. In this case the phase transition occurs at H_{c2} which is different from H_{c2}^o and the transition becomes first order.

We introduce H_{c1}^o by $H_{c1}^o = H(n=0)$. Then $H_{c1} = H_{c1}^o$ for the second-order transition, while for the first-order transition the magnetization at H around H_{c1} behaves as in Fig. 9.2. In this case H_{c1} is slightly different from H_{c1}^o. In this paper we approximately use H_{c1}^o for H_{c1}. The first-order phase transition at $H = H_{c1}$ is a consequence of an attractive vortex-vortex interaction.

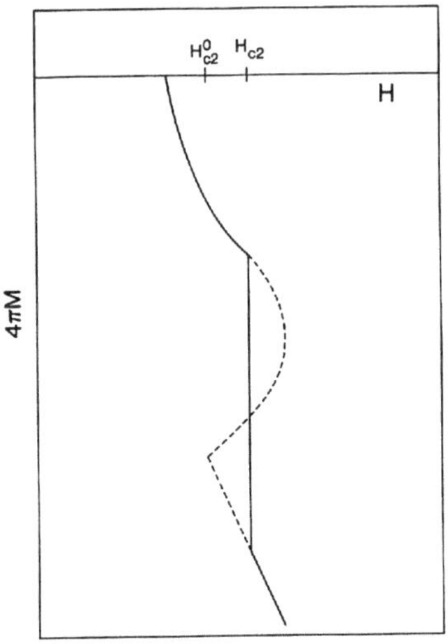

Fig. 9.1. Schematic illustration of the magnetization curve in the neighborhood of H_{c2} for a type $II_{i,1}$ superconductor

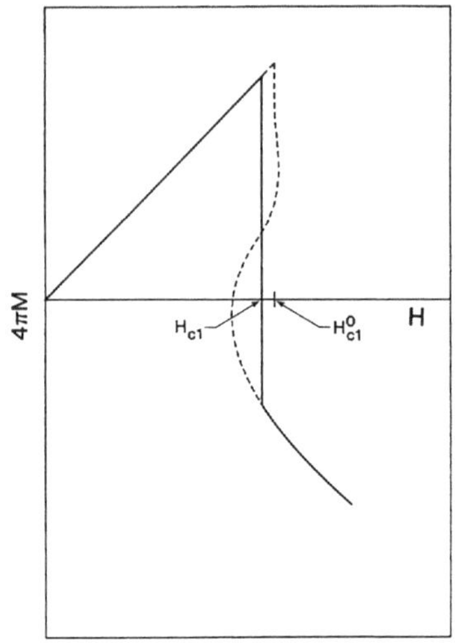

Fig. 9.2. Schematic illustration of the magnetization curve in the neighborhood of H_{c1} for a type $II_{1,j}$ superconductor

First-order transitions at H_{c1} or H_{c2} seem to occur frequently in the case of magnetic superconductors at low temperature. The situation regarding the nature of the transition at H_{c1} may be understood roughly as follows. The quantization of the magnetic flux associated with a single vortex means that an increase in local magnetization density M must be accompanied by a corresponding decrease in the magnetic field H, since B, M and H are related through $B = H + 4\pi M$. Thus, if the polarization of the localized moments in the vicinity of a vortex is sufficiently strong, then the magnetic field may reverse its direction, inducing an attractive interaction between the vortices.

We can understand this in another way. The expression (7.7) for $\vec{B}$ given by the Maxwell equation differs from that for nonmagnetic superconductors only by the presence of $[1+4\pi\chi]$ factor. Thus, $\vec{B}(\vec{x})$ behaves effectively as the nonmagnetic one with λ_L being replaced by $\lambda_L/\sqrt{1+4\pi\chi}$. This suggests that the magnetic induction B in a magnetic superconductor will behave in a manner similar to that of a nonmagnetic superconductor with effective κ given by $\kappa/[1+4\pi\chi]^{1/2}$. As a matter of fact, it was shown in ref. [18] (see also ref. [19]) that, when d-f interaction is disregarded, the magnetic induction is effectively obtained by the replacement $\kappa \to \kappa[1+4\pi\chi]^{-1/2}$ provided T is not too close to T_p or T_{c2}. Since χ increases with decreasing T, κ_{eff} becomes smaller at lower temperature, causing the first-order transition at H_{c1} (i.e. sudden drop in B at H_{c1}). On the other hand, as we will demonstrate later, the first-order nature of the transition at H_{c2} seems to be strongly related to the effect of the polarization of the electron spins by the localized moments through the d-f interaction.

When the first-order transition occurs only at H_{c1}, we call the system type II/1-2. If on the other hand the system shows a first-order transition only at H_{c2} then we refer to it as type II/2-1. These names were given by Crabtree et al. in ref. [20]. Extending this nomenclature, we define type

II/i-j (i,j = 1,2) in the following way; i = 1 (j = 1) means the first-order transition at H_{C1} (H_{C2}) while i = 2 (j = 2) means the second-order transition at H_{C1} (H_{C2}). We will return to this point in later sections.

10. Quantitative Analysis of the Magnetic Shielding, Spin Fluctuations and Electron Polarization

As has been emphasized at several points in the course of our discussion the behaviors of magnetic superconductors may be understood in terms of the following three mechanisms.

(a) The shielding of localized spins by the persistent current. This changes the exchange coupling from γ to $\tilde{\gamma}$, and creates the spin-periodic phase in $ErRh_4B_4$ and $HoMo_6S_8$. As it was pointed out earlier, provided T is not close either to T_D or T_{C2}, the system behaves as a nonmagnetic one with $\kappa_{eff} = \kappa/\sqrt{1+4\pi\chi}$ when the d-f interaction is ignored. This κ_{eff} becomes smaller when χ increases at lower temperature. This may tend to make the system type II_{1-j} (j = 1 or 2) or type I. This usually creates the transition type II_{2-j} → type II_{1-j} → type I with decreasing temperature. However, type I → type II_{1-j} is also possible in situations where the saturation effect reduces the susceptibility χ in applied field at lower temperature. The d-f interaction can, however, modify these situations.

The pure electromagnetic effect is obtained in our formalism by putting $s(t,H) \to 1$ and $\mu \to 0$ where s(t,H) is the scaling factor (6.14) and μ is the energy gap due to polarization of the electron spin. In the pure electromagnetic situation, T_C should be the same as that obtained in the nonmagnetic case. Furthermore $B_{C2} = n_C\phi$ vs T curve is also roughly the same as that obtained in the case of a nonmagnetic superconductor, although H_{C2} differs from that obtained in the case of a nonmagnetic superconductor, because $H_{C2} = B_{C2} - 4\pi M_{C2}$. This tends to result in a bell-shape curve if H_{C2} is plotted as a function of T [19].

Since in the high temperature regime ($T \simeq T_C$), s(t,H) approaches unity and μ decreases, the pure electromagnetic situation may be realized to a good approximation. Therefore, in this domain $\kappa_{eff} = \kappa/[1+4\pi\chi]^{1/2}$ may be a very good approximation. This provides us with a useful way for determining κ. Using this method Berhoozi et al. determined κ for a single crystal of $ErRh_4B_4$; κ was found to be about 4 [21].

(b) The localized spin fluctuation effect. In our formalism this effect was expressed by the scaling factor s(t,H). Thus, this effect will be referred to as the scaling effect. The situation in which the electromagnetic and scaling effect is dominant over the polarization effect is obtained by putting $\mu = 0$. This situation will be called the scaling dominant case.

In Fig. 10.1 we present a curve showing the temperature dependence of the condensation energy $H_C^2(t;0)$. The upper and lower critical fields H_{C2} and H_{C1} calculated in the scaling dominant limit for the easy axis are shown in Fig. 10.2 and the upper critical field for the hard axis in Fig. 10.3. The parameters are chosen in order to obtain a reasonable degree of agreement with the experimental values of H_{C2} measured along the easy axis. Fig. 10.4 shows a typical magnetization curve and it should be noted that it does not give any indication of a 1st order phase transition at H_{C2}.

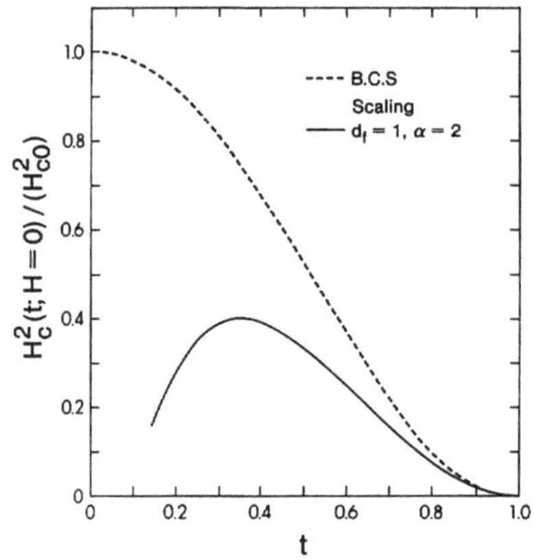

Fig. 10.1. The condensation energy $H_c^2(t;n=0)/H_{c0}^2$ calculated, including the effect of the scaling, as a function of reduced temperature t for $d_f = 1$, $\alpha = 2$ and the parameters shown in Table 2, together with the B.C.S. result

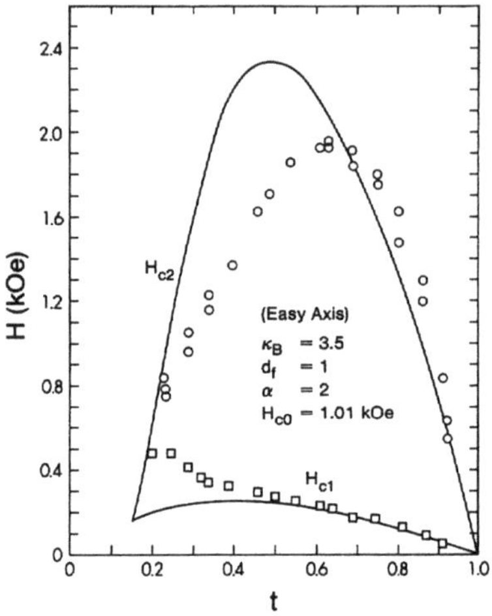

Fig. 10.2. The easy axis upper and lower critical fields, H_{c_2} and H_{c_1}, calculated including the effect of the scaling and the electromagnetic interaction but neglecting the effect of the spin splitting of the electrons (i.e., $\mu = 0$) as a function of temperature. Here the parameters used are those given in Table 2 together with $d_f = 1$, $\alpha = 2$, $(I^2/g(T_c) = 0.9067)$, $\kappa_B = 3.5$ and $H_{co} = 1.01$ KOe. Also shown are the experimental measurements after Crabtree et al.[22]

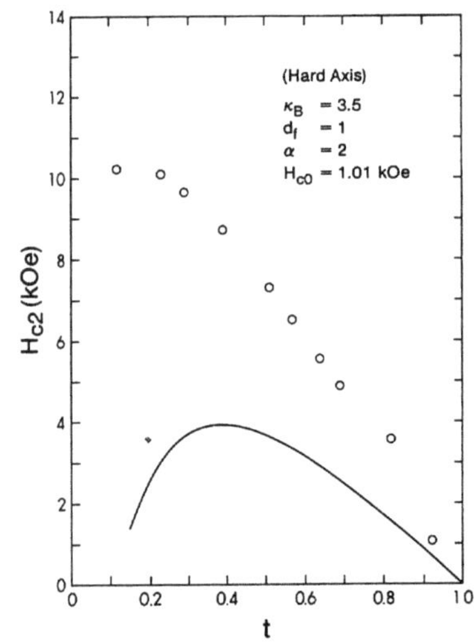

Fig. 10.3. The hard axis upper critical field, H_{c_2}, calculated as a function of temperature, including the effect of the scaling and the electromagnetic interaction but neglecting the effect of the spin splitting of the electrons (i.e. $\mu = 0$). Here the parameters used are those given in Table 2 together with $d_f = 1$, $\alpha = 2$, $(I^2/g(T_c) = 0.9067)$, $\kappa_B = 3.5$ and $H_{co} = 1.01$ KOe. Also shown are the experimental measurements after Crabtree et al.[22]

The comparison of these curves with the single-crystal $ErRh_4B_4$ measurements indicates that the magnetic properties of this material may not be explained solely on the basis of the scaling effect and that we need to consider a considerable amount of electron spin polarization.

(c) The electron spin polarization. This effect arises from the nonvanishing of the spin-splitting μ. When the effect of μ dominates over that induced by the scaling $s(t;H)$, the situation is referred to as the polarization

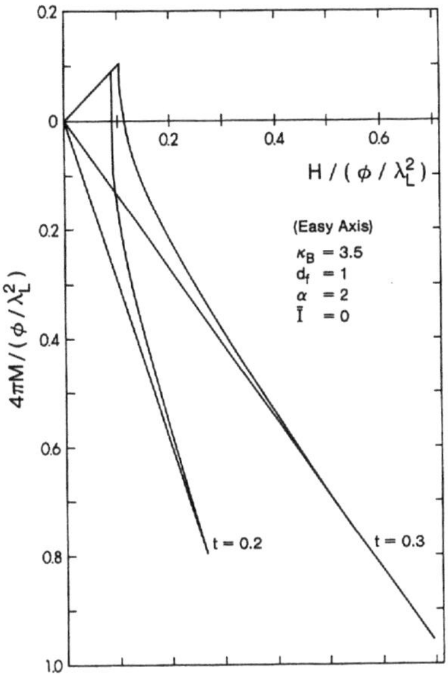

Fig. 10.4. Typical easy axis magnetization curves calculated, including the effect of the scaling and the electromagnetic interaction but neglecting the spin splitting of the electrons (i.e., $\mu = 0$). Here the parameters used are those presented in Table 2 together with $d_f = 1$, $\alpha = 2$, $(I^2/g(T_c) = 0.9067)$, $\kappa_B = 3.5$ and $H_{co} = 1.01$ KOe

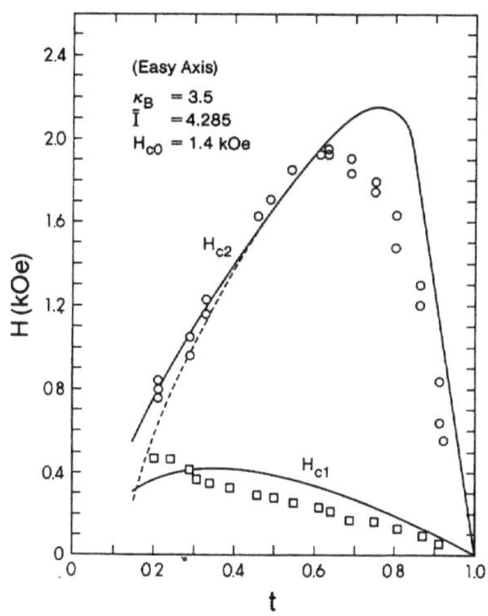

Fig. 10.5. The easy axis upper and lower critical fields, H_{c2} and H_{c1}, calculated as a function of temperature, including the effect of the spin splitting of the electron and the electromagnetic interaction but neglecting the scaling effect (i.e., $s = 1$). Here the parameters used are those presented in Table 2, together with $\bar{I} = 4.285$, $\kappa_B = 3.5$ and $H_{co} = 1.4$ KOe. The dotted portion represents H^0_{c2}. Also shown are the experimental measurements after Crabtree et al. [22]

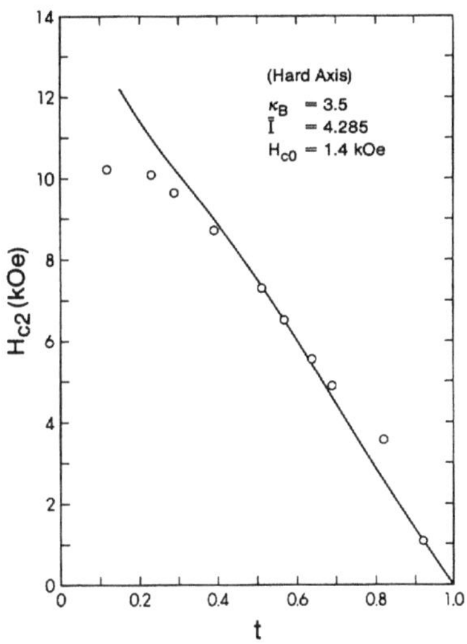

Fig. 10.6. The hard axis upper critical fields, H_{c2}, calculated as a function of temperature including the effect of the spin splitting of the electrons and the electromagnetic interaction but neglecting the scaling effect (i.e., $s = 1$). Here the parameters chosen are those presented in Table 2 together with $\bar{I} = 4.285$, $\kappa_B = 3.5$ and $H_{co} = 1.4$ KOe. Also shown are the experimental measurements after Crabtree et al. [22]

dominant case. This situation is given by putting $s(T;H) = 1$. The upper and lower critical fields H_{c1} and H_{c2} calculated in the polarization dominant case are shown for the easy axis in Fig. 10.5 and for the hard axis in Fig. 10.6. Fig. 10.7 shows several typical magnetization curves. The observed magnetization curves for $ErRh_4B_4$ seems to be similar to those presented in this limit.

11. Application to $ErRh_4B_4$

In this section, we employ the method of analysis outlined in the preceding sections to consider the magnetic properties of single-crystal $ErRh_4B_4$. Let us begin with a discussion of the parameters. In Table 1 we list the available experimental values. The values given within parentheses should be regarded as temporary. The values of C and $4\pi M$ are calculated from the observed lattice constant [23] for $RERh_4B_4$ together with J and g_J which are assumed to have the same value as the Er-free ion. The Curie temperatures $T_m^{(a)}$ and $T_m^{(c)}$ are obtained from an extrapolation of the inverse susceptibility measurements measured in the normal state with respect to the a axis (easy) and c axis (hard) respectively [24]. The term $dT_c(x)/dx$ refers to the reduction of the superconducting transition temperature for the pseudo-ternary compound $Er_{1-x}Y_xRh_4B_4$ [25]. The value of T_p is identified by an extrapolation of the satellite intensity distribution obtained from the neutron scattering [8],neglecting the tail part.

For the theoretical analysis we require the nondimensional parameters listed in Table 2; these may be calculated from the parameters presented in Table 1. The value of d is estimated from T_p, $T_m^{(a)}$ and C by means of the

Table 1. Experimental values for $ErRh_4B_4$. The values in parentheses are temporary.

T_{c1} = 8.7 K	$J = 15/2$
T_{c2} = 0.7 K	$g_J = 6/5$
T_m^a = (1.0 K)	$C = 0.184$ K
T_m^c = (-20.0 K)	$4\pi M = 10.11$ KOe
T_p = (0.8 K)	$\left.\frac{dT_c(x)}{dx}\right\|_{x=o} = (0.14)$

Table 2. Nondimensional parameters used in the theoretical calculation

$t_m^a = (T_m^a/T_c) = 0.115$	$g(T_c)N(0) = 0.3$
$t_m^c = (T_m/T_c) = -2.300$	$d = (\frac{D}{T_m\lambda_L^2}) = 0.452 \times 10^{-2}$
$t_p = (T_p/T_c) = 0.092$	$J = 7.5$
$c = (4\pi C/T_m^a) = 2.312$	

Parameters d_f, u and $\bar{I}$ are defined as

$$d_f = 4Dk_f^2/T_m^a$$

$$u = M/(\phi/\lambda_{Lo}^2)$$

$$\bar{I} = IM/\Delta_0 = \left(\frac{8\pi^5}{3}\, g(T_c)N(0)\, \frac{I^2}{g(T_c)}\right)^{1/2} \frac{u}{\kappa_B}$$

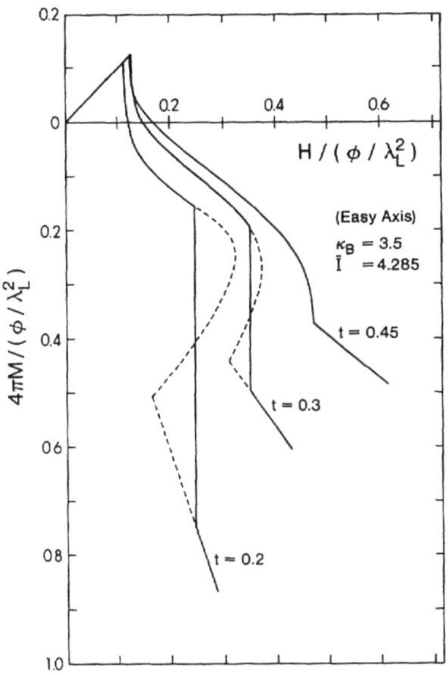

Fig. 10.7. Typical easy axis magnetization curves calculated including the effect of the spin splitting of the electrons and the electromagnetic interaction but neglecting the effect of the scaling (i.e., s = 1). The solid line denotes the physical magnetization curve while the dotted line represents the unstable super-cooled portion. Here the parameters chosen are those presented in Table 2 together with $\bar{I} = 4.285$, $\kappa_B = 3.5$ and $H_{co} = 1.4$KOe.

result obtained in the London limit that $T_p = T_m^{(a)}[1 + d - 2\sqrt{dc}]$ and $g(T_c)N(0)$ is chosen to have the value 0.3 which is the typical value for weak super-conductors.

When α, d_f and u are fixed, then $I^2/g(T_c)$ may be calculated from $(dT_c(x)/dx)/T_c$ and $\bar{I}$ may be obtained for a given value of κ_B. The parameter u (and therefore ϕ/λ_{Lo}^2) is estimated from the thermodynamic critical field, H_{co}, of $LuRh_4B_4$, which is reported to be 1.85 KOe [26]. Using the relation $T_c(Lu)/T_c(Er) = H_{co}(Lu)/H_{co}(Er)$, we have a value of $H_{co}(Er) = 1.40$ KOe. By means of the relation $H_{co}^2 = \frac{3}{2}\frac{\kappa_B^2}{\pi^4}\left(\frac{\phi}{\lambda_{Lo}^2}\right)^2$ we obtain for a given κ_B an estimate of ϕ/λ_{Lo}^2 and therefore u. The parameters α and d_f cannot as yet be determined from experimental values and are therefore adjusted to provide a reasonable fit to the data.

In this analysis we assume that the observed anisotropy of the magnetic properties in the mixed state may be attributed completely to the anisotropic nature of the localized spin system. We therefore try to explain the various experimental measurements such as the condensation energy, the upper and lower critical fields measured with respect to both the hard and easy axes and the easy axis magnetization curves with a single set of parameters which are determined in the way mentioned above.

The temperature dependence of the condensation energy is presented in Fig. 11.1 for $d_f = 20$ and $\alpha = 5$ together with the B.C.S. result. The results clearly show the suppression of the condensation energy due to the spin fluctuations. It is worth noting that our calculations show that once α and d_f are larger than a certain value ($\sim$1 say), then their effect on the scaling behavior is determined largely through the combination of parameters used to determine $(dT_c(x)/dx)_{x=o}/T_c$ and hence since this is taken as an input parameter in our calculation the dependence on α and d_f is rather mild. This

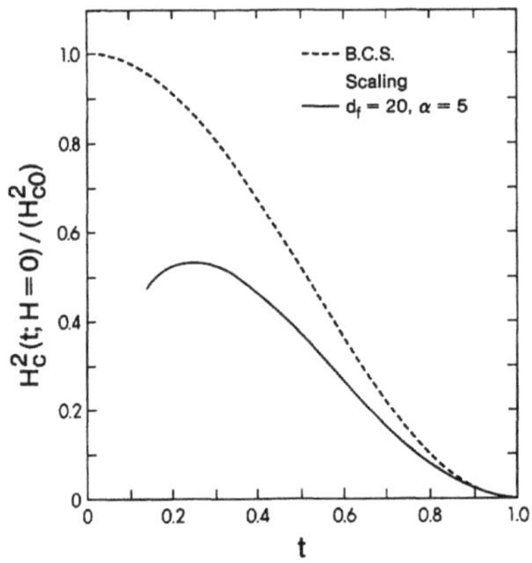

Fig. 11.1. The condensation energy $H_c^2(t;n=0)/H_{c0}^2$ calculated, including the effect of the scaling, as a function of the reduced temperature t for $d_f=20$, $\alpha=5$, $(I^2/g(Tc)=10.1466)$, together with the parameters presented in Table 2. Also shown is the B.C.S. result

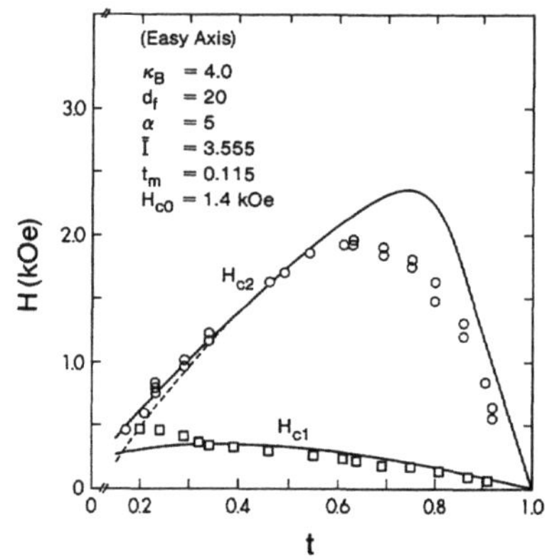

Fig. 11.2. The easy axis upper and lower critical fields, H_{C2} and H_{C1}, calculated as a function of temperature, including the effect of both the scaling and the electron spin polarization together with the effect of the electromagnetic interaction. Here the parameters chosen are those presented in Table 2 together with $\kappa_B = 4.0$, $d_f = 20$, $\alpha = 5$, $(I^2/g(Tc) = 10.1466)$ and $H_{CO} = 1.4$ KOe. The dotted portion represents H_{C2}^0. Also shown are the experimental measurements after Crabtree et al. [22]

means that the main effect of increasing α and d_f beyond a certain point is the subsequent change induced in $\bar{I}^2/g$ (and therefore $\bar{I}$) in order that the calculated value of $(dT_C(x)/dx|_{x=0})/T_C$ be consistent with that estimated from the experiment.

Presented in Fig. 11.2 are the upper and lower critical fields calculated with respect to the easy axis. The parameters used are those presented in Table 2 together with $\kappa_B = 4$; $\alpha = 5$, $d_f = 20$ (which gives $\bar{I} = 3.555$) and $H_{CO} = 1.4$ KOe. The corresponding results for the hard axis are presented in Fig. 11.3. Also shown in Figs. 11.2 and 11.3 are the recent experimental measurements of Crabtree et al. [22] on single-crystal $ErRh_4B_4$. The agreement with the calculated values is reasonably good. With the same parameters several easy axis magnetization curves are illustrated in Fig. 11.4. Using the nomenclature of Crabtree et al. [20] the theoretical curves show the following sequence of phase transitions as the temperature is lowered,

type $II_{2,2}$ $\rightarrow$ type $II_{1,2}$ $\rightarrow$ type $II_{1,1}$ $\rightarrow$ type I.

The calculated jump in the magnetization at H_{C_2} $(=\Delta M_{II})$ is shown in the lower curve in Fig. 11.5 together with several experimental points [20]. We see that while the temperature at which the jump at H_{C2} appears seems to agree with experiment the overall magnitude is rather low.

It is interesting to compare and contrast the results presented in this section with those presented in the preceding section. From the results presented in Fig. 10.2 we see that while it is possible to account quali-

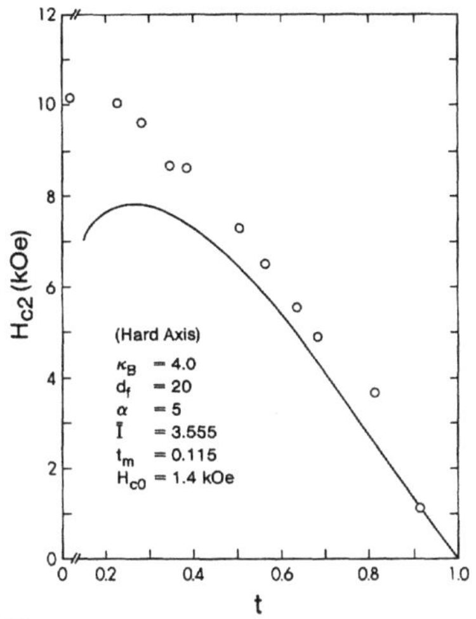

Fig. 11.3. The hard axis upper critical fields H_{c2}, calculated as a function of temperature including the effect of both the scaling and the electron spin polarization together with the effect of the electromagnetic interaction. Here the parameters chosen are those presented in Table 2 together with $\kappa_B = 4.0$, $d_f = 20$, $\alpha = 5$, $(I^2/g(T_c) = 10.1466)$, and $H_{co} = 1.4$ KOe. Also shown are the experimental measurements after Crabtree et al. [22]

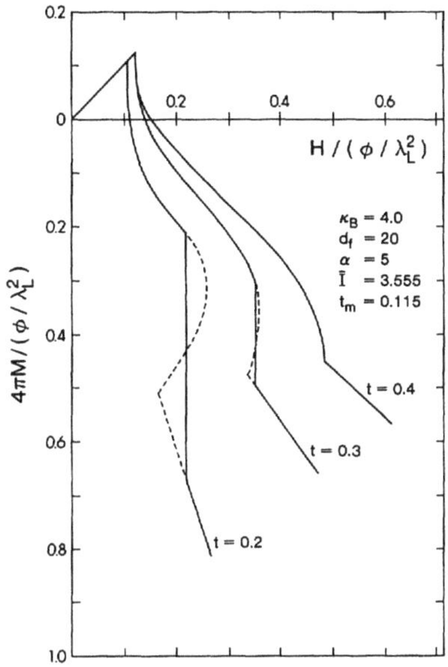

Fig. 11.4. Typical easy axis magnetization curves calculated including the effect of both the scaling and the electron spin polarization together with the effect of the electromagnetic interaction. Here the parameters chosen are those presented in Table 2 together with $\kappa_B = 4.0$, $d_f = 20$, $\alpha = 5$, $(I^2/g(T_c) = 10.1466)$ and $H_{co} = 1.4$ KOe. The solid line denotes the physical magnetization curve while the dotted line shows the unstable supercooled portion

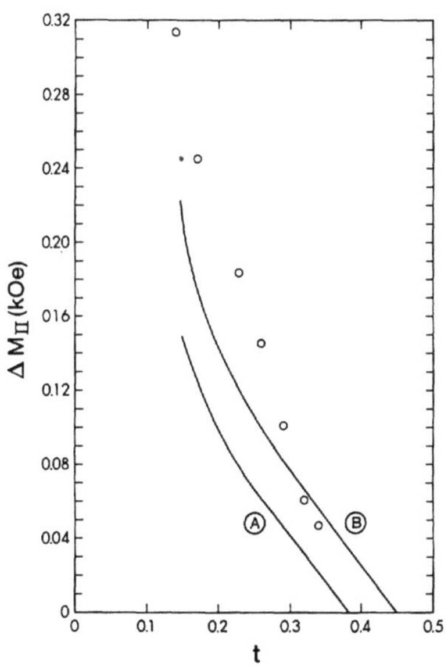

Fig. 11.5. The jump in the magnetization at H_{c2}, ΔM_{II}, for $\kappa_B = 4.0$, $d_f = 20$, $\alpha = 5$, $(I^2/g(T_c) = 10.1466)$ and $H_{co} = 1.4$ KOe including scaling, electron spin polarization and the electromagnetic interaction (lower curve A). Also ΔM_{II} calculated for $\kappa_B = 3.5$, $d_f = 30$, $\alpha = 5$, $(I^2/g(T_c) = 14.8417)$ and $H_{co} = 1.40$ KOe including the effect of the electron spin polarization and the electromagnetic interaction but neglecting the effect of the scaling (i.e., $s(t;h) = 1$) (upper curve B). Circles denote the experimental results of Behroozi et al. [20]

tatively at least for the easy axis critical field curves $H_{C1}^{(a)}$ and $H_{C2}^{(a)}$ purely on the basis of the scaling and the electromagnetic interaction, the hard axis upper critical field curve bears little resemblance to that observed experimentally (Fig. 10.3). Furthermore, while the magnetization curves show a 1st order transition at H_{C1} the transition to the normal state at H_{C2} is continuous (Fig. 10.4), indicating that the sequence of the transitions as the temperature is lowered is

type $II_{2,2}$ $\rightarrow$ type $II_{1,2}$ $\rightarrow$ type I.

The results show that if we completely neglect the effect of the scaling, a slight modification of the parameters allows us to obtain reasonably good agreement with the experimentally observed upper and lower critical fields in both the hard and easy axes (Figs. 10.5 and 10.6). Furthermore, the magnetization curves show the appearance of a 1st order transition at H_{C2}, while the transition at H_{C1} is now of 2nd order, suggesting that the sequence of transitions in this case to be

type $II_{2,2}$ $\rightarrow$ type $II_{2,1}$ $\rightarrow$ type I.

The jump in the magnetization calculated using the parameters of Fig. 10.5 is shown by the upper curve in Fig. 11.5.

This comparison together with other numerical computations allow us to draw the following conclusions regarding the nature of the d-f interaction.

(1) Both the scaling effect and the induced splitting of the electron energies can considerably alter the upper and lower critical field curves. Their effect with respect to the anisotropy of the localized spin system is however quite different.

(2) The appearance of the 1st order transition at H_{C1} arising from the electromagnetic interaction is enhanced by the effect of the scaling. The role of the electron spin splitting is minor.

(3) The splitting of the electron energy by the polarization of the localized spins through the d-f interaction can give rise to a first-order transition to the normal state around H_{C2} [27]. The role of the scaling is minor.

(4) A quantitative analysis of $ErRh_4B_4$ reveals that most of its properties may be adequately described if the d-f interaction is included [28], the dominant effect being the splitting of the electron energy due to the polarization of the localized moments.

12. Summary

We have presented the theory for ferromagnetic superconductors, in which the dipole interaction and d-f interaction are taken into account. Since both interactions work to suppress the superconductivity, the modification of the magnetic properties in the mixed state arising from those effects show, in many cases, a similar qualitative tendency. However, their effects of temperature and field dependence of the superconducting properties are quite distinct, and overall consistency can reveal the dominant mechanism in each material. The present analysis shows that in the $ErRh_4B_4$ case, dipole effect and the conduction electron polarization effect with small fluctuation effect can explain all the magnetic properties of this compound.

Acknowledgements

The authors would like to acknowledge the valuable discussions with Drs. M. Tachiki and G.W. Crabtree. Also we would like to thank Mr. R. Teshima for his fine computer work. This work was supported by the Natural Sciences and Engineering Research Council, Canada and by the Dean of the Faculty of Science, The University of Alberta, Edmonton, Alberta, Canada. One of the authors (J.W.) is supported by the Alberta Heritage Scholarship fund through a Ralph Steinhauer Award.

References

1. See for example the review articles by M. Tachiki, Physica 109-110B&C, 1699 (1982); J. Magn. & Magn. Mater. 31-34, pt.2, 484 (1983); H. Matsumoto and H. Umezawa, Cryogenics 23, 37 (1983).

2. W.A. Fertig, D.C. Johnston, L.E. DeLong, R.E. McCallum, M.B. Maple and B.T. Matthias, Phys. Rev. Letters 38, 987 (1977); M. Ishikawa and Ø. Fisher, Solid State Commun. 23, 37 (1977).

3. See for example "Ternary Superconductors", G.K. Shenoy, B.D. Dunlap and F.Y. Fradin, Eds. (North Holland, 1981).
4. P.W. Anderson and H. Suhl, Phys. Rev. 116, 898 (1959).
5. M. Tachiki, A. Kotani, H. Matsumoto and H. Umezawa, Solid State Commun. 31, 927 (1979).
6. H. Matsumoto, H. Umezawa and M. Tachiki, Solid State Commun. 31, 157 (1979).
7. D.E. Moncton, J. Appl. Phys. 50, 1880 (1979); D.E. Moncton, D.C. McWhan, P.H. Schmidt, G. Shirane, W. Thomlinson, M.B. Maple, H.B. MacKay, L.D. Woolf, Z. Fisk and D.C. Johnston, Phys. Rev. Letters 45, 2060 (1980).
8. S.K. Sinha, G.W. Crabtree, D.G. Hinks and H. Mook, Phys. Rev. Letters 48, 950 (1982).
9. J.W. Lynn, G. Shirane, W. Thomlinson and R.N. Shelton, Phys. Rev. Letters 46, 368 (1981); J.W. Lynn, G. Shirane, W. Thomlinson, R.M. Shelton and D.E. Moncton, Phys. Rev. B24, 3817 (1981).
10. E.I. Blount and C.M. Varma, Phys. Rev. Letters 42, 1079 (1979); H.S. Greenside, E.I. Blount and C.M. Varma, Phys. Rev. Letters 46, 49 (1981).
11. H. Suhl, in Felix Bloch Festschrift (1980).
12. S. Maekawa and M. Tachiki, Phys. Rev. B18, 4688 (1978).
13. Here we did not consider the dynamical effect on the gap equation. As for this effect, see L. Coffey, K. Levin and G.S. Grest, Phys. Rev. B27, 2740 (1983).
14. G. Sarma, J. Phys. Chem. Solids 24, 1029 (1963).
15. H. Matsumoto and H. Umezawa, F. Phys. 24, 357 (1977); H. Matsumoto, M. Tachiki and H. Umezawa, F. Phys. 25, 273 (1977).
16. H. Umezawa, H. Matsumoto and M. Tachiki, "Thermo Field Dynamics and Condensed States" (North Holland, 1982).
17. I. Shapira, M.N. Shah and H. Umezawa, Physica 84B, 213 (1976).
18. H. Matsumoto, M. Tachiki and H. Umezawa, Phys. Rev. B25, 6633 (1982).
19. H. Matsumoto, R. Teshima, H. Umezawa and M. Tachiki, Phys. Rev. B27, 158 (1983).
20. F. Behroozi, G.W. Crabtree, S.A. Cambell and D.G. Hinks, Phys. Rev. B27, 6849 (1983).
21. F. Behroozi, Private Communication (1983).
22. G.W. Crabtree, F. Behroozi, S.A. Cambell and D.G. Hinks, Phys. Rev. Letters 49, 1342 (1982).

23. J.M. Vanderberg and B.T. Matthias, Natl. Acad. Sci. USA 74, 1336 (1977).
24. G.W. Crabtree, Private Communication (1983).
25. K. Okuda, Y. Nakakura and K. Kadowaki, Solid State Commun. 32, 185 (1979).
26. L.D. Woolf, D.C. Johnston, H.B. MacKay, R.W. McCallum and M.B. Maple, J. Low Temp. Phys. 35, 651 (1979).
27. This is in agreement with the recent calculations of Koyama et al. which show the magnetization curve to have convex curvature around H_{c_2}. However, their calculation did not cover the low temperature domain where the first-order transition at H_{c_2} takes place. T. Koyama, S. Maekawa and M. Tachiki, J. Phys. Soc. of Japan 52, 1750 (1983).
28. This conclusion is also supported by the independent calculations of Sakai et al. O. Sakai, M. Suzuki, S. Maekawa, M. Tachiki, G.W. Crabtree and F. Behroozi, J. Phys. Soc. of Japan 52, 1341 (1983).

Superconductivity and Magnetism in Rare Earth and Actinide Compounds

Masashi Tachiki

The Research Institute for Iron, Steel and Other Metals
Tohoku University, Katahira, Sendai 980, Japan

In rare earth and actinide compounds, the interaction between the magnetic moments of f electrons and superconducting electrons causes various unusual properties. The interaction generates anomalous temperature dependence of H_{c2} and magnetic field dependence of magnetization in the mixed state of ferromagnetic superconductors such as $ErRh_4B_4$. In the magnetic phase diagrams of ferromagnetic superconductors, various phases and critical points appear. Recently, superconductivity was found in $CeCu_2Si_2$ and UBe_{13}, in which the specific heat, the electric resistivity, etc.,behave like those in the Kondo lattice. In relation to the phenomena, physical properties in the Kondo lattice and the stability of superconducting state in the lattice were theoretically studied.

1. Introduction

The f electrons in rare earth and actinide compounds exhibit various characteristic properties. The f electrons in rare earth compounds are relatively localized at the lattice sites and carry magnetic moments. In many rare earth compounds such as (RE)Mo_6S_8 and (RE)Rh_4B_4, RE being rare earth ions, magnetic orders coexist with superconductivity [1]. In rare earth compounds, Ce compounds are special. The hybridization between the f- and conduction electron states is strong and the compounds are in valence-fluctuating states or in Kondo lattice state, depending on the depth of the f-electron level in the Fermi sea. The temperature dependence of the specific heat and the resistivity of $CeCu_2Si_2$ shows the Kondo lattice behavior, nevertheless the compound becomes superconductive below 0.6 K [2-6]. An actinide compound UBe_{13} was recently found to be the same type of superconductor as $CeCu_2Si_2$ [7]. The superconducting transition temperature of the compound is 0.86 K.

The following two topics are chosen for discussions in the present paper. One is the phases and critical points appearing in ferromagnetic superconductors and the other is superconductivity in the Kondo lattice.

2. Phases and Critical Points in Ferromagnetic Superconductors

The rare earth compound $ErRh_4B_4$ is a typical ferromagnetic superconductor [8]. The superconducting transition temperature T_{c1} of the compound is 8.7 K. On lowering the temperature, superconducting phase is preserved down to the reentrance phase transition temperature T_{c2}, below which the compound becomes the normal ferromagnetic state. The temperature T_{c2} has a value between 0.7 K and 0.9 K depending on the samples. A new sinusoidal magnetic order phase appears in the superconducting state between T_{c1} and a temperature T_Q which is 1 K - 1.2 K depending on the samples [9,10]. The easy axis of magnetization is the a axis in the tetragonal crystal and the hard axis of magnetization is the c axis. The upper and lower critical fields were measured by CRABTREE et al.[11] and the results are shown in Fig. 1. The temperature dependence of the upper critical field H_{c2} above T_{c2} is similar to that in usual non-magnetic superconductors. However, when an external magnetic field is applied along the a axis the dependence is very different from that in usual superconductors, as seen in Fig.1. The curve of H_{c2} is bell-shaped which has a maximum at 5.5 K. The continuous curve of H_{c2} above T = 3.3 K indicates the second-order phase transition between the vortex and normal phases, and the dashed curve indicates the first-order phase transition. Therefore, the point indicated by t_2 in Fig. 1 corresponds to a tricritical point. Another tricritical point has been predicted at t_1 on the line of the lower critical fields H_{c1}. In the following, the origin of the unusual behavior in the phase diagram of $ErRh_4B_4$ will be discussed.

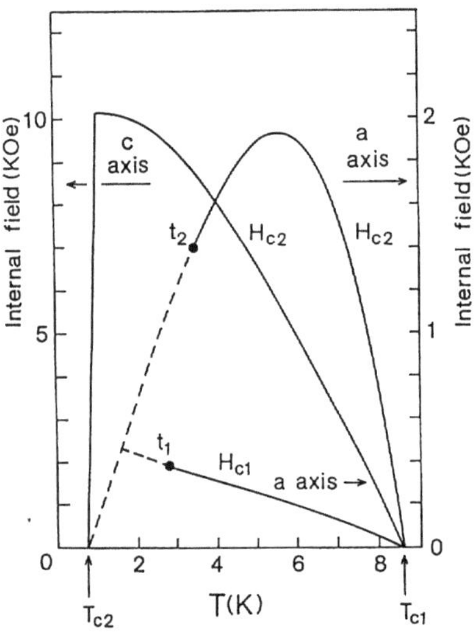

Fig. 1 Phase diagram in single crystal of $ErRh_4B_4$. Continuous curves and dashed curves indicate the second-and first-order phase transition, respectively. The tricritical point t_2 was experimentally obtained and the tricritical point t_1 was theoretically predicted. After CRABTREE et al. [11]

The two kinds of dominant interaction are considered between the superconducting electrons and the rare earth magnetic moments. One is the exchange type interaction or the so-called c-f interaction expressed by the form $I\vec{\sigma}\cdot\vec{J}$, where I is the intensity of the interaction, $\vec{\sigma}$ the spin of conduction electron and $\vec{J}$ the total angular momentum of rare earth ion. The other is the interaction between the rare earth magnetic moments and the superconducting current which is induced by the magnetic moments. The former interaction causes the anomalous temperature dependence of H_{c2} when an external magnetic field is applied along the a axis. The magnetic field penetrates into the sample in the vortex state and induces a large magnetization at low temperatures. The magnetization produces the strong molecular field acting on conduction electrons through the c-f interaction. In the molecular field the up and down spin electrons in the singlet Cooper pair have the different Zeeman energies. The strength of the singlet Cooper pairs

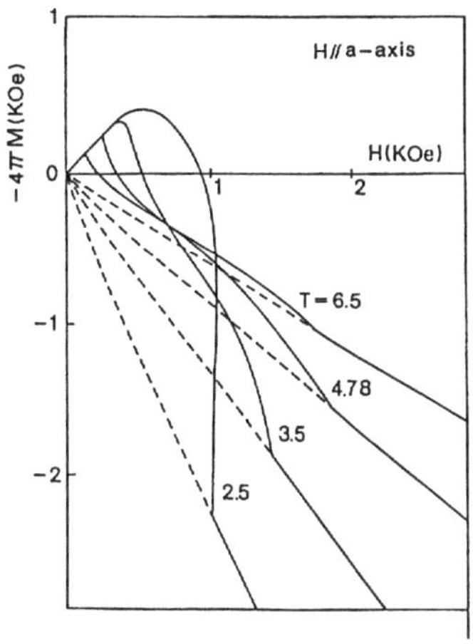

Fig. 2 Magnetization curves of single crystal of $ErRh_4B_4$. After BEHROOZI et al. [18]

is weakened by this energy difference. This weakening effect of the Cooper pairs causes the drastic decrease of H_{c2} at low temperatures [11-16].

The figure 2 shows the anomalous magnetization curves observed by BEHROOZI et al. [17] using single crystal of $ErRh_4B_4$. When the temperature is above T_{t2} the magnetization curve near H_{c2} has a convex curvature which is opposite to that of non-magnetic superconductors. With decreasing temperature, the curvature increases and the magnetization jump occurs at H_{c2} below T_{t2} = 3.3 K. Therefore, t_2 corresponds to the tricritical point. When the temperature decreases further, the magnitude of the magnetization jump increases and H_{c2} decreases, and finally the sample becomes a type I superconductor.

Another tricritical point is predicted to occur at the point labelled t_1 on the H_{c1} line in Fig. 1. Below T_{t1} the jump of magnetization occurs at H_{c1}. The jump originates from the following mechanism [18]. Let us consider a single vortex in the sample in the superconducting state. If the magnetic induction, the magnetic field induced by the vortex current, and the rare earth magnetization are written as $b(\vec{x})$, $h(\vec{x})$, and $m(\vec{x})$, respectively, the condition of flux quantization is expressed as

$$\int b(\vec{x})d^2x = \int h(\vec{x})d^2x + 4\pi\int m(\vec{x})d^2x = \phi_0 \quad , \tag{1}$$

where $\phi_0 = hc/2e$ is the unit flux. The magnetic field induces the rare earth magnetization in the vortex. Since the magnetization becomes very large near the magnetic phase transition temperature, $\int h(\vec{x})d^2x$ decreases and $h(\vec{x})$ develops a negative tail, as shown schematically in Fig. 3. When there are two vortices a distance x apart, the interaction between them is proportional to $h(x)$. For $T > T_{t1}$, the interaction is repulsive and the transition between the Meissner and vortex states is continuous. However, below T_{t1}, $h(x)$ develops an attractive tail and the vortices first appear at the distance corresponding to the minimum in $h(x)$. This leads to a jump of the magnetization curve as shown schematically in Fig. 4. The details of the critical behavior at the tricritical point t_1 and t_2 have been discussed by CRABTREE et al. [19].

Fig. 3 Magnetic field induced by the vortex current

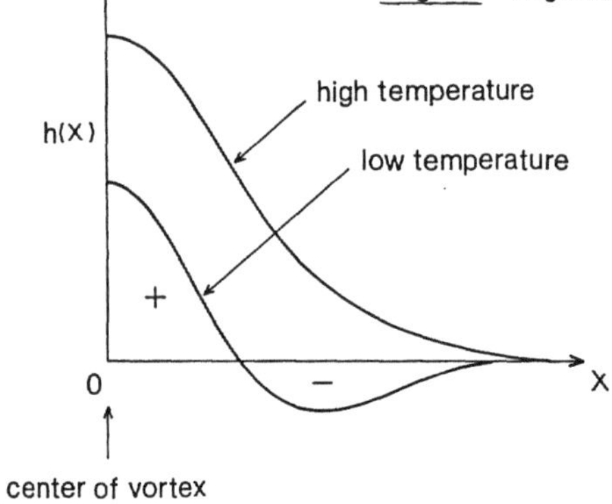

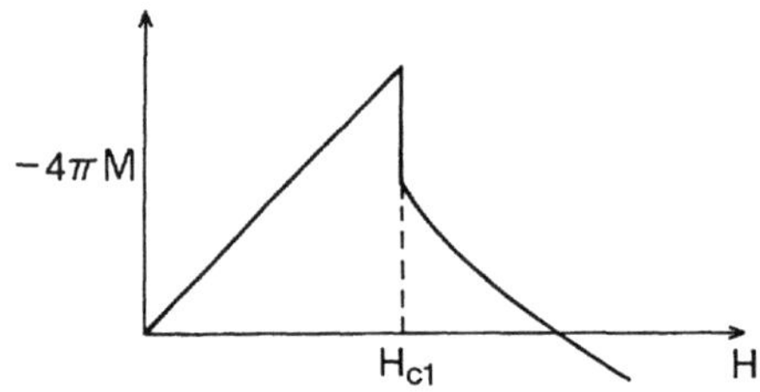

Fig. 4 Magnetization jump at H_{c1}

Let us discuss the origin of the sinusoidal magnetic order which appears in the superconducting state between T_{C2} and T_Q in $ErRh_4B_4$ [20,21]. If we write the persistent current as $\vec{j}(\vec{x})$, and the magnetic field induced by the current as $\vec{h}(\vec{x})$, both the quantities are connected with each other by the Maxwell equation

$$\vec{\nabla}\times\vec{h}(\vec{x}) = (4\pi/c)\vec{j}(\vec{x}) \quad . \tag{2}$$

Writing the rare earth magnetization as $\vec{m}(\vec{x})$, we define the magnetic induction by

$$\vec{B}(\vec{x}) = \vec{h}(\vec{x}) + 4\pi\vec{m}(\vec{x}) \quad . \tag{3}$$

The persistent current is given by

$$\vec{j}(\vec{x}) = -(c/4\pi\lambda_L{}^2)c(-i\vec{\nabla})\vec{a}(\vec{x}) \quad , \tag{4}$$

where λ_L is the London penetration depth, $\vec{a}(\vec{x})$ is the vector potential, and $c(-i\vec{\nabla})$ is a characteristic function which is defined by $c(-i\vec{\nabla})\exp(i\vec{k}\cdot\vec{x}) = c(\vec{k})\exp(i\vec{k}\cdot\vec{x})$. The $c(\vec{k})$ is approximately expressed by

$$c(\vec{k}) = \exp(-\xi^2k^2/2) \quad , \tag{5}$$

ξ being the coherence length. The magnetic induction is connected with the vector potential by

$$\vec{B}(\vec{x}) = \vec{\nabla}\times\vec{a}(\vec{x}) \quad . \tag{6}$$

Using (2), (3), (4), and (6), we have

$$[\lambda_L{}^2\nabla^2 - c(-i\vec{\nabla})]\vec{h}(\vec{x}) = 4\pi c(-i\vec{\nabla})\vec{m}(\vec{x}) \quad , \tag{7}$$

where $\vec{m}(\vec{x})$ stands for a transverse magnetization. We assume the following sinusoidal forms for $\vec{m}(\vec{x})$ and $\vec{h}(\vec{x})$

$$\vec{m}(\vec{x}) = \vec{m}\sin\vec{q}\cdot\vec{x} \quad , \tag{8}$$

$$\vec{h}(\vec{x}) = \vec{h}\sin\vec{q}\cdot\vec{x} \tag{9}$$

where $\vec{m}$ and $\vec{h}$ are perpendicular to $\vec{q}$. From (7), we have

$$\vec{h} = -F(q)\vec{m} \quad , \tag{10}$$

with

$$F(q) = \frac{4\pi c(q)}{\lambda_L{}^2 q^2 + c(q)} \quad . \tag{11}$$

The interaction energy between the sinusoidal magnetization and the magnetic field induced by the magnetization is given by

$$-(1/2)\int \vec{h}(\vec{x})\cdot\vec{m}(\vec{x})d^3x = (1/4)F(q)m^2 \quad , \tag{12}$$

where the expressions (8), (9), and (10) were used. The energy (12) is positive, because the direction of the magnetic field is opposite to that of the magnetization. The energy (12) is the largest for a uniform magnetization. When the wave number q of the sinusoidal wave increases, the energy monotonically decreases.

If (8) is used, the magnetic ordering energy is expressed as

$$-(1/2)\int \vec{m}(\vec{x})\gamma(-i\vec{\nabla})\vec{m}(\vec{x})d^3x = -(1/4)\gamma(q)m^2 \quad , \tag{13}$$

where $\gamma(q)$ is the Fourier component of the exchange interaction constant. We assume that if the system is in the normal state, the ferromagnetic state is stable. Then, the energy (13) has a minimum value at $q = 0$, and the energy increases as the wave number increases. In the superconducting state, the energy of the system is given by the sum of the energies (12) and (13), which has a minimum at a finite wave number $q = Q$. This fact indicates that the sinusoidal magnetic order with wave number Q is stable rather than the ferromagnetic order in the superconducting state. In magnetically isotropic systems, the component of the magnetization perpendicular both to $\vec{m}$ and $\vec{q}$ produces the same energies as those of (12) and (13). In this case a spiral magnetic order with wave number Q is realized. In magnetically anisotropic systems, the magnetic anisotropy energy stabilizes the sinusoidal magnetic order compared with the spiral magnetic order.

Neutron diffraction experiments using single crystals of $ErRh_4B_4$ reveal that the sinusoidal magnetic order with wave length 100 Å appears in the temperature range between 0.7 K and 1 K, and the magnetization is perpendicular to the propagation vector [10,22], consistent with the present mechanism. In addition to the sinusoidal order, neutron diffraction experiments observed a ferromagnetic order in the temperature range. The currently accepted view is that superconductivity and ferromagnetism do not coexist microscopically, but occur in separate domains. It has recently been suggested that the occurrence of the ferromagnetic domains is an artifact caused by inhomogeneous strains in the crystal [23,24]. Recent neutron diffraction experiments also show that a sinusoidal or spiral magnetic order exists in the superconducting state of $HoMo_6S_8$ [25] and $HoMo_6Se_8$ [26]. The temperature range in which the periodic order appears is 0.7 K - 1.0 K in $HoMo_6S_8$ and its wave length is 200 Å. In $HoMo_6Se_8$, the periodic order appears below 0.53 and its wave length is 100 Å.

3. Superconductivity in the Kondo Lattice

First the experimental results on $CeCu_2Si_2$ [2-6, 27] and UBe_{13} [7, 28] are summarized. (1) These compounds have extremely large electronic specific heat at low temperatures. When the specific heat is expressed as $C = \gamma T$, the magnitude of γ is of the order of 1 Joule $mole^{-1}$ K^{-2}, which is several hundred times as large as those of usual metals. (2) The superconducting transition temperatures T_c of $CeCu_2Si_2$ and UBe_{13} are 0.6 K and 0.86 K, respectively. The transition temperature of $CeCu_2Si_2$ strongly depends on samples. (3) The jump of the specific heat ΔC at the superconducting transition temperature is extremely large, but the ratio $\Delta C/\gamma T_c$ is of the order of unity, which is comparable to the value of the BCS theory 1.4. (4) The electric resistivity has a maximum at a certain temperature above T_c. (5) The temperature gradient of the upper critical field $|\partial H_{c2}/\partial T|$ at T_c is extremely large. The upper critical field at low temperatures is considerably high for relatively low T_c. (6) The temperature dependence of the electric resistivity and the line width of quasi elastic scattering of neutrons show that these compounds are in the Kondo lattice state rather than in the mixed valence state. (7) T_c is much lower than the Kondo temperature T_K.

Let us first try to analyze the experimental results on the basis of the theory of independent Kondo impurities. If we write the electronic specific heat of the system as C, the specific heat of the conduction electrons as C_c, and the density of states of conduction electrons at Fermi level as ρ, C/C_c is given by $1/4\rho T_K$ [29-31]. The experimental value of C/C_c is of the order of 1000. The attractive interaction acting on the electrons in the Cooper pairs is expressed as $g_{eff} = g_{BCS} - \Gamma_{\uparrow\downarrow}$, where g_{BCS} is the attractive interaction through phonons and $\Gamma_{\uparrow\downarrow}$ is the repulsive interaction through the spin fluctuation of Kondo impurities. When the temperature is sufficiently lower than T_K, ρg_{eff} is given by [32,33]

$$\rho g_{eff} = \rho g_{BCS} - \frac{1}{4T_K\rho} . \tag{14}$$

Since the magnitude of ρg_{BCS} is of the order of unity and $1/4T_K\rho$ is of the order of 1000, the repulsive interaction overcomes the attractive interaction and the resultant interaction becomes repulsive. This fact implies that the superconducting state is not stable in the system in contradiction to the experimental results. To explain the electric resistivity below T_K in the normal state, the coherence between Kondo ions was crucial. The coherence may also be important to understand the superconducting state in the Kondo lattice [34].

We use a periodic Anderson model for the Kondo lattice system. In this case there may be two ways to take account of the coherence mentioned above. One way is that first we solve a single-ion Kondo problem by taking account of the correlation between f electrons, and after that the coherence between ions is considered. The other way is that the coherence is first considered in the non-magnetic state of the system by the Hartree-Fock approximation, and then the correlation between f electrons is taken into account. In the present paper we will take the latter way according to the following reasons. In $CeSn_3$, the results of the band theoretical calculation can explain the Fermi surface determined by the experiments of the de Haas-van Alphen effect, although the theory cannot interpret the effective mass [35]. The spread of the Kondo state of single ion is very wide and many Kondo ions are included in the spread.

Since Ce ions in $CeCu_2Si_2$ and U ions in UBe_{13} are well separated in the crystals, there is almost no overlapping of the f-electron wave functions. Therefore, we assume a dispersionless f band. For simplicity we neglect the orbital degeneracy of the f electron. The assumption may be reasonable for these compounds, because the ground levels of Ce and U in the crystalline field are the Kramers doublet. If we assume a single conduction band which hybridizes the f band and two electrons per atom, the paramagnetic state of the system becomes an insulating state. For the system being in a metallic state, we should consider two conduction bands at least. For simplicity, we introduce two conduction bands with the same dispersion which are expressed by ε_k+a and ε_k-a. We assume the symmetric Anderson model or $\varepsilon_f = -U/2$, ε_f and U being the f-electron energy and the correlation energy of the f electrons, respectively. Then, the Hamiltonian for the model is written as

$$H = \sum_{k\sigma} \varepsilon_{1k} c_{1k\sigma}^{+} c_{1k\sigma} + \sum_{k\sigma} \varepsilon_{2\sigma} c_{2k\sigma}^{+} c_{2k\sigma}$$
$$+ \sum_{k\sigma} V(c_{1k\sigma}^{+} f_{k\sigma} + f_{k\sigma}^{+} c_{1k\sigma}) + \sum_{k\sigma} V(c_{2k\sigma}^{+} f_{k\sigma} + f_{k\sigma}^{+} c_{2k\sigma})$$
$$+ \sum_{i} U(f_{i\uparrow}^{+} f_{i\uparrow} - \tfrac{1}{2})(f_{i\downarrow}^{+} f_{i\downarrow} - \tfrac{1}{2}) \quad , \tag{15}$$

with $\varepsilon_{1k} = \varepsilon_k+a$ and $\varepsilon_{2k} = \varepsilon_k-a$, where k and i express the wave number and the lattice site of ion, respectively. In (15), $c_{1k\sigma}$, $c_{2k\sigma}$, and $f_{k\sigma}$ are the electron operators for the two conduction bands and the f-electron band, respectively, V is the hybridization parameter which is assumed to be common to the two conduction bands and independent of k.

The figure 5 shows the energy levels of the system in a non-magnetic Hartree-Fock state. We assumed three electrons per atom. The continuous curves show the energy levels filled with up and down spin electrons, and the dashed curves show the empty energy levels. The Hartree-Fock state includes the following configurations of electrons at the i site; no electron, one electron, and two electrons occupy the site. The correlation energy which is expressed by the last term of the Hamiltonian (15) is -U/4 for one-electron occupancy at the i site and U/4 for no electron occupancy and two-electron occupancy. Since U is very large, the energies of no electron and two-electron occupancy states are much larger than that of one-electron occupancy state. Therefore, the correlation effect of the f electrons is taken into account, the one-electron occupancy state predomi-

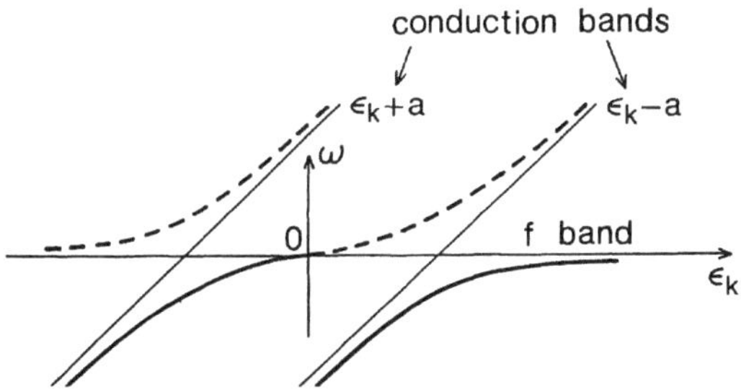

Fig. 5 Energy level of the system in a non-magnetic Hartree-Fock state. The thick continuous curves indicate the energy levels filled with up and down spin electrons and the dashed curves indicate the energy levels

nates over other states. In this many-body state, there are large spin fluctuations and small charge fluctuations of f electrons.

The f-electron self-energy in the many-body state mentioned above is given by

$$\Sigma(\omega_n) = -A(i\omega_n) - iB[(i\omega_n)^2 + (\pi T)^2] + \cdots \tag{16}$$

with $\omega_n = \pi T(2n+1)$, n being integers. We use units such that $\hbar = k_B = 1$. YOSHIMORI and KASAI used a single site approximation for obtaining the self-energy, when they discussed the electric resistivity in the normal state of Kondo lattice [36]. If we use their approximation, A and B are expressed as functions of U and other physical parameters by applying the results of YOSIDA and YAMADA's theory for single impurity problem [31].

With the use of the self-energy (16), the Green's functions of f electron is given by

$$G_f(k, \omega_n) = [i\omega_n - \Sigma(\omega_n) - V^2(i\omega_n - \varepsilon_{1k})^{-1} - V^2(i\omega_n - \varepsilon_{2k})^{-1}]^{-1} . \tag{17}$$

For small value of $|\omega_n|$, (17) is reduced to

$$G_f(k, \omega_n) = \frac{1}{\tilde{\gamma} + \alpha} \times \frac{1}{i\omega_n - E_k} , \tag{18}$$

with

$$E_k = \frac{\alpha}{1 + \alpha} \varepsilon_k , \tag{19}$$

$$\alpha = \frac{1}{\tilde{\gamma}} \times \frac{2V^2}{a^2} , \tag{20}$$

$$\tilde{\gamma} = 1 - \left.\frac{\partial\Sigma(\omega_n)}{i\omega_n}\right|_{\omega_n=0} = 1 + A . \tag{21}$$

The quantity E_k is the energy level of quasi-fermion near the Fermi level. The effective mass of the quasi-fermion is given by $m^* = m(1+(1/\alpha))$, m being the mass of conduction electron. The effective mass is very heavy, because $\alpha \ll 1$. The life time of the particle at the Fermi level is infinity. Therefore, this quasi-fermion system has a character of heavy Fermi liquid.

Due to the heavy effective mass, we expect a large specific heat. Using the excitation energy (19), we calculated the specific heat in the normal state. The result is

$$C_N = \frac{2\pi^2}{3} \rho_{eff} T , \tag{22}$$

with the effective density of states

$$\rho_{eff} = (1 + \frac{1}{\alpha})\rho . \tag{23}$$

The calculation shows that $\rho_{eff}/\rho = 1/8T_K\rho$. The value which was obtained from the specific heat experiments is of the order of 1000. This large value may be understood, if we consider that $T_K \sim 10$ K and $\rho \sim 10^{-4}K^{-1} - 10^{-5}K^{-1}$. This large enhancement of the density of states mainly comes from the spin functions of f electrons.

4. Superconducting Properties of the Kondo Lattice

Let us consider the singlet Cooper pairs of the quasi-fermions. The quasi-fermions are mainly composed of the f electrons. Therefore, the attractive interaction acting between the f electrons plays an important role in the formation of the Cooper pairs [34]. If we write the attractive interaction constant as λ and use the weak coupling approximation, we have an equation

$$1 = \lambda T_c \sum_n \sum_k G_f(k, \omega_n) G_f(-k, -\omega_n) \quad . \tag{24}$$

The superconducting transition temperature T_c is obtained by solving the equation (24). Inserting (18) into (24) we have

$$1 = \lambda T_c \times \frac{1}{\tilde{\gamma}^2(1+\alpha)^2} \sum_n \sum_k \frac{1}{\omega_n^2 + E_k^2} \quad . \tag{25}$$

The summation over k is replaced by the integral over ε_k. The cut-off energy T_K for the summation over ω_n is introduced, since the states with $\omega_n > T_K$ cannot contribute to superconductivity due to the strong depairing effect. Then, T_c is calculated from (25) as

$$T_c = 1.13\, T_K \exp\left[-\frac{\alpha(1+\alpha)\tilde{\gamma}^2}{\lambda\rho}\right] \quad . \tag{26}$$

The effective attractive interaction constant λ is written as

$$\lambda = g - \Gamma_{\uparrow\downarrow} \quad , \tag{27}$$

where g is the attractive interaction constant and $\Gamma_{\uparrow\downarrow}$ is the renormalized Coulomb repulsive interaction constant acting between the f electrons with up and down spins. The attractive interaction constant g was introduced as a phenomenological parameter. At present we do not know whether the attractive interaction originates from the interaction through phonons or an interaction through some exotic excitations characteristic in the periodic Anderson system. Even if the attractive interaction originates from phonons, the Migdal theorem might be violated, because the effective mass of the quasi-fermions is comparable to the ion masses. In the quasi-fermion state, the correlation energy is strongly reduced. The reduced correlation energy constant $\Gamma_{\uparrow\downarrow}$ is calculated as

$$\Gamma_{\uparrow\downarrow} = \frac{1}{4T_K\rho^2}\left(\frac{V}{a}\right)^4 \quad . \tag{28}$$

The repulsive interaction constant (28) is much smaller than that in the independent impurity model discussed in section 3, since (29) includes the very small factor $(V/a)^4$. Therefore, the attractive interaction constant g can be larger than $\Gamma_{\uparrow\downarrow}$ and the system possibly becomes superconductive.

The difficulty of the instability of the superconducting state in an independent Kondo impurity model mentioned in section 3 may be removed. The compounds $LaCu_2Si_2$ and $ThBe_{13}$ have the same crystal structures as $CeCu_2Si_2$ and UBe_{13}. The experimental results show that $LaCu_2Si_2$ [2] and $ThBe_{13}$ [37] are not superconductive. Since the f electrons are responsible for superconductivity in the present model, the experimental results mentioned above may be understood.

Using the present model we calculated the thermodynamic potential difference between the superconducting state and the normal state $\Omega_S - \Omega_N$. The specific heat jump at T_c is obtained from the potential difference as

$$\Delta C = - T_c \left.\frac{d^2(\Omega_S - \Omega_N)}{dT^2}\right|_{T \to T_c + 0^-} = \frac{8\pi^2}{7\zeta(3)} \rho_{eff} T_c \quad , \tag{29}$$

where $\zeta(3)$ is the Riemann ζ function. The expression (29) is the same as that in the BCS theory, if ρ_{eff} is replaced by ρ. The ratio of ΔC and the specific heat of the normal state at T_c is estimated from (22) and (29) as

$$\Delta C/C_N(T_N) = 12/7\zeta(3) = 1.426 \quad , \tag{30}$$

which is the BCS value of the ratio. The order of the magnitude of the ratio is in agreement with the experimental results in $CeCu_2Si_2$ and UBe_{13}.

The thermodynamic critical field H_c at 0 K is calculated from the relation $\Omega_S - \Omega_N = -H_c^2/8\pi$ as

$$H_c = 1.76\, T_c(4\pi\rho_{eff})^{1/2} \quad . \tag{31}$$

If the critical field (31) is compared to the BCS value $H_c = 1.76T_c(4\pi\rho)^{1/2}$, it is seen that H_c in the system is strongly enhanced by ρ_{eff}. Therefore, the superconducting state is very stable and the system is suggested to have a very high upper critical field.

Acknowledgments

The author wishes to thank Dr. G.W. Crabtree, Prof. F. Behroozi, Dr. R.K. Kalia, Prof. M.B. Maple. Prof. H. Umezawa, Prof. H. Matsumoto, Prof. A. Kotani, Prof. S. Maekawa, Dr. T. Koyama, and Dr. S. Takahashi for valuable discussions on the studies presented in this paper.

References

1. For a review, see Topics in Current Physics, ed. by Ø. Fischer and M.B. Maple (Springer-Verlag, New York, 1982) Vols. 32 and 34
2. F. Steglich, J. Aarts, C.D. Bredl, W. Leike, D. Meschede, W. Franz, and H. Schäfer: Phys. Rev. Lett. 43, 1892 (1979)
3. U. Rauchschwalbe, W. Leike, C.D. Bredl, F. Steglich, J. Aarts, K.M. Martini, and A.C. Mota: Phys. Rev. Lett. 49, 1448 (1982)
4. W. Leike, U. Rauchschwalbe, C.D. Bredl, F. Steglich, J. Aarts, and F.R. de Boer: J. Appl. Phys. 53, 2111 (1982)
5. W. Assumus, M. Herrmann, U. Rauchschwalbe, S. Riegel, W. Leike, H. Spille, S. Horn, G. Weber, F. Steglich, and G. Cordier: to be published

6. M. Ishikawa, H.F. Braun, J.L. Jorda: Phys. Rev. B27, 3092 (1983)
7. H.R. Ott, H. Rudigier, Z. Fisk, and J.L. Smith: Phys. Rev. Lett. 50, 1595 (1983)
8. W.A. Fertig, D.C. Johnston, L.D. DeLong, R.W. McCallum, M.B. Maple, and B.T. Matthias: Phys. Rev. Lett. 38, 987 (1979)
9. D.E. Moncton, D.B. McWhan, P.H. Schmidt, G. Shirane, W. Thomlinson, M.B. Maple, H.B. MacKay, L.D. Woolf, Z. Fisk, and D.C. Johnston: Phys. Rev. Lett. 45, 2060 (1980)
10. S.K. Sinha, G.W. Crabtree, D.G. Hinks, H.A. Mook: Phys. Rev. Lett. 48, 14 (1982)
11. G.W. Crabtree, F. Behroozi, S.A. Campbell, and D.G. Hinks: Phys. Rev. Lett. 49, 1342 (1982)
12. S. Maekawa and M. Tachiki: Phys. Rev. B18, 4688 (1978)
13. K. Machida and D. Youngner: J. Low Temp. Phys. 35, 449 (1979); ibid. 35, 561 (1979)
14. C.Y. Huang, S.E. Kohn, S. Maekawa, and J.L. Smith: Solid State Commun. 32, 929 (1979)
15. O. Sakai, M. Suzuki, S. Maekawa, M. Tachiki, G.W. Crabtree, and F. Behroozi: J. Phys. Soc. Jpn. 52, 1341 (1983)
16. T. Koyama, S. Maekawa, and M. Tachiki: J. Phys. Soc. Jpn. 52, 1750 (1982)
17. F. Behroozi, G.W. Crabtree, S.A. Campbell, and D.G. Hinks: Phys. Rev. B27, 6849 (1983) and private communication
18. M. Tachiki, H. Matsumoto, and H. Umezawa: Phys. Rev. B20, 1915 (1979)
19. G.W. Crabtree, R.K. Kalia, F. Behroozi, and M. Tachiki: to be published
20. E.I. Blount and C.M. Varma: Phys. Rev. Lett. 42, 1079 (1979)
21. H. Matsumoto, H. Umezawa, and M. Tachiki: Solid State Commun. 31, 157 (1979); M. Tachiki, A. Kotani, H. Matsumoto, and H. Umezawa: Solid State Commun. 31, 927 (1979)
22. S.K. Sinha, G.W. Crabtree, D.G. Hinks, and H.A. Mook: Physica 109 & 110B, 1693 (1982)
23. L.N. Bulaevskii, A.I. Buzdin, D.I. Khomskii, and P.V. Panjukov: Solid State Commun. 46, 133 (1983)
24. M. Tachiki, B.D. Dunlap, and G.W. Crabtree: Phys. Rev. B28, 5342 (1983)
25. J.W. Lynn, G. Shirane, W. Thomlinson, and R.N. Shelton: Phys. Rev. B24, 3817 (1981); J.W. Lynn, J.L. Ragazzoni, R. Pynn, and J. Joffrin: J. de Phys. Lett. 42, L-45 (1981); J.W. Lynn, R. Pynn, J. Joffrin, J.L. Ragazzoni,and R.N. Shelton: Phys. Rev. B27, 581 (1983)
26. J.W. Lynn, J.A. Gotaas, R.W. Erwin, R.A. Ferrell, J.K. Bhattacharjee, R.N. Shelton, and P. Klavins: Phys. Rev. Lett. 52, 133 (1984)
27. S. Horn, E. Holland-Moritz, M. Loewenhaupt, F. Steglich, E. Scheuer, A. Beneit, and J. Floquet: Phys. Rev. B23, 3171 (1981)
28. J.W. Chen, S.E. Lambert, and M.B. Maple: private communication
29. K.G. Wilson: Rev. Mod. Phys. 47, 773 (1975)
30. P. Nozieres: J. Low Temp. Phys. 17, 31 (1974)
31. K. Yamada: Prog. Theor. Phys. 53, 970 (1975); ibid. 54, 316 (1975); K. Yosida and K. Yamada: Prog. Theor. Phys. 53, 1286 (1975)
32. T. Matsuura, S. Ichinose, and Y. Nagaoka: Prog. Theor. Phys. 57, 717 (1977)
33. A. Sakurai: Phys. Rev. B17, 1195 (1978)
34. M. Tachiki and S. Maekawa: Phys. Rev. B29, No. 5 (1984)
35. W.R. Johanson, G.W. Crabtree, D.D. Koelling, A.S. Edelstein, and O.D. McMasters: J. Appl. Phys. 52, 2134 (1981); W.R. Johanson, G.W. Crabtree, A.S. Edelstein, and O.D. McMaster: J. Magn. Magn. Mater. 31-34, 377 (1983)
36. A. Yoshimori and H. Kasai: J. Magn. Magn. Mater. 31-34, 475 (1983)
37. E. Bucher, J.P. Maita, G.W. Hull, R.C. Fulton, and A.S. Cooper: Phys. Rev. B11, 440 (1973)

Electromagnetic Properties of Ferromagnetic Superconductors

A. Kotani and K. Okada

Department of Physics, Faculty of Science, Osaka University
Toyonaka 560, Japan

Both the static and dynamic normal modes of ferromagnetic superconductors are obtained theoretically as the coupled modes of spin, photon and electric current. With the use of these normal modes, we study, from a unified viewpoint, the static and dynamic electromagnetic properties of the bulk, surface and film of ferromagnetic superconductors. We discuss the oscillation of the field around a vortex, the oscillation of the field penetrating from the surface, the spontaneous surface magnetization, the interaction between a vortex and the surface, the magnetic order in the film, and the reflectance for an electromagnetic wave.

1. Introduction

In the interplay of magnetism and superconductivity in ferromagnetic superconductors, $ErRh_4B_4$ and $HoMo_6S_8$, the electromagnetic interaction between the rare earth magnetic moment and the persistent current plays an essentially important role [1-3]. Through the electromagnetic interaction, the long-range part of the interaction between spins is screened by the persistent current. In this sense, we can regard the ferromagnetic superconductor as a coupled system among the spin, photon and electric current [4, 5]. We study here both the static and dynamic normal modes of this coupled system. When we obtain the normal modes, any field inside the ferromagnetic superconductor can be expanded in terms of the normal modes, and thereby we will discuss the static and dynamic properties of the bulk, surface and film of ferromagnetic superconductors from a unified viewpoint.

The d-f exchange interaction between the rare earth spin and the conduction electron is less important than the electromagnetic interaction in $ErRh_4B_4$ and $HoMo_6S_8$, so that we disregard the depairing effect originating from the d-f interaction.

2. Normal Mode Equation

Basic equations of our coupled system are the Maxwell equations

$$\vec{\nabla} \times \vec{H}(\vec{r},t) = \frac{4\pi}{c}\vec{J}(\vec{r},t) + \frac{1}{c}\frac{\partial}{\partial t}\vec{E}(\vec{r},t), \tag{1}$$

$$\vec{\nabla} \times \vec{E}(\vec{r},t) = -\frac{1}{c}\frac{\partial}{\partial t}\vec{B}(\vec{r},t), \tag{2}$$

and the current representation

$$\vec{J}(\vec{r},t) = -\frac{c}{4\pi\lambda_L^2}\int d\vec{r}'dt'K(\vec{r}-\vec{r}', t-t')\vec{A}(\vec{r}', t'), \tag{3}$$

where $\vec{H}$ is the magnetic field, $\vec{J}$ the current, $\vec{E}$ the electric field, $\vec{B}$ the magnetic induction and $\vec{A}$ the vector potential. λ_L represents the London penetration depth and the kernel K represents the nonlocal effect of the current. We also use the relations

$$\vec{\nabla} \times \vec{A}(\vec{r},t)= \vec{B}(\vec{r},t)= \vec{H}(\vec{r},t) + 4\pi\vec{M}(\vec{r},t), \tag{4}$$

where $\vec{M}$ is the rare earth magnetization. By eliminating $\vec{J}$, $\vec{E}$, $\vec{B}$ and $\vec{A}$ from (1) - (4), we obtain the relation between H and M in the following form:

$$[q^2 + \frac{K(\vec{q},\omega)}{\lambda_L{}^2} - (\frac{\omega}{c})^2]H(\vec{q},\omega)= 4\pi[- \frac{K(\vec{q},\omega)}{\lambda_L{}^2} + (\frac{\omega}{c})^2]M(\vec{q},\omega). \tag{5}$$

Here, we have taken the Fourier transform

$$\vec{f}(\vec{r},t)= \int \frac{d\vec{q}d\omega}{(2\pi)^4} \vec{f}(q,\omega)\ e^{i(\vec{q}\cdot\vec{r}-\omega t)} \tag{6}$$

with $\vec{f}= \vec{H}$, $\vec{M}$ and so on, and confined ourselves to the transverse mode where the field direction is perpendicular to the wave vector $\vec{q}$.

We introduce the dynamical susceptibility $\chi(\vec{q},\omega)$ of the spin system in the paramagnetic phase by the definition

$$\vec{M}(\vec{q},\omega)= \chi(\vec{q},\omega)\vec{H}(\vec{q},\omega). \tag{7}$$

Then from (5) and (7) we obtain the equation which describes the relation between the frequency ω and the wave number q of the normal mode:

$$q^2 + [1 + 4\pi\chi(\vec{q},\omega)][\frac{K(\vec{q},\omega)}{\lambda_L{}^2} - (\frac{\omega}{c})^2]= 0 . \tag{8}$$

This equation is denoted by the <u>normal mode equation</u>. Hereafter, we assume, for simplicity, that the <u>current is local, so</u> that we put $K(\vec{q},\omega)= 1$.

3. Static Limit

3.1 Normal Modes

First we study the static limit with $\omega= 0$. For the static susceptibility of the paramagnetic spin system, we use the expression

$$\chi(\vec{q},0)= \frac{C}{T - T_m+Dq^2}, \tag{9}$$

where T_m and C are the Curie temperature and the Curie constant of the spin system without coupling to the persistent current, and D is the exchange stiffness constant. Then the normal mode equation is written in the following form [4]:

$$(\lambda_L{}^2q^2 + 1)(Dq^2 + T - T_m) + 4\pi C= 0, \tag{10}$$

and its solution for q is obtained as follows:

(i) For $t > t_0 \equiv 1 + d + 2\sqrt{d\bar{c}}$,

$$q\lambda_L = \pm i\gamma_1, \pm i\gamma_2, \quad \text{where} \tag{11}$$

$$\binom{\gamma_1}{\gamma_2} = \left[\frac{1}{2d} \{ t + d - 1 \mp \sqrt{(t-d-1)^2 - 4d\bar{c}} \} \right]^{1/2}.$$

(ii) $t_0 > t > t_p \equiv 1 + d - 2\sqrt{d\bar{c}}$,

$$q\lambda_L = \pm k_1 \pm ik_2, \quad \pm k_1 \mp ik_2 , \quad \text{where} \tag{12}$$

$$\binom{k_1}{k_2} = \left[\left(\frac{t+\bar{c} - 1}{4d} \right)^{1/2} \mp \frac{t+d - 1}{4d} \right]^{1/2} .$$

(iii) $t = t_p$,

$$q\lambda_L = \pm k_1. \tag{13}$$

Here we have defined $t = T/T_m$, $\bar{c} = 4\pi C/T_m$ and $d = D/\lambda_L{}^2 T_m$. It is to be emphasized that the nature of the normal modes is characterized by the two characteristic temperatures t_0 and t_p; for $t > t_0$, q's of the normal modes are purely imaginary, while for $t < t_0$ they are complex numbers and at $t = t_p$ they become real.

We show in Fig. 1 a typical example of the temperature dependences of γ_1, γ_2, k_1 and k_2 [4]. In the numerical results shown in this paper, we always use the parameter values of $\bar{c} = 2.0$ and $d = 0.01$ (then $t_0 = 1.29$ and $t_p = 0.73$). The oscillatory behavior of the normal modes below t_0 is an important feature of ferromagnetic superconductors, and its physical origin is understood as follows. Through the electromagnetic interaction, the long-range part of the spin-spin interaction is screened by the persistent current, so that the interaction between spins becomes strongest for a finite wave number. As a result, the normal modes also oscillate with the finite wave number. With increasing temperature above t_0, the contribution of the spin system to the normal mode becomes smaller, and the oscillation of the normal modes vanishes.

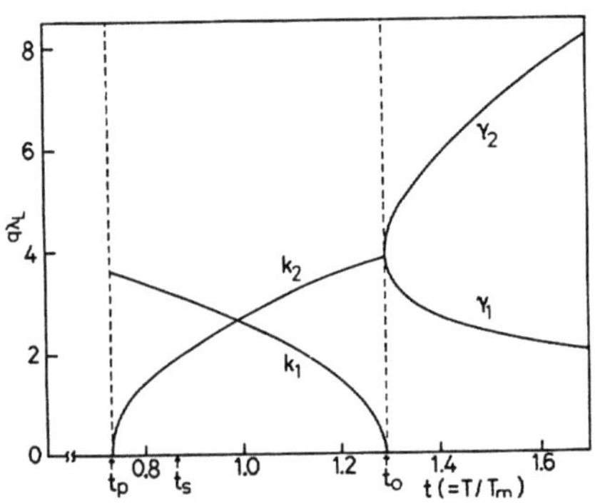

Fig. 1. Damping constants γ_1, γ_2 and k_2, and the wave number of oscillation k_1 of the static normal modes as a function of temperature

Now we can express any field in the bulk, surface and film of superconductors as a linear combination of the normal modes obtained here. In the following, we first give a brief comment on the bulk problem, and then discuss in more detail the surface and film problems.

3.2 Bulk System

(i) Field oscillation around a vortex

Let us consider the magnetic field $H(r)$ around a single vortex located in the bulk system. $H(r)$ is expressed as [4]

$$H(r)=\frac{\phi}{2\pi\lambda_L^2}\left[\frac{\tau+q_1^2}{q_1^2-q_2^2}K_0(-iq_1r/\lambda_L)+\frac{\tau+q_2^2}{q_2^2-q_1^2}K_0(-iq_2r/\lambda_L)\right], \qquad (14)$$

where τ is given by $(t-1)/d$, ϕ is the unit flux, K_0 is the modified Bessel function, and q_1 and q_2 are the wave numbers of the normal modes (decaying modes), i.e.,

$$q_1=i\gamma_1,\ q_2=i\gamma_2 \qquad \text{for } t>t_0\ , \qquad (15)$$

$$q_1=k_1+ik_2,\ q_2=-k_1+ik_2 \qquad \text{for } t<t_0.$$

Therefore, $H(r)$ decays monotonically for $t>t_0$, whereas it decays oscillatorily for $t<t_0$. When we consider two vortices separated by r, the interaction between them is proportional to $H(r)$, so that it is attractive when r is in the region of negative $H(r)$. In this case, by applying an external field, a finite density of vortices is introduced into the superconductor at the lower critical field through the first-order phase transition. Therefore, for $t<t_0$ the ferromagnetic superconductor becomes the type II/1 or type I superconductor, although it behaves as the usual type II/2 superconductor for $t>t_0$.

(ii) Spin-sinusoidal instability

We have shown that the damping constant k_2 of the normal mode vanishes at the characteristic temperature t_0. Therefore, at this temperature the static normal mode with a finite wave number k_1 can exist without decay. This corresponds to the onset of the spin-sinusoidal order, which has been observed experimentally [6].

3.3 Surface of Semi-Infinite System

(i) Field penetration

We assume that an external magnetic field H_0 is applied parallel to the surface (at $x=0$) of a semi-infinite ($x>0$) ferromagnetic superconductor, and use the boundary conditions $H=H_0$ and $dM/dx=0$ at $x=0$. Then we can express the magnetic field in the superconductor as the following linear combination of the normal modes [4] :

$$H(x)=\frac{H_0}{q_2(1+q_2^2)-q_1(1+q_1^2)}\left[q_2(1+q_2^2)e^{iq_1x/\lambda_L}-q_1(1+q_1^2)e^{iq_2x/\lambda_L}\right], \qquad (16)$$

where q_1 and q_2 are given in (15). Therefore, the field penetrating from the surface decays monotonically for $t > t_0$, while it makes an oscillatory damping for $t < t_0$.

In Fig. 2, we plot H(x) as a function of x/λ_L with t as a parameter. It is found that H(x) oscillates for $t < t_0$ and the amplitude of the oscillation increases remarkably with decreasing temperature. The denominator of (16), $q_2(1+q_2^2) - q_1(1+q_1^2)$, vanishes at a critical temperature

$$t_s \equiv 1 + \frac{d}{2} - \frac{1}{2}\sqrt{d^2+4d\bar{c}} \ .$$

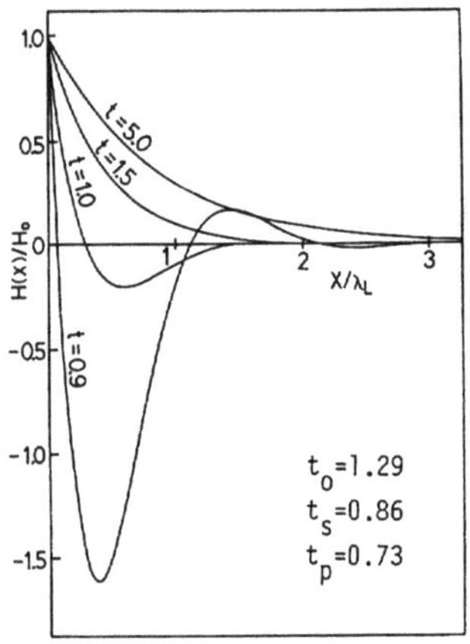

Fig. 2. Spatial variation of magnetic field penetrating from the surface at x= 0. Temperature is taken as a parameter

Thus H(x) diverges at t_s (t_s= 0.86 with our parameter values), and this indicates that the spontaneous magnetic order bound near the surface occurs below t_s. The temperature t_s is lower than t_0, but higher than t_p. For the temperature between t_s and t_p, as studied in [7], the spontaneous surface magnetization decays with an oscillation as we go into the bulk crystal. For $t < t_p$, the spontaneous surface magnetization does not decay, but is connected smoothly to the spin sinusoidal order of the bulk system.

The physical reason of the occurrence of the spontaneous surface magnetization is understood as follows. When a rare earth spin is located well inside the surface, the magnetic field around the spin is well screened by the persistent current, which extends spatially in the range of λ_L around the spin. When the spin is located within λ_L from the surface, the screening current is prevented by the surface, so that the screening effect is weakened. Therefore the interaction between spins near the surface is stronger than that in the bulk, and thus the spontaneous magnetic order near the surface occurs at the temperature (t_s) higher than that in the bulk system (t_p).

(ii) A vortex in the vicinity of the surface

Let us consider a system where H_0 (in the z direction) is applied at the surface (x= 0) and a vortex is introduced at a distance R from the surface. Here we assume the boundary condition M= 0 at x= 0, instead of dM/dx= 0, to make the calculation easier. Then the magnetic field of the system is given by

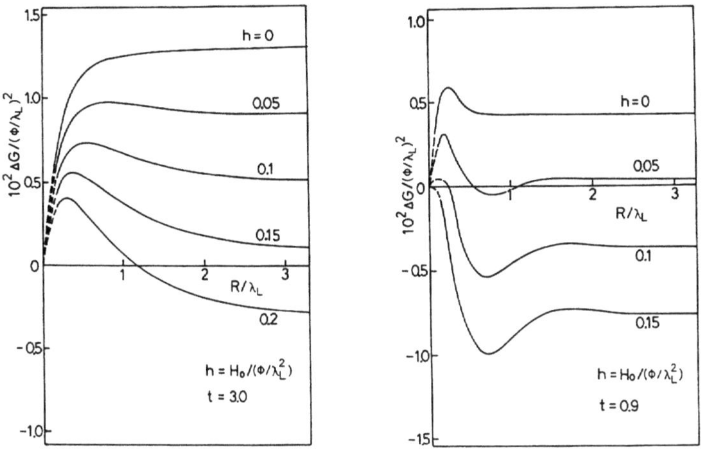

Fig. 3. Gibbs free energy ΔG of a vortex near the surface as a function of its position R (the distance from the surface) with the external field strength as a parameter. Temperature is taken as (a) t= 3.0 and (b) t= 0.9

$$H(x, y)= H_{ex}(x) + H_v(\sqrt{(x-R)^2+y^2}) + H_{iv}(\sqrt{(x+R)^2+y^2}) , \qquad (17)$$

where H_{ex} is the field penetrating from the surface, and H_v and H_{iv} are the fields around the vortex and an image vortex, respectively. The expression of $H_{ex}(x)$ is similar to (16) but slightly different due to the different boundary condition. $H_v(r)$ is the same as the expression (14), and $H_{iv}(r)= -H_v(r)$. It is to be noted that the interaction between the vortex and the penetrating field is proportional to $H_{ex}(R)$, while the interaction between the vortex and the image vortex is proportional to $H_{iv}(2R)$. For $t > t_0$, $H_{ex}(R)$ is always positive and $H_{iv}(2R)$ is always negative, whereas for $t<t_0$ both of $H_{ex}(R)$ and $H_{iv}(2R)$ oscillate as R changes.

By using H(x,y) in (17), we calculate the free energy of the system. In Fig. 3(a) and (b) we show the change of the Gibbs free energy ΔG by the introduction of the single vortex, as a function of the vortex position with the external field strength as a parameter. In the case of $t > t_0$, as shown in Fig. 3(a), ΔG for $H_0= 0$ increases monotonically with R, resulting from the attractive interaction between the vortex and the image vortex. For finite H_0, an energy barrier of ΔG appears near the surface due to the repulsive interaction between the vortex and the penetrating field. Above a critical strength (lower critical field) of H_0, ΔG becomes negative, and the stable position of the vortex is found to be well inside the surface ($R \to \infty$). However, in the case of $t < t_0$ as shown in Fig. 3(b), we have an energy barrier near the surface even for $H_0= 0$ due to the repulsive part of the interaction between the vortex and the image vortex. Furthermore, above the critical field, the stable position of the vortex is not at $R \to \infty$ but near the surface, originating from the attractive interaction between the vortex and the penetrating field.

The present calculation shows that the pinning of vortices near the surface occurs quite differently, according as the temperature $t \gtrless t_0$. In order to know further details, the calculation should be extended to the case of finite number of vortices especially for $t < t_0$, where our system is the type II/1 or type I superconductor as shown before.

3.4 Film

We study the magnetic field in the film of thickness 2a, where the external field H_0 is applied parallel to the film surfaces at $x = \pm a$. The boundary conditions are taken as $H(-a) = H(a) = H_0$ and $dM/dx = 0$ at $x = \pm a$. The magnetic field $H(x)$, which is calculated in the case of $a = \lambda_L$ as a linear combination of the normal modes, is plotted in Fig. 4 for various values of t [8]. At low temperatures ($t \lesssim t_0$), $H(x)/H_0$ is found to become negative. When t is lowered further, $H(x)$ diverges at a certain temperature, below which the spontaneous magnetic order occurs. In Fig. 5 the critical temperature where $H(x)$ diverges is plotted, with the solid curve, as a function of the film thickness. On the other hand, when we apply the external field antisymmetrically such as $H(-a) = -H(a) = H_0$, we find the different critical temperature which is plotted with the dashed curve. Therefore, the true critical temperature of the spontaneous magnetic order should be the higher one of the solid and dashed curves. Below the true critical temperature of the solid curve, the spontaneous magnetization with the symmetric pattern is stabilized, whereas below the dashed curve that with the antisymmetric pattern is stabilized, as shown in Fig. 5 schematically [9, 10].

Fig. 4

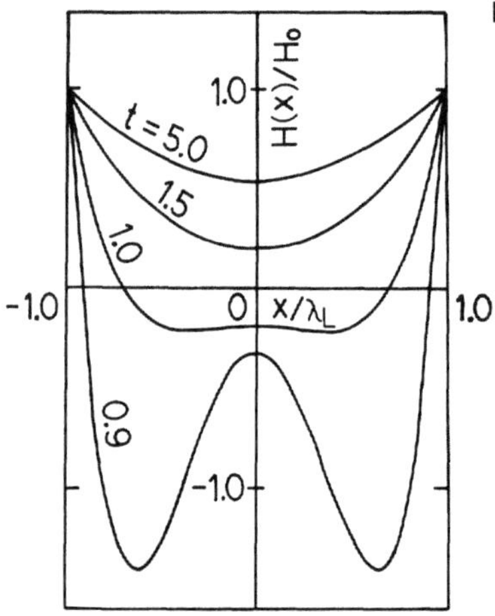

Fig. 5

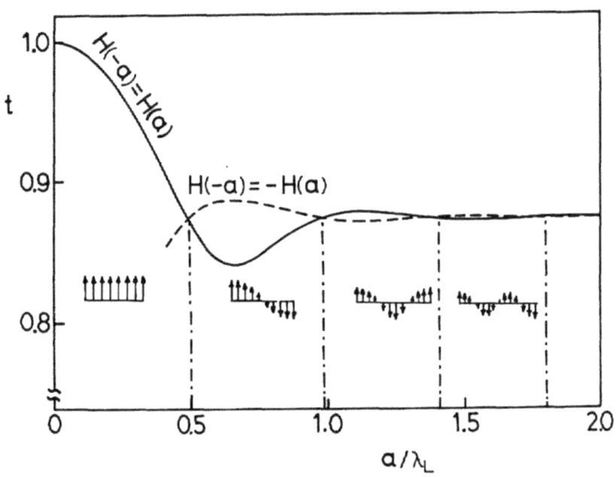

Fig. 4. Spatial variation of magnetic field inside the film of thickness $2\lambda_L$ for various values of temperature. External field is assumed to be applied as $H(-a) = H(a) = H_0$

Fig. 5. Critical temperature of the spontaneous magnetic order in the film as a function of the film thickness. The critical temperature for symmetric spin order is shown with the solid curve and that for antisymmetric spin order with the dashed curve; the true critical temperature is the higher one of them. The patterns of the stable magnetic order are also shown schematically

4. Dynamical Case

4.1 Normal Modes

The dynamical susceptibility is assumed to be

$$\chi(\vec{q},\omega)=\chi(\vec{q},0)\frac{\Gamma}{\Gamma-i\omega}, \tag{18}$$

where Γ is the damping constant of the spin system and taken as

$$\Gamma=b/\chi(\vec{q},0) \tag{19}$$

with a constant b [2, 5]. Then the normal mode equation is written in the form

$$q^2+[1+\frac{4\pi bC}{b(T-T_m+Dq^2)-i\omega C}][\frac{1}{\lambda_L^2}-(\frac{\omega}{c})^2]=0, \tag{20}$$

from which q^2 is expressed as functions of ω and t by

$$(q\lambda_L)^2=\frac{1}{2d}[-(t-1-i\frac{\bar{\omega}}{\bar{b}})-d(1-\bar{\omega}^2)$$
$$\pm\{[t-1-i\frac{\bar{\omega}}{\bar{b}}-d(1-\bar{\omega}^2)]^2-4d\bar{c}(1-\bar{\omega}^2)\}^{1/2}]. \tag{21}$$

Here $\bar{\omega}$ and $\bar{b}$ are defined by

$$\bar{\omega}=\omega\lambda_L/c \quad\text{and}\quad \bar{b}=\lambda_L bT_m/cC,$$

respectively.

In this way $q\lambda_L$ is given by the form

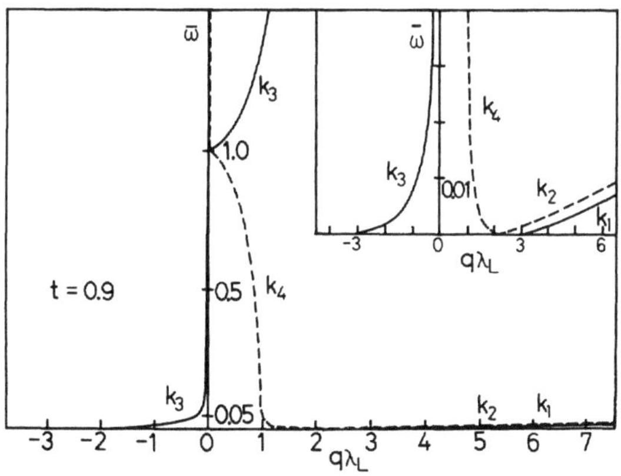

Fig. 6. Frequency ($\bar{\omega}$) vs wave number (k_1 and k_3) of the dynamic normal modes at the temperature t= 0.9, plotted with the solid curves. The dampings (k_2 and k_4) are plotted with the dashed curves. The inset shows the result in the low frequency region

$$q\lambda_L = \pm(k_1+ik_2), \quad \pm(k_3+ik_4) \quad ,$$

and the relations between the frequency ω and wave numbers $k_1 \sim k_4$ are shown in Fig. 6 for a fixed temperature t= 0.9 [5]. Here $\bar{b}$ is taken as 0.01. When $\bar{\omega}$ is larger than 1.0, namely when $\bar{\omega}$ is larger than the plasma frequency c/λ_L, the $\bar{\omega}$ versus k_3 relation represents the dispersion of the transverse plasmon, which is a coupled mode of the photon and the current. In such a high frequency region, the spin system gives no effect on the normal mode because it cannot follow the rapid motion of the plasmon. When $\bar{\omega}$ becomes smaller than 1.0, the plasmon changes to a monotonically decaying mode with vanishing k_3. However, when $\bar{\omega}$ becomes small enough, the spin system couples with the plasmon, and as a result the wave number k_3 reappears with the negative value. On the other hand, the mode consisting of k_1 and k_2 originates mainly from the spin motion, and is modified by the coupling with the damped plasmon. In this way, we obtain the coupled modes of the spin, photon and current in the low frequency region, as plotted in the inset with the extended scale of $\bar{\omega}$.

4.2 Reflectance for Electromagnetic Wave

(i) Semi-infinite system

When the electromagnetic wave is incident normally on the surface of the semi-infinite system, the electromagnetic wave inside the superconductor is expressed as a linear combination of the dynamical normal modes obtained above. By using the boundary conditions that E and H are continuous and dM/dx vanishes at the surface, we can calculate the amplitudes of the normal modes and the reflected electromagnetic wave. In Fig. 7, we show the reflectance for the electromagnetic wave calculated in this way, as a function of t with $\bar{\omega}$ as a parameter. The deviation of R from 1.0 originates from the energy loss of the electromagnetic wave inside the superconductor due to the energy absorption by the spin system. The energy absorption occurs when the motion of the spin system has a finite delay behind the driving electromagnetic wave. When $\bar{\omega}$ is sufficiently small (as the case $\bar{\omega}= 10^{-5}$) the spins can follow the electromagnetic field, so that we have $R \sim 1.0$. When $\bar{\omega}$ increases (as the cases $\bar{\omega}= 10^{-4}$ and 10^{-3}), the spin system can still follow the driving electromagnetic wave at high temperatures, but it

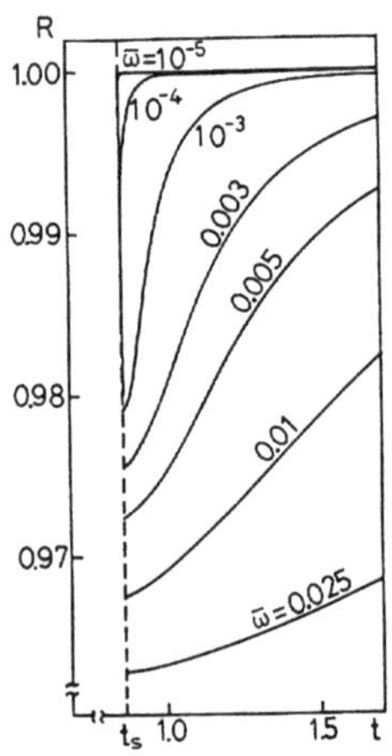

Fig. 7. Reflectance of the semi-infinite system for a normally incident electromagnetic wave as a function of the temperature with the frequency as a parameter. The temperature t_s indicates the critical temperature of the spontaneous surface magnetization

has a finite delay near the critical temperature t_s of the spontaneous surface magnetization, because of the critical slowing down of the spin motion. Thus, we have a sharp decrease of R in the neighborhood of t_s. The energy loss of the electromagnetic wave near t_s will also be reflected in the surface impedance [5].

(ii) Film

We discuss here the reflectance R of the film for a normally incident electromagnetic wave. When temperature is changed, R shows a sharp drop at the critical temperature of the spontaneous magnetic order [11], in a similar way to the case of semi-infinite system. Moreover, R also changes characteristically as a function of the film thickness 2a. In Fig. 8, R is plotted as a function of a/λ_L with $\bar{\omega}$ as a parameter [11]. Here the temperature is fixed at 0.9, and then we have the magnetic order when the thickness 2a becomes smaller than a critical value $2a_c$ ($a_c/\lambda_L \sim 0.35$, see also Fig. 5). It is found that R shows the oscillatory behavior as a function of the film thickness. This is in sharp contrast with R in the usual nonmagnetic superconductors, where R decreases monotonically with decrease of the film thickness, as shown here with the thin dashed curves for comparison.

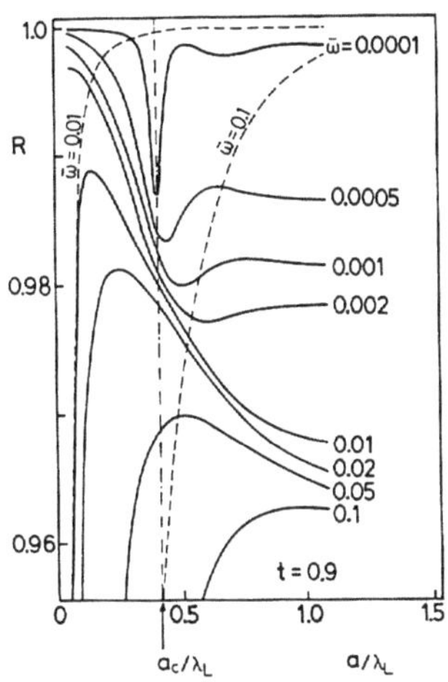

Fig. 8. Reflectance of the film for a normally incident electromagnetic wave as a function of the film thickness with the frequency as a parameter. Temperature is taken as t=0.9 and the critical thickness of the spontaneous magnetic order is indicated with the chain line. The reflectance of the nonmagnetic superconductor is also shown with thin dashed curves for comparison

5. Concluding Remarks

We have obtained both the static and dynamic normal modes of ferromagnetic superconductors as the coupled modes of spin, photon and electric current. With the use of the normal modes, we have discussed various electromagnetic properties of the bulk, surface and film of ferromagnetic superconductors. Many unusual properties, especially for the surface and the film, have been predicted theoretically, and they are expected to be observed experimentally in the near fututre.

We have confined ourselves to the local electric current, for simplicity. However, we expect that the most essential features in this paper will not be changed by the nonlocal current effect. The effect of the nonlocal persistent current on the static field penetration from the surface was already studied by one of the present authors and his coworkers [12], and the essential features, i.e.,the oscillation of the penetrating field and the occurrence of the spontaneous surface magnetization, were still found in the nonlocal case. More recently, KLEIN [13] also studied the same problem. When we take account of the nonlocal current effect, an infinite number of normal modes occurs. However, KLEIN took account only of a small number of them (the same number as that in the local case), so that his calculation is not accurate. On the other hand, our calculation [12] is accurate enough.

We have mainly used the boundary condition for the magnetization in the form dM/dx= 0. It is not well established what boundary condition is most appropriate for M. The effect of the change of the boundary condition on the field penetration was also studied in our paper [12]. The penetrating field depends quantitatively on the boundary condition, but the essential features are unchanged. When a vortex is in the vicinity of the surface (the problem of 3.3 (ii) in this paper), the free energy of the system depends sensitively on the boundary condition for M, and the details will be published elsewhere.

Part of this paper is based on the collaborations with Dr. S. Takahashi, Dr. T. Koyama, Prof. M. Tachiki, Prof. H. Matsumoto and Prof. H. Umezawa, to whom the authors would like to express their sincere thanks.

References

1 H. Matsumoto, H. Umezawa and M. Tachiki: Solid State Commun. 31, 157 (1979).
2 M. Tachiki, A. Kotani, H. Matsumoto and H. Umezawa: Solid State Commun. 31, 927 (1979).
3 E. I. Blount and C. M. Varma: Phys. Rev. Lett. 42, 1079 (1979).
4 A. Kotani, M. Tachiki, H. Matsumoto, H. Umezawa and S. Takahashi: Phys. Rev. B23, 5960 (1981).
5 A. Kotani and K. Okada: Solid State Commun. 46, 811 (1983).
6 S. K. Sinha, G. W. Crabtree, D. G. Hinks and H. Mook: Phys. Rev. Lett. 48, 950 (1982).
7 T. Koyama, A. Kotani, S. Takahashi, M. Tachiki, H. Matsumoto and H. Umezawa: J. Phys. Soc. Jpn. 51, 3469 (1982).
8 A. Kotani, S. Takahashi, M. Tachiki, H. Matsumoto and H. Umezawa: Solid State Commun. 37, 619 (1981).
9 A. Kotani, S. Takahashi, M. Tachiki, H. Matsumoto and H. Umezawa: Physica 108 B+C, 1043 (1981).
10 S. Takahashi, A. Kotani, M. Tachiki, H. Matsumoto and H. Umezawa: J. Phys. Soc. Jpn. 52, 989 (1983).
11 K. Okada: thesis, Osaka University (1983), unpublished.
12 A. Kotani, S. Takahashi, T. Koyama, M. Tachiki, H. Matsumoto and H. Umezawa: J. Phys. Soc. Jpn. 50, 3254 (1981).
13 U. Klein: Phys. Rev. B27, 1925 (1983).

Josephson Effect in Ferromagnetic Superconductors

T. Koyama and M. Tachiki

The Research Institute for Iron, Steel and Other Metals
Tohoku University, Sendai 980, Japan

1. Introduction

In the experiments for the Josephson effect in the ferromagnetic superconductors such as $ErRh_4B_4$, the following anomalous properties have been observed [1,2].

(1) The dc Josephson current disappears at low temperature slightly above the reentrant transition temperature.
(2) In the external field dependence of the maximum current (I_c vs. H_{ex} curve) the central peak of the Fraunhofer pattern shifts and splits on cooling just before the Josephson current is quenched.

In this note we develop a theory for the Josephson effect, taking account of the effect of the localized spin magnetization of the rare earth ions. We assume that the electromagnetic interaction dominantly works between the superconducting current and the localized spin magnetization of the rare earth ions. One of the roles of the localized spin magnetization is to change the magnetic flux around the junction through the interaction. This effect leads to a modulation of the spatial distribution of the tunneling current and a decrease in the Josephson penetration depth λ_J. Another important effect of the localized spin magnetization comes from its critical fluctuations near the magnetic transition temperature. As shown in a later section, the critical spin fluctuation can cause the fluctuations of the phase difference across the junction, so that the noise current is generated by the critical spin fluctuations. Thus, the characteristics of the Josephson junction are expected to deviate from the usual ones in non-magnetic superconductors by the existence of the localized spin magnetization at low temperatures.

In the following sections we derive basic equations for our system and calculate the temperature and external field dependence of the maximum Josephson current. The experimental results mentioned above are interpreted on the basis of the present calculation.

2. Basic Equations

Let us consider a tunneling junction made by two ferromagnetic superconductors. The equation for the phase difference $\gamma(x, t)$ can be derived from the Maxwell equations in the usual manner by noting the relation

$$\vec{b}(\vec{r}, t) = \vec{h}(\vec{r}, t) + 4\pi\vec{m}(\vec{r}, t) \quad , \tag{2.1}$$

where $\vec{b}$, $\vec{h}$ and $\vec{m}$ are the magnetic induction field, the magnetic field related to the superconducting current, and the localized spin magnetization, respectively. In the following we take the junction in the x-y plane at z = 0 and assume that $\vec{b}$, $\vec{h}$ and $\vec{m}$ are parallel to the y axis and uniform in this direction. In this case the equation is obtained [3] as

$$\partial_x^2\gamma(x, t) - \frac{1}{\bar{c}^2}\partial_t^2\gamma(x, t) - \frac{g}{\bar{c}^2}\partial_t\gamma(x, t) - \frac{8\pi^2}{\phi_0}\int_{-\infty}^{\infty} dz e^{-|z|/\lambda_L}\partial_x m(x, z, t)$$
$$= \frac{1}{\lambda_J^2}\sin\gamma(x, t) \quad , \tag{2.2}$$

where $\bar{c}$ is light velocity in the junction, g is the normalized conductance, ϕ_0 is the unit flux (hc/2e), λ_L is the London penetration depth, and λ_J is the Josephson penetration depth.

In order to solve eq. (2.2), we need one more equation for the localized spin magnetization m(x, z, t). The equation is set up as follows, based on the time-dependent Ginzburg-Landau equation

$$\partial_t m(x, z, t) = -\beta\frac{\delta F[\gamma, m]}{\delta m(x, z, t)} \quad , \tag{2.3}$$

with β being a positive constant. In the above equation $F[\gamma, m]$ is the free energy of our system and is expressed in the form

$$F[\gamma, m] = F_{el}[\gamma] + F_{spin}[m] - \frac{\phi_0}{4\pi\lambda_L}\int dx\partial_x\gamma(x)\int dz e^{-|z|/\lambda_L} m(x, z, t) \quad , \tag{2.4}$$

where $F_{el}[\gamma]$ ($F_{spin}[m]$) represents the part which depends only on $\gamma(x, t)$ ($m(x, z, t)$). Therefore, $F_{el}[\gamma]$ should have the same form as the one for the non-magnetic system. Assuming that $F_{spin}[m]$ can be expanded in powers of m and keeping the lowest order term, we have

$$F_{spin}[m] = \int dx\int dz \frac{1}{2} m(x, z, t)\tilde{\alpha}(T, \partial_z)m(x, z, t) \quad . \tag{2.5}$$

In the following we confine ourselves to the paramagnetic phase. The Fourier component of the coefficient $\tilde{\alpha}(T, \partial_z)$ in (2.5) is taken as

$$\tilde{\alpha}(T, q) = \frac{1}{\chi_0(T, q)} + \frac{4\pi\lambda_L^{-2}}{q^2 + \lambda_L^{-2}} \quad , \tag{2.6}$$

where $\chi_0(T, q)$ is the susceptibility in the normal state. The second term comes from the diamagnetic screening effect due to the superconducting current. This choice of $\tilde{\alpha}(T, q)$ is based on the theory by BLOUNT and VARMA [4] and MATSUMOTO et al. [5]. Therefore, $\tilde{\alpha}(T, q)$ vanished at T_p, the transition temperature of the periodic phase observed in the ferromagnetic superconductors. Now substituting (2.4) into (2.3), we have an equation

$$\partial_t m(x, z, t) = -\beta\tilde{\alpha}(T, \partial_z)m(x, z, t) + \frac{\beta\phi_0}{4\pi\lambda_L}\partial_x\gamma(x, t)e^{-|z|/\lambda_L} \quad . \tag{2.7}$$

Eqs. (2.2) and (2.7) provide the basic equations for the Josephson effect in the ferromagnetic superconductors.

Let us consider the effect of fluctuations. When the temperature is close to the magnetic transition temperature, the critical fluctuations of the localized spin magnetization arise. The fluctuations can couple to the phase difference γ through eqs. (2.2) and (2.7). This fact indicates that the fluctuations of the phase difference γ result from the critical spin fluctuations near the magnetic transition temperature. In the following we take account of the fluctuations of γ and m in eqs. (2.2) and (2.7), based on the stochastic theory. The equations can be rewritten in the Langevin type stochastic equations if we look on the functions $\gamma(x, t)$, $\eta(x, t) \equiv \partial_t\gamma(x, t)$ and $m(x, z, t)$ as stochastic functions, introducing random forces:

$$\text{(i)}\ \frac{1}{\bar{c}^2}\,\partial_t\eta(x, t) = -\frac{g}{\bar{c}^2}\,\eta(x, t) + \partial_x^{\,2}\gamma(x, t) - \frac{8\pi^2}{\phi_0}\int dz e^{-|z|/\lambda_L}\,\partial_x m(x, z, t) - \frac{1}{\lambda_J^{\,2}}\sin\gamma(x, t) + \frac{1}{\lambda_J^{\,2}}R_1(x, t)/j_0\ , \tag{2.8}$$

$$\text{(ii)}\ \partial_t\gamma(x, t) = \eta(x, t)\ , \tag{2.9}$$

$$\text{(iii)}\ \partial_t m(x, z, t) = -\beta\tilde{\alpha}(T, \partial_z)m(x, z, t) + \frac{\phi_0}{4\pi\lambda_L}\,\partial_x\gamma(x, t)e^{-|z|/\lambda_L} + R_2(x, z, t)\ , \tag{2.10}$$

where $R_1(x, t)$ and $R_2(x, z, t)$ represent random forces, respectively, due to the resistive flow of the quasi-electron current and thermal fluctuations of the localized spin magnetization. It is assumed in the following that $R_1(x, t)$ and $R_2(x, z, t)$ are statistically independent of one another and their distributions are given by a Gaussian with white spectrum, for simplicity. We have, then, the following correlation functions for the random forces:

$$\langle R_1(x, t)R_1(x', t')\rangle = \frac{2}{RL}\,k_BT\delta(x - x')\delta(t - t')\ , \tag{2.11}$$

$$\langle R_2(x, z, t)R_2(x', z', t')\rangle = 2\beta k_BT\delta(x - x')\delta(z - z')\delta(t - t')\ , \tag{2.12}$$

$$\langle R_1(x, t)R_2(x', z', t')\rangle = 0\ . \tag{2.13}$$

Here R is the resistance of the junction and L is its extent.

Let us introduce the probability density functional $P[\gamma, \eta, m, t]$. The Fokker-Planck equation for $P[\gamma, \eta, m, t]$ can be derived from eqs. (2.8) - (2.10) with the use of (2.11) - (2.13). We have

$$\begin{aligned}\partial_t P[\gamma, \eta, m, t] = \int dx\Big[&-\frac{\delta}{\delta\gamma(x, t)}\,\eta(x, t) + \frac{\delta}{\delta\eta(x, t)}\,\{g\eta(x, t) - \bar{c}^2(\partial_x^{\,2}\gamma(x, t) \\ &- \frac{8\pi^2}{\phi_0}\int_{-\infty}^{\infty} dz e^{-|z|/\lambda_L}\,\partial_x m(x, z, t) - \frac{1}{\lambda_J^{\,2}}\sin\gamma(x, t)) - g\cdot\frac{32\pi^3\lambda_L}{\phi_0}\,k_BT\,\frac{\delta}{\delta\eta(x, t)}\} \\ &+ \int dz\,\frac{\delta}{\delta m(x, z, t)}\,\{\beta\tilde{\alpha}(T, \partial_z)m(x, z, t) - \frac{\beta\phi_0}{4\pi\lambda_L}\,\partial_x\gamma(x, t)e^{-|z|/\lambda_L} \\ &+ \beta k_BT\,\frac{\delta}{\delta m(x, z, t)}\}\Big]\,P[\gamma, \eta, m, t]\ .\end{aligned} \tag{2.14}$$

The noise averaged values of any physical quantities can be obtained, using the solution to eq. (2.14), from

$$\langle A\rangle_{av.} \equiv \int \delta\gamma\delta\eta\delta m A P[\gamma, \eta, m, t] \quad . \tag{2.15}$$

Here the integration on the right-hand side is understood to be a functional integration by $\gamma(x)$, $\eta(x)$ and $m(x)$. In the next section, we calculate the averaged value of the Josephson current, solving (2.14) in the dc case and investigate the dc Josephson effect near the magnetic transition temperature.

3. dc Josephson Effect

Before calculating the Josephson current we examine the equilibrium state without a tunneling current to see how the critical spin fluctuations induce the fluctuations of γ. In this case, setting $\partial_t P[\gamma, \eta, m, t] = 0$ in eq. (2.14), we have the following probability density functional:

$$\begin{aligned} P[\gamma, \eta, m] = C \exp[- \frac{1}{k_B T} \frac{\phi_0{}^2 \ell}{32\pi^3 \lambda_L} \int dx \{ \frac{1}{2\bar{c}^2} \eta^2(x) + \frac{1}{2} (\partial_x \gamma(x))^2 - \frac{8\pi^2}{\phi_0} \partial_x \gamma(x) \\ \times \int dz e^{-|z|/\lambda_L} m(x, z) + \frac{1}{\lambda_J{}^2} (1 - \cos\gamma(x)) \} \\ - \frac{\ell}{k_B T} \int dx dz \frac{1}{2} m(x, z) \tilde{\alpha}(T, \partial_z) m(x, z) \}] \quad , \end{aligned} \tag{3.1}$$

where C is a normalization constant and ℓ is a characteristic length in the y direction. The role of the critical spin fluctuations on the characteristic of the junction can be visualized from (3.1) if we eliminate $\eta(x)$ and $m(x, z)$ by performing the functional integration as follows:

$$\begin{aligned} P[\gamma] &\equiv \int \delta\eta\delta m P[\gamma, \eta, m] \\ &= C \exp[- \frac{1}{k_B T} \frac{\phi_0{}^2 \ell}{32\pi^3 \lambda_L} \int dx \{ \frac{1}{2} S(T) (\partial_x \gamma(x))^2 + \frac{1}{\lambda_J{}^2} (1 - \cos\gamma(x)) \}] \quad , \end{aligned} \tag{3.2}$$

with

$$S(T) = 1 - \frac{2\pi}{\lambda_L} \int_{-\infty}^{\infty} dz e^{-|z|/\lambda_L} \tilde{\alpha}(T, \partial_z) e^{-|z|/\lambda_L} \quad . \tag{3.3}$$

The effect of the spin fluctuations thus appears through the function $S(T)$. This function has the following properties. By the previous assumption that $\tilde{\alpha}(T, \partial_z)$ vanishes at a temperature T_p, the inverse function $\tilde{\alpha}^{-1}(T, \partial_z)$ diverges at T_p. This fact means that $S(T)$ decreases with decreasing temperature and becomes zero at a temperature near and above T_p. We denote this temperature by T_s. This result indicates that the inhomogeneous fluctuations of γ are enhanced near T_s because the term including $(\partial_x\gamma)^2$ in (3.2) decreases as $T \to T_s$ as a result of the fluctuations of the localized spin magnetization. It can be shown that the temperature T_s is equal to the transition temperature at which the spontaneous magnetization bound to the surface of the ferromagnetic superconductors appears [6-9].

Since a Josephson junction may be regarded as a kind of surface, the surface magnetization is expected to appear around the junction at T_S. The temperature dependence of S(T) is shown by the dot-dashed line in Fig. 1. In the calculation we used the values of the parameters in the susceptibility $\chi_0(T, q)$, $\bar{c} = 4\pi C/T_m = 2$ and $d = D/\lambda_L{}^2 T_m = 0.01$, where C, D and T_m are the Curie constant, the magnetic stiffness constant and the Curie temperature in the normal state, respectively. The above choice of the parameters leads to the transition temperatures, $T_S/T_m = 0.761$ and $T_p/T_m = 0.745$.

Let us consider the case in which the dc Josephson current flows through the junction and an external field is applied in the y direction. In this case $\gamma(x, t)$ and $m(x, z, t)$ can be separated into their average values, $\gamma_0(x)$ and $m_0(x, z)$, and their fluctuations, $\tilde{\gamma}(x, t)$ and $\tilde{m}(x, z, t)$, respectively as

$$\gamma(x, t) = \gamma_0(x) + \tilde{\gamma}(x, t) \quad , \tag{3.4}$$

$$m(x, z, t) = m_0(x, z) + \tilde{m}(x, z, t) \quad . \tag{3.5}$$

Substituting (3.4) and (3.5) into eqs. (2.8) - (2.10) and separating the equations into the ones for the average values and for the fluctuations, we obtain the following equations:

$$\left\{\begin{aligned} &\text{(i)} \quad \partial_x{}^2\gamma_0(x) - \frac{8\pi^2}{\phi_0}\int_{-\infty}^{\infty} dz e^{-|z|/\lambda_L}\partial_x m_0(x, z) = \frac{1}{\lambda_J{}^2}\sin\gamma_0(x) \quad , && (3.6)\\ &\text{(ii)} \quad \tilde{\alpha}(T, \partial_z)m_0(x, z) - \frac{\phi_0}{4\pi\lambda_L}\partial_x\gamma_0(x)e^{-|z|/\lambda_L} = 0 \quad , && (3.7)\end{aligned}\right.$$

and

$$\left\{\begin{aligned} &\text{(i)} \quad \frac{1}{\bar{c}^2}\partial_t \eta(x, t) = -\frac{g}{\bar{c}^2}\eta(x, t) + \partial_x{}^2\tilde{\gamma}(x, t)\\ &\qquad - \frac{8\pi^2}{\phi_0}\int_{-\infty}^{\infty} dz e^{-|z|/\lambda_L}\partial_x\tilde{m}(x, z, t)\\ &\qquad - \frac{1}{\lambda_J{}^2}\sin(\gamma_0(x) + \tilde{\gamma}(x, t)) + \frac{1}{\lambda_J{}^2}\sin\gamma_0(x)\\ &\qquad + \frac{1}{\lambda_J{}^2}R_1(x, t)/j_0 \quad , && (3.8)\\ &\text{(ii)} \quad \partial_t\tilde{\gamma}(x, t) = \tilde{\eta}(x, t) \quad , && (3.9)\\ &\text{(iii)} \quad \partial_t\tilde{m}(x, z, t) = -\beta\tilde{\alpha}(T, \partial_z)\tilde{m}(x, z, t) + \frac{\phi_0}{4\pi\lambda_L}\partial_x\tilde{\gamma}(x, t)e^{-|z|/\lambda_L}\\ &\qquad + R_2(x, z, t) \quad . && (3.10)\end{aligned}\right.$$

First we consider eqs. (3.6) and (3.7). These coupled equations lead to the following equation for $\gamma_0(x)$.

$$S(T)\partial_x{}^2\gamma_0(x) = \frac{1}{\lambda_J{}^2}\sin\gamma_0(x) \quad . \tag{3.11}$$

Here, the function S(T) is the same as (3.3). Eq. (3.11) tells that the Josephson penetration depth λ_J should be scaled into $\lambda_J{}^{eff} = S^{1/2}(T)\lambda_J$ in the magnetic superconductors. The effective Josephson penetration depth $\lambda_J{}^{eff}$ therefore shortens with decreasing temperature and vanishes at T_s. This result implies that the self-field effect [10] becomes important as the temperature approaches T_s even in the case that λ_J is larger than the junction width.

The probability density functional for the fluctuations in this case can be obtained from eqs. (3.8) - (3.10):

$$P[\tilde{\gamma}, \tilde{n}, \tilde{m}; \gamma_0, m_0] = C\exp[-\frac{1}{k_BT}(\frac{\phi_0{}^2\ell}{32\pi^3\lambda_L})\int dx\{\frac{1}{2\bar{c}^2}\tilde{n}^2(x) + \frac{1}{2}(\partial_x\tilde{\gamma}(x))^2$$

$$-\frac{8\pi^2}{\phi_0}\partial_x\tilde{\gamma}(x)\int_{-\infty}^{\infty} dz e^{-|z|/\lambda_L}\tilde{m}(x, z) - \frac{1}{\lambda_J{}^2}[\cos(\gamma_0(x) + \tilde{\gamma}(x)) - \cos\gamma_0(x)$$

$$+\sin\gamma_0(x)\cdot\tilde{\gamma}(x) - \frac{16\pi^3}{\phi_0{}^2}\int dz\tilde{m}(x, z)\tilde{\alpha}(T, \partial_z)\tilde{m}(x, z)\} \quad . \tag{3.12}$$

This result can be derived also by substituting (3.4) and (3.5) into (3.1) with the use of eqs. (3.6) and (3.7). Now, let us calculate the total DC Josephson current. The current I is obtained by averaging $j_0\sin\gamma(x)$ and summing over the junction width as

$$I = j_0\int_{-L/2}^{L/2} dx\int\delta\tilde{\gamma}\delta\tilde{n}\delta\tilde{m}\,\sin(\gamma_0(x) + \tilde{\gamma}(x))P[\tilde{\gamma}, \tilde{n}, \tilde{m}; \gamma_0, m_0]$$

$$= j_0\int_{-L/2}^{L/2} dx\int\delta\tilde{\gamma}\,\sin(\gamma_0(x) + \tilde{\gamma}(x))P[\tilde{\gamma}; \gamma_0] \quad , \tag{3.13}$$

where

$$P[\tilde{\gamma}; \gamma_0] = C\exp[-\frac{1}{k_BT}(\frac{\phi_0{}^2\ell}{32\pi^3\lambda_L})\int dx\{\frac{S(T)}{2}(\partial_x\tilde{\gamma}(x))^2$$

$$-\frac{1}{\lambda_J{}^2}(\sin\gamma_0(x)\cdot\tilde{\gamma}(x) + \cos(\gamma_0(x) + \tilde{\gamma}(x)) - \cos\gamma_0(x))\}] \quad . \tag{3.14}$$

Because the functional integration in (3.13) can not be performed without a certain approximation, we make the following approximation. Introducing the Gaussian functional

$$P_0[\tilde{\gamma}; \gamma_0] = \exp[-\frac{1}{k_BT}(\frac{\phi_0{}^2\ell}{32\pi^3\lambda_L})\int dx\{\frac{S(T)}{2}(\partial_x\tilde{\gamma}(x))^2$$

$$+ \frac{1}{2\lambda_J^2} \cos\gamma_0(x)\cdot\tilde{\gamma}^2(x)\}]$$

$$\equiv \exp[- \frac{1}{2} \int dxdy\tilde{\gamma}(x)K(x, y)\tilde{\gamma}(y)] \tag{3.15}$$

with

$$K(x, y) = \frac{\phi_0^2 \ell}{32\pi^3\lambda_L} \frac{1}{k_BT} [-S(T)\partial_x^2 + \frac{1}{\lambda_J^2} \cos\gamma_0(x)]\delta(x-y) \quad , \tag{3.16}$$

we rewrite (3.14) in the form

$$P[\tilde{\gamma}; \gamma_0] = C\ P_0[\tilde{\gamma}; \gamma_0]\exp[\frac{1}{k_BT} (\frac{\phi_0^2\ell}{32\pi^3\lambda_L}) \frac{1}{\lambda_J^2} \int dx\{\cos(\gamma_0(x) + \tilde{\gamma}(x))$$

$$+ \sin\gamma_0(x)\cdot\tilde{\gamma}(x) + \frac{1}{2} \cos\gamma_0(x)\cdot\gamma^2(x) - \cos\gamma_0(x)\}] \tag{3.17}$$

$$= C\ P_0[\tilde{\gamma}; \gamma_0][1 + \frac{\phi_0^2\ell}{32\pi^3\lambda_L k_BT} \frac{1}{\lambda_J^2} \int dx\{\cos(\gamma_0(x) + \tilde{\gamma}(x))$$

$$+ \sin\gamma_0(x)\cdot\tilde{\gamma}(x) + \frac{1}{2} \cos\gamma_0(x)\cdot\tilde{\gamma}^2(x) - \cos\gamma_0(x)\} + \cdots] . \tag{3.18}$$

In deriving (3.18) we expanded the exponential factor in (3.17) into Taylor's series and kept only the terms up to first order. The normalization constant C is determined by the condition

$$\int \delta\tilde{\gamma}\ P[\tilde{\gamma}; \gamma_0] = 1 \quad , \tag{3.19}$$

which leads to

$$C = \exp[- \frac{1}{2} \int dx \ell n K^{-1}(x, x)]\{1 + \frac{\phi_0^2\ell}{32\pi^3\lambda_L k_BT} \frac{1}{\lambda_J^2} \int dx[\cos\gamma_0(x)e^{- \frac{1}{2} K^{-1}(x, x)}$$

$$+ \frac{1}{2} \cos\gamma_0(x)\cdot K^{-1}(x, x) - \cos\gamma_0(x)]\}^{-1} . \tag{3.20}$$

The inverse function $K^{-1}(x, x)$ in (3.20) is the solution of the equation

$$(\frac{\phi_0^2\ell}{32\pi^3\lambda_L k_BT})[-S(T)\partial_x^2 + \frac{1}{\lambda_J^2} \cos\gamma_0(x)]K^{-1}(x, y) = \delta(x-y) \quad . \tag{3.21}$$

Now the functional integration (3.13) can be performed by the use of (3.18). The result is given by

$$I = j_0\tilde{C} \int_{-L/2}^{L/2} dx \exp[- \frac{1}{2} K^{-1}(x, x)]\{\sin\gamma_0(x)$$

$$+ \left(\frac{\phi_0{}^2 \ell}{32\pi^3 \lambda_L k_B T} \right) \frac{1}{\lambda_J{}^2} \int_{-L/2}^{L/2} dx' [\sin\gamma_0(x) \cos\gamma_0(x') \{ e^{-K^{-1}(x,x) - \frac{1}{2}K^{-1}(x',x')}$$

$$+ \frac{1}{2} [K^{-1}(x, x') - (K^{-1}(x, x'))^2] - 1\}$$

$$+ \sin\gamma_0(x') \cos\gamma_0(x) K^{-1}(x, x')]\} \quad , \quad \text{where} \tag{3.22}$$

$$\tilde{C} = \{1 + \frac{\phi_0{}^2 \ell}{32\pi^3 \lambda_L k_B T} \frac{1}{\lambda_J{}^2} \int_{-L/2}^{L/2} dx \cos\gamma_0(x) [e^{-\frac{1}{2}K^{-1}(x,x)} + \frac{1}{2} K^{-1}(x,x) - 1]\}^{-1} . \tag{3.23}$$

It is noted that the Josephson current is reduced by the factor exp $[-K^{-1}(x,x)/2]$. To see the behavior of the factor we put $\gamma_0(x)$ to be a constant. This circumstance is realized in the case of small junction without an applied field. Setting $\gamma_0(x) = \gamma_0$ in eq. (3.21), we have the solution

$$K^{-1}(x, x) = \text{const.} = \nu t / 2\sqrt{S(T)\cos\gamma_0} \tag{3.24}$$

with $\nu = (32\pi^3 \lambda_L \lambda_J k_B T_m / \phi_0{}^2 \ell)$ and $t = T/T_m$. This result shows that the factor $\exp[-K^{-1}(x, x)/2]$ vanishes at temperature T_s as a consequence of the critical spin fluctuations. In Fig. 1 numerical results obtained from (3.22) - (3.24) for the temperature dependence of the maximum current are presented for the parameters, ν = 0.05, 0.1 and 0.2. In the calculation γ_0 was chosen so that the current takes a maximum value. As seen in the figure, the maximum current shows a sharp decrease just above T_s. Such a behavior of the temperature variation has been observed at low temperatures by ROWELL et al. [2], and UMBACH and GOLDMAN [1]. Therefore, we conclude that the dc Josephson effect in the ferromagnetic superconductors is quenched by the critical spin fluctuations at T_s, the temperature at which surface magnetization coexisting with superconductivity appears.

Next we show the numerical results for the external field dependence of the maximum Josephson current. The procedure of the calculation is as follows. From eq. (3.11) we first obtain

$$\gamma_0(x) = -\sin^{-1}[2\mathrm{sn}(\frac{x - x_0}{k\lambda_J{}^{\mathrm{eff}}} |k^2) \mathrm{cn}(\frac{x - x_0}{k\lambda_J{}^{\mathrm{eff}}} |k^2)] \ , \tag{3.25}$$

where sn and cn denote the elliptic functions. Then, substituting (3.25) into (3.21), we get numerically the function $K^{-1}(x, y)$. The total current can be calculated from (3.22), using the result for $K^{-1}(x, y)$. The value of the parameters k and x_0 in (3.25) were chosen in such a way that the total current I takes a maximum value under an given external field. In Fig. 2(a) we plotted two branches of the I_c vs. H_{ex} curves at several temperatures. The branches located in the weaker (stronger) field region correspond to the mode with less than one flux quantum ϕ_0 (more than ϕ_0 but less than $2\phi_0$) in the junction. As seen in the figure, the I_c vs. H_{ex} curve deviates from the usual Fraunhofer-like pattern as the temperature is lowered and the overlap of the two modes appears. This overlap is caused by the self-field effect which comes from the shortening of the effective Josephson penetration depth $\lambda_J{}^{\mathrm{eff}}$ near T_s. It is also seen that the Josephson current

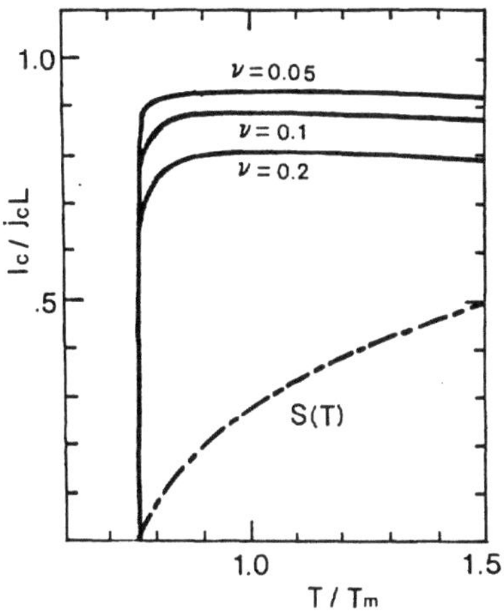

Fig. 1 Temperature dependence of the maximum Josephson current

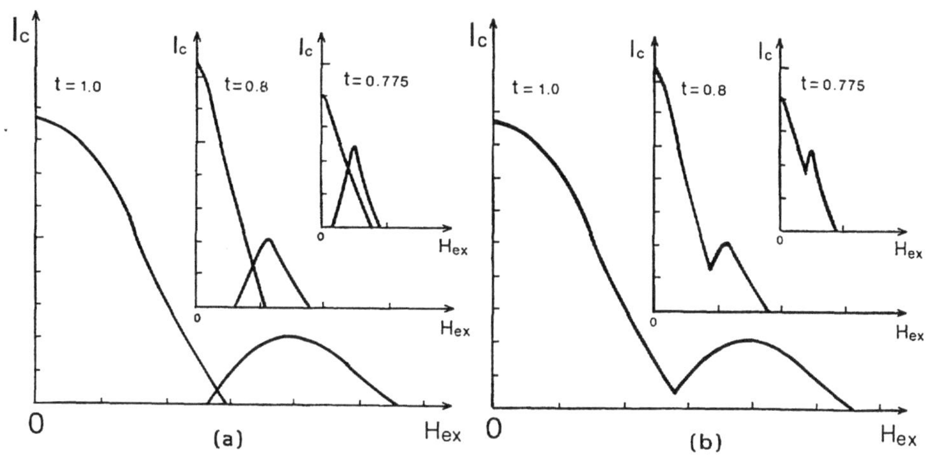

Fig. 2 External field dependence of the Josephson current

in the smaller flux mode is greatly reduced with decreasing temperature. This is due to the critical fluctuations of the localized spin magnetization. In the experiments we always measure the maximum tunneling current, so that the temperature dependence of the I_c vs. H_{ex} curve shown in Fig. 2(b) is expected to be observed from our theory. It is seen in the figure that, as a consequence of the overlap of the two modes and the reduction of the current in the smaller flux mode, a side peak appears and grows with decreasing temperature in the I_c vs. H_{ex} curve. The central peak in the Fraunhofer pattern thus seems to split just before the Josephson current is quenched, as observed by UMBACH and GOLDMAN [1] in $ErRh_4B_4$-Lu_xO_y-In junctions.

Finally, we briefly discuss the effect of polycrystalline electrode with a magnetic anisotropy on the I_c vs. H_{ex} curve. When the uniform tunneling current flows through the junction, the current creates a magnetic field $h_0(x) = 4\pi I_0 x/c$ with I_0 being the tunneling current density. The magnetic field induces, in turn, an irregular localized spin magnetization $m_0(x)$ because of random orientation of the easy directions of grains in the electrode. This fact implies that the total magnetic flux induced by the tunneling current does not vanish even when no external field is applied, i.e.,

$$\int_{-L/2}^{L/2} dx(h_0(x) + 4\pi m_0(x)) = \int_{-L/2}^{L/2} dx 4\pi m_0(x) \neq 0 \quad . \qquad (3.26)$$

The existence of the flux without an applied field causes a shift of the Fraunhofer pattern. Thus, the shift of the I_c vs. H_{ex} curves observed in the polycrystalline $ErRh_4B_4$-Lu_xO_y-In junctions may be understood as the effect of the polycrystalline electrode.

References

1. C.P. Umbach and A.M. Goldman, Phys. Rev. Lett. 48, 1433 (1982)
2. J.M. Rowell, R.C. Dynes and P.A. Schmit, in Superconductivity in d- and f-Band Metals, edited by H. Suhl and M.B. Maple (Academic, New York 1980) p. 409
3. T. Koyama and M. Tachiki, preprint
4. E.I. Blount and C.M. Varma, Phys. Rev. Lett. 42, 1079 (1979)
5. H. Matsumoto, H. Umezawa and M. Tachiki, Solid State Commun. 31, 157 (1979)
6. A. Kotani, M. Tachiki, H. Matsumoto, H. Umezawa and S. Takahashi, Phys. Rev. B23, 5960 (1981)
7. M. Tachiki, A. Kotani, S. Takahashi, T. Koyama, H. Matsumoto and H. Umezawa, Solid State Commun. 37, 113 (1981)
8. A. Kotani, S. Takahashi, T. Koyama, M. Tachiki, H. Matsumoto and H. Umezawa, J. Phys. Soc. Japan 50, 3254 (1981)
9. T. Koyama, A. Kotani, S. Takahashi, M. Tachiki, H. Matsumoto and H. Umezawa, J. Phys. Soc. Japan 51, 3469 (1982)
10. C.S. Owen and D.J. Scalapino, Phys. Rev. 164, 538 (1967)

An Inhomogeneous Superconducting State in a Ferromagnetic Superconductor

Hiizu Nakanishi

Department of Physics, Faculty of Science and Technology, Keio University Yokohama 223, Japan

Kazushige Machida

Department of Physics, Kyoto University, Kyoto 606, Japan

An inhomogeneous superconducting state under a ferromagnetic molecular field is investigated theoretically. This problem is solved exactly within the mean field approximation for a one-dimensional electron band. It is found that the solution gives a spin modulated structure associated with a spatially inhomogeneous superconducting order parameter and a two-gap structure in the energy spectrum. Experiments on $ErRh_4B_4$ are interpreted by this solution.

1. Introduction

The study of competition and coexistence of magnetism and superconductivity has a long history since the work of GINZBURG [1]. In the early days of the study, magnetic impurities were doped in superconductors to investigate the competition of the magnetism and superconductivity. But that ended up destroying the superconductivity far before the appearance of the magnetic order [2].

The situation changed totally when ternary compounds such as (RE)Mo_6S_8, (RE)Mo_6Se_8 and (RE)Rh_4B_4(RE: rare earth atom) were synthesized [3]. Most of these compounds undergo the superconducting transition although they have magnetic ions as constituents. Besides, some of them (RE = Er, Tb, Dy, Gd for (RE)Mo_6S_8 and RE = Sm, Nd for (RE)Rh_4B_4) show antiferromagnetic order and others ($HoMo_6S_8$, $ErRh_4B_4$) show ferromagnetic order. One of the reasons why these ternary compounds are able to show both orders is that the third atoms (S, Se, B) separate the magnetic atoms and metallic atoms from each other. Therefore the exchange interaction both among the local spins on the magnetic ions and between the local spins and conduction electrons becomes very weak.

As for the case of antiferromagnetic superconductors, both the long-range orders coexist very clearly and theories which take into account the effects of the antiferromagnetic order on the superconductivity as a mean field explain experiments well. On the other hand, in the case of the ferromagnetic superconductors, the situation is rather controversial. In this case, both the orders coexist much more difficultly because of an electromagnetic interaction even if the s-f interaction (the interaction between conduction electrons and local spins) is very weak. Some experimental data, however, which indicate a certain "intermediate phase" between the superconducting phase (at the upper temperature) and the ferromagnetic phase (at the lower temperature), have been accumulating [4-6]. Especially the neutron scattering experiment on a single crystal of $ErRh_4B_4$(ERB) done by SINHA et al.[5] seems to indicate that there should be a coexistence phase of superconductivity and ferromagnetism.

Their results are summarized as in the following: ERB becomes superconducting at $T_{C_1} \cong 8.7K$ but a ferromagnetic component of the neutron scattering appears at $T_M \cong 1.2K$. At the same time a transverse linearly polarized modulated structure of spins develops as evidenced by satellite peaks and their intensities are almost proportional to that of the ferromagnetic component. At $T_{C_2} \cong 0.71K$ in the cooling case, this structure disappears suddenly with loss of the superconductivity and the system becomes a normal ferromagnet. The wave length of the modulated phase is about 100Å and almost temperature independent. The structure is coherent over 2400Å at least. No higher harmonics were found in that phase. The transition at T_{C_2} shows a hysteresis behavior which implies the first-order transition.

These data suggest strongly a coexisting phase, but a neutron scattering experiment cannot tell whether or not a superconductivity and a ferromagnetism stay at the same part of a sample. That caused controversies among us [7-11]. Some suggest a domain structure consisting of ferromagnetic regions and superconducting regions where a spiral spin phase is stabilized, and others point out possibility of a spontaneous vortex state [11]. These theories do not suppose a genuine coexistence phase of ferromagnetism and superconductivity because such a state is likely to have higher electromagnetic energy.

Here we try to find out the effects of the ferromagnetism on the superconductivity in the coexistence phase. If we deal with the effect of the ferromagnetism through the mean field, the problem becomes the one considered by FULDE and FERRELL [12], and LARKIN and OVCHINNIKOV [13], where they have shown that an inhomogeneous superconducting state is stabilized under a uniform field acting on electron spins. TAKADA and IZUYAMA [14] have made detailed study of a helical solution. One of the authors has examined the experimental data on ERB in terms of the spatially modulated superconducting state [15,16].

In this paper, we are going to show that the Fulde-Ferrell problem in one-dimension can be solved exactly within the mean field approximation and study physical properties of the solution. Then we try to compare them with the experiments on ERB.

The model is set up in §2, and former theories for this system are reviewed in §3. In §4, we show that the problem in one dimension corresponds to that of the one-dimensional Peierls system, and solve it. The physical properties of the solution are investigated in §5 and §6, and with them the experiments on ERB are interpreted in §7. We give a summary and some discussions in §8.

2. Model

The system consists of localized spins which undergo a ferromagnetic transition and conduction electrons which undergo a superconducting transition. There should be the BCS attraction among the conduction electrons, and an exchange interaction between the conduction electrons and the localized spins. Other than that the localized spins may also interact on each other through an exchange interaction and there is the electromagnetic interaction. The total Hamiltonian will be

$$\mathcal{H} = \mathcal{H}_{BCS} + \mathcal{H}_{s-f} + \mathcal{H}_{f-f} + \mathcal{H}_{em} \tag{2.1}$$

$$\mathcal{H}_{BCS} = \int d^3x \Big[\sum_\sigma \psi^+(\vec{x})\Big(\frac{p^2}{2m} - \mu\Big)\psi_\sigma(\vec{x}) - \frac{1}{2}g \sum_{\sigma,\sigma'} \psi^+_\sigma(\vec{x})\psi^+_{\sigma'}(\vec{x})\psi_{\sigma'}(\vec{x})\psi_\sigma(\vec{x})\Big] \tag{2.2}$$

$$\mathcal{H}_{s\text{-}f} = \sum_{j,\mu,\nu} \int d^3x J(\vec{R}_j - \vec{x})\psi_\mu^+(\vec{x})\vec{\sigma}_{\mu\nu}\psi_\nu(\vec{x})\cdot\vec{S}_j \qquad (2.3)$$

where $\psi(\vec{x})$ denotes a conduction electron field, σ's are Pauli's spin matrices, and $\vec{S}_j$ and $\vec{R}_j$ represent a local spin and its position respectively.

We are going to study the effects of the ferromagnetism on the conduction electrons by treating the local spins as external entities. Then the effects of the electromagnetic interaction is much smaller than that of the s-f interaction, thus we will neglect them. If we consider the stability of the state, those effects are important.

Treating the effects of the local spins as a mean field below the ferromagnetic transition temperature, we get the Hamiltonian for the conduction electrons

$$\mathcal{H} = \mathcal{H}_{BCS} + \int d^3x \vec{h}(\vec{x})\cdot\psi_\mu^+(\vec{x})\vec{\sigma}_{\mu\nu}\psi_\nu(\vec{x}) \quad : \quad \vec{h}(\vec{x}) = \sum_{j,\mu,\nu} J(\vec{x}-\vec{R}_j)\langle\vec{S}_j\rangle \ . \qquad (2.4)$$

If we approximate that $\vec{h}(\vec{x})$ is uniform, we will have

$$\mathcal{H} = \int d^3x[\sum_\sigma \psi^+(\vec{x})(\frac{p^2}{2m} - \mu + h\sigma_z)\psi(\vec{x}) - \frac{1}{2}g\sum_{\sigma,\sigma'} \psi_\sigma^+(\vec{x})\psi_{\sigma'}^+(\vec{x})\psi_{\sigma'}(\vec{x})\psi_\sigma(\vec{x})] \qquad (2.5)$$

where we have assumed that $\langle\vec{S}_j\rangle$ is parallel to the z axis, the last approximation is not essential because the periodicity of $h(x)$ is the same as that of the Bloch function for the crystalline system which we are considering here.

3. Fulde-Ferrell State

The superconducting states for the Hamiltonian [(2.5)] have been studied by FULDE and FERRELL, and others [12-14]. We review them in this section.

The Gor'kov equations for the Hamiltonian are given by

$$[-\frac{\partial}{\partial\tau} - \frac{p^2}{2m} + \mu - h\sigma_z]\ G(\tau,\vec{x},\vec{x}') + \Delta(x)F^+(\tau,\vec{x},\vec{x}') = \delta(\vec{x}-\vec{x}')$$

$$[\frac{\partial}{\partial\tau} - \frac{p^2}{2m} + \mu - h\sigma_3]\ F^+(\tau,\vec{x},\vec{x}') - \Delta^+(x)G(\tau,\vec{x},\vec{x}') = 0$$

$$\Delta^+(\vec{x}) = gF^+(0+,\vec{x}) \quad . \qquad (3.1)$$

The superconducting order parameter $\Delta(\vec{x})$ has been introduced. $\Delta(\vec{x})=0$ is always the solution of this set of equations. $\Delta(\vec{x})$=const and nonzero is also a solution if $h<\delta_0$ for zero temperature, where δ_0 is a gap Δ for h=0 and T=0. If we consider only these homogeneous solutions, the phase diagram would look like the one in Fig. 1.a.

What FULDE and FERRELL have shown is [12], however, that although the coefficients in the system of equations are homogeneous and isotropic functions of the coordinates, the system may have solutions which Δ depend on the coordinates. Namely the Cooper pairs which have nonzero pairing momenta may appear. Physically this is because the time-reversed states are not degenerate if $h\neq0$. This is illustrated in Fig. 2.

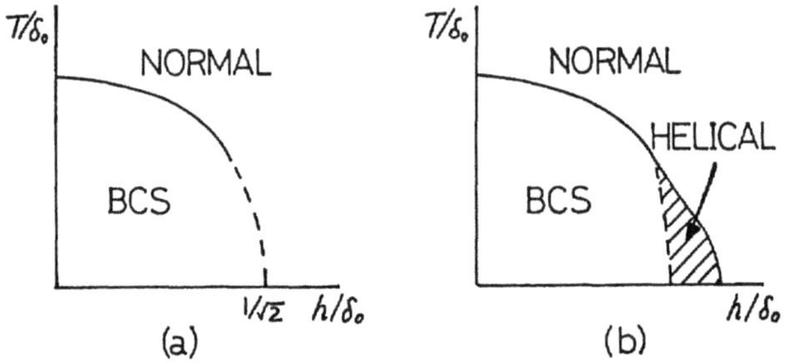

Fig.1. The schematic phase diagrams in the temperature (T) vs. field (h), (a) considering only a homogeneous BCS state, (b) considering also a helical state. The solid lines denote the second-order transition and the dashed lines denote the first-order transition. T and h are normalized by δ_0

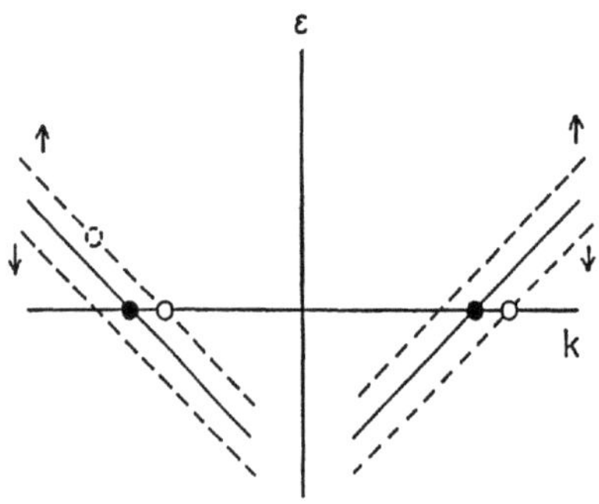

Fig.2. The schematic diagram of energy spectra for h=0 (solid lines) and h≠0 (dashed lines) to explain the pairing scheme. For h=0 time reversed pairs are degenerate (●), but for h≠0 pairs with finite momenta (○) are more favorable than time reversed pairs (◌ and right ○)

If we assume the helical solution $\Delta(x)=\Delta e^{i\vec{q}\vec{x}}$, (3.1) can be solved easily and the phase diagram would be like the one shown in Fig. 1.b. The helical state becomes stable in the shaded region. The pitch q for that state is determined by minimizing the free energy.

On the other hand, LARKIN and OVCHINNIKOV [13] expanded the gap equation and studied several possible configurations. They have shown that the sinusoidal solution is more stable than the helical solution at least near the transition line between the normal state and the inhomogeneous superconducting state.

There are some differences between the two states (helical and sinusoidal). The energy spectrum for the sinusoidal state should have the mid gap states. The spin density for this state is not homogeneous but that for the helical state is homogeneous. However, it is not easy to solve for the sinusoidal state generally because the higher harmonics will come into the problem inevitably and the two-wave approximation may not be suited the present case: $q \ll q_F$. We are going to show that the system of (3.1) can be solved exactly for the one-dimension case.

4. The Fulde-Ferrell Problem in One Dimension

In this section, we will show that (3.1) can be solved exactly for one dimension by mapping the problem to that of the Peierls system where the exact solutions have been found within the mean field approximation[17-19].

4.1 Formulation

We denote a right (left) going electron field by ψ_σ (ϕ_σ). Then the Hamiltonian for the problem is

$$\mathcal{H} = \int dx[\psi_\sigma^+(x)v_F(p-p_F)\psi_\sigma^+(x) + \phi_\sigma^+(x)v_F(-p-p_F)\phi_\sigma(x)$$
$$+ h\{\psi_\mu^+(x)\sigma_{\mu\nu}^z\psi_\nu(x) + \phi_\mu^+(x)\sigma_{\mu\nu}^z\phi_\nu(x)\} - \frac{1}{2}g\{\psi_\mu^+(\psi_\nu^+\cdot\psi_\nu)\psi_\mu + \phi_\mu^+(\phi_\nu^+\cdot\phi_\nu)\phi_\mu +$$
$$+ \psi_\mu^+(\phi_\nu^+\cdot\phi_\nu)\psi_\mu + \phi_\mu^+(\psi_\nu^+\cdot\psi_\nu)\phi_\mu + \phi_\mu^+(\psi_\nu^+\cdot\phi_\nu)\psi_\mu + \psi_\mu^+(\phi_\nu^+\cdot\psi_\nu)\phi_\mu\}] \qquad (4.1)$$

where v_F and p_F are the Fermi velocity and momentum respectively. To elliminate p_F from the kinetic parts we put $\psi \to \psi\cdot\exp(ip_Fx/h)$, $\phi \to \phi\cdot$ $\cdot\exp(-ip_Fx/h)$ and employ the Nambu notation: $\Psi^+(x) = (\psi_\uparrow^+(x),\phi_\downarrow(x))$, $\Phi^+(x) = (\phi_\uparrow^+(x), \psi_\downarrow(x))$.

As for the interaction term, we adopt the mean field approximation and define the order parameters

$$\Delta_I(x) = g\langle\psi_\uparrow(x)\phi_\downarrow(x)\rangle \quad , \qquad \Delta_{II}(x) = g\langle\phi_\uparrow(x)\psi_\downarrow(x)\rangle \quad , \qquad (4.2)$$
$$\Delta(x) = \Delta_I(x) + \Delta_{II}(x) = \Delta_r(x) + i\Delta_i(x) \quad . \qquad (4.3)$$

Then the interaction terms of (4.1) becomes

$$-\int dx\{\Delta^*(x)(\psi_\uparrow(x)\phi_\downarrow(x) + \phi_\uparrow(x)\psi_\downarrow(x)) + h.c.\}$$
$$= \int dx\{\Psi^+(x)\cdot(\Delta_r(x)\sigma_x - \Delta_i(x)\sigma_y)\cdot\Psi(x) + \Phi^+(x)\cdot(\Delta_r(x)\sigma_x - \Delta_i(x)\sigma_y)\cdot\Phi(x)\} .$$

Thus the total Hamiltonian for the mean field approximation is

$$\mathcal{H} = \int dx\{\Psi^+(x)\cdot(v_Fp\sigma_z + h + \Delta_r(x)\sigma_x - \Delta_i(x)\sigma_y)\cdot\Psi(x)$$
$$+ \Phi^+(x)\cdot(-v_Fp\sigma_z + h + \Delta_r(x)\sigma_x - \Delta_i(x)\sigma_y)\cdot\Phi(x)\} \qquad (4.4)$$

and the self-consistent equations are given by

$$\Delta_r(x) = -\frac{1}{2}g\{\langle\Psi^+(x)\cdot\sigma_x\cdot\Psi^+(x)\rangle + \langle\Phi^+(x)\cdot\sigma_x\cdot\Phi(x)\rangle\}$$
$$\Delta_i(x) = +\frac{1}{2}g\{\langle\Psi^+(x)\cdot\sigma_y\cdot\Psi(x)\rangle + \langle\Phi^+(x)\cdot\sigma_y\cdot\Phi(x)\rangle\} \quad . \qquad (4.5)$$

To diagonalize this Hamiltonian, we consider the following Bogoliubov-de Gennes (BdG) equations [20] for Ψ and Φ fields.

$$\begin{cases} (v_Fp + h)u(x) + \Delta(x)v(x) = \varepsilon u(x) \\ (-v_Fp + h)v(x) + \Delta^*(x)u(x) = \varepsilon v(x) \end{cases} \qquad (4.6)$$

$$\begin{cases} (-v_Fp + h)u'(x) + \Delta(x)v'(x) = \varepsilon' u'(x) \\ (v_Fp + h)v'(x) + \Delta^*(x)u'(x) = \varepsilon' v'(x) \quad . \end{cases} \qquad (4.7)$$

If $\Delta(x)$ is real, we can see that (4.6) and (4.7) are equivalent by taking the complex conjugate. Since we are seeking for a sinusoidal solution, we assume $\Delta(x)$ is real hereafter.

We denote the eigenvalues and eigenfunctions by ε_ν, ε_ν' and $u_\nu(x)$, $v_\nu(x)$, $u_\nu'(x)$, $v_\nu'(x)$ respectively and expand $\Psi(x)$ and $\Phi(x)$ by them

$$\Psi(x) = \sum_\nu \alpha_\nu \begin{pmatrix} u_\nu(x) \\ v_\nu(x) \end{pmatrix} \qquad \Phi(x) = \sum_\nu \beta_\nu \begin{pmatrix} u_\nu'(x) \\ v_\nu'(x) \end{pmatrix} = \sum_\nu \beta_\nu \begin{pmatrix} u^*(x) \\ v_\nu^*(x) \end{pmatrix} \tag{4.8}$$

where α_ν and β_ν are annihilation operators. The Hamiltonian, BdG equations, and self-consistent equation can be expressed as

$$\mathcal{H} = \sum_\nu \varepsilon_\nu (\alpha_\nu^+ \alpha_\nu + \beta_\nu^+ \beta_\nu) \tag{4.9}$$

$$\begin{cases} v_F p\, u_\nu(x) + \Delta(x) v_\nu(x) = (\varepsilon_\nu - h) u_\nu(x) \\ -v_F p\, v_\nu(x) + \Delta(x) u_\nu(x) = (\varepsilon_\nu - h) v_\nu(x) \end{cases} \tag{4.10}$$

$$\Delta(x) = -g \sum_\nu (v_\nu^*(x) u_\nu(x) + v_\nu(x) u_\nu^*(x)) f(\varepsilon_\nu) \tag{4.11}$$

where $f(\varepsilon_\nu)$ is the Fermi distribution function.

To simplify the expression we introduce the following notations:

$$\tilde{\varepsilon}_\nu = \varepsilon_\nu - h, \quad \tilde{x} = x/v_F\hbar, \quad \tilde{\partial} = \partial/\partial\tilde{x}, \quad U_\nu^t(x) = (u_\nu(x), v_\nu(x)).$$

Then (4.10) and (4.11) become

$$(-\sigma_z i\tilde{\partial} + \sigma_x \Delta(\tilde{x}))\cdot U_\nu(\tilde{x}) = \tilde{\varepsilon}_\nu U_\nu(\tilde{x}), \quad \Delta(\tilde{x}) = -g\sum_\nu U_\nu^+(\tilde{x})\cdot\sigma_x\cdot U_\nu(\tilde{x}) f(\varepsilon_\nu). \tag{4.12}$$

Transforming $U_\nu(\tilde{x})$ to $F_\nu(\tilde{x})$ which is defined by

$$F_\nu(\tilde{x}) = \begin{pmatrix} f_{\nu-}(\tilde{x}) \\ f_{\nu+}(\tilde{x}) \end{pmatrix} = T\cdot U_\nu(\tilde{x}) \qquad T = \frac{1}{\sqrt{2}}(\sigma_y + \sigma_z) \quad , \tag{4.13}$$

we finally obtain

$$(-\sigma_y i\tilde{\partial} - \sigma_x \Delta(\tilde{x}))\cdot F_\nu(\tilde{x}) = \tilde{\varepsilon}_\nu F_\nu(\tilde{x}), \quad \Delta(\tilde{x}) = g\sum_\nu F_\nu^+(\tilde{x})\cdot\sigma_x\cdot F_\nu(\tilde{x}) f(\varepsilon_\nu). \tag{4.14}$$

These equations are the same as those for the one-dimensional Peierls system [21] where the exact solutions are known. We will sketch how we can get the exact solution in the next subsection. We especially refer to the paper by MERTSCHING et al. [18].

4.2 Exact solution

Suppose that the self-consistent solutions of (4.14) give

$$\Delta(\tilde{x}) = \Delta_1 k_1 \,\mathrm{sn}(\Delta_1 \tilde{x}, k_1) \tag{4.15}$$

where $\mathrm{sn}(x, k)$ is the Jacobi elliptic function with the modulus k. The two parameters Δ_1 and k_1 are to be determined self-consistently and characterized the superconducting state. By eliminating $f_{\nu-}(\tilde{x})$ from (4.14), we obtain

$$(\tilde{\partial}^2 + \tilde{\varepsilon}_\nu^{\,2} - Q(\tilde{x}))f_{\nu+}(\tilde{x}) = 0, \quad Q(\tilde{x}) = \Delta^2(\tilde{x}) - \tilde{\partial}\Delta(\tilde{x}) \quad . \tag{4.16}$$

It is not difficult to see [18] that (4.16) can be written in terms of the Weierstrass function [22] $\wp(x)$ as $Q(\tilde{x}) = e_1 + 2\wp(\tilde{x} + \tilde{\omega}_3)$ where the three parameters associated with the Weierstrass function are given by

$$e_1 = \frac{1}{6}\Delta_1^2(1 + k_1^2)$$

$$e_2 = -\frac{1}{12}\Delta_1^2(1 - 6k_1 + k_1^2), \quad e_3 = -\frac{1}{12}\Delta_1^2(1 + 6k_1 + k_1^2). \tag{4.17}$$

The fundamental periodicities $2\tilde{\omega}_1$ and $2\tilde{\omega}_3$ of the elliptic function are

$$\tilde{\omega}_1 = K(k)/\sqrt{e_1 - e_3}, \quad \tilde{\omega}_3 = iK(k')/\sqrt{e_1 - e_3} \tag{4.18}$$

and the modulus k and complementary modulus k' are given by

$$k^2 = (e_2 - e_3)/(e_1 - e_3), \quad k'^2 = 1 - k^2 \tag{4.19}$$

with $K(k)$ being the complete elliptic integral of the first kind. We have used the same notation for the periodicity ω as for the x. Therefore the eigenvalue equation (4.16) is reduced to solving the Lamé equation [22]:

$$(\tilde{\partial}^2 + \tilde{\varepsilon}_\nu^{\,2} - e_1 - 2\wp(\tilde{x} + \tilde{\omega}_3))f_{\nu+}(\tilde{x}) = 0 \quad . \tag{4.20}$$

This is nothing but the one-dimensional Bloch problem in a periodic potential. It can be shown that the periodic solution of the Lamé equation is given by

$$f_{\nu+}(\tilde{x}) = \left(\frac{\wp(\tilde{x}+\tilde{\omega}_3) - e}{2L(\bar{\wp} - e)}\right)^{1/2} \exp[iC(\tilde{\varepsilon})\int_0^{\tilde{x}} \frac{dx'}{\wp(x'+\tilde{\omega}_2) - e}] \tag{4.21}$$

with the proper normalization [18], where

$$e = e_1 - \tilde{\varepsilon}_\nu^{\,2}, \quad C(\tilde{\varepsilon}) = \pm\tilde{\varepsilon}\sqrt{(\tilde{\varepsilon}^2 - \tilde{\varepsilon}_2^{\,2})(\tilde{\varepsilon}^2 - \tilde{\varepsilon}_3^{\,2})}, \quad \tilde{\varepsilon}_i^{\,2} = e_1 - e_i$$

$$\bar{\wp} = \frac{1}{\tilde{\omega}_1}\int_0^{\tilde{\omega}_1} dx\,\wp(x + \tilde{\omega}_3) = e_1 - (e_1 - e_3)\frac{E(k)}{K(k)} \tag{4.22}$$

with $E(k)$ being the complete elliptic integral of the second kind and L being the system length. Note that for $\tilde{\varepsilon}_2^{\,2} < \tilde{\varepsilon}_\nu^2 < \tilde{\varepsilon}_2^{\,3}$, $C(\tilde{\varepsilon})$ becomes complex which means that the energy gaps appear in the spectrum.

By applying the Bloch theorem: $f_{\nu+}(\tilde{x} + 2\tilde{\omega}_1) = e^{2i\tilde{q}\tilde{\omega}_1} f_{\nu+}(\tilde{x})$ we can determine the $v_F\hbar$ times wave number q as

$$\tilde{q}(\varepsilon) \equiv \hbar v_F q(\varepsilon) = \pm\frac{1}{K(k)}\sqrt{\frac{(e_1 - e)(e_2 - e)}{e_3 - e}} \cdot \Pi\left(\frac{e_2 - e_3}{e_3 - e}, k\right) \tag{4.23}$$

where $\pi(\rho, k)$ is the complete elliptic integral of the third kind. The density of states $N(\varepsilon)$ is derived through $N(\varepsilon) = 1/\pi|dq/d\varepsilon|$ as

$$N(\varepsilon) = \frac{|\tilde{\varepsilon}^2 + \wp - e_1|}{\pi\hbar v_F \sqrt{(\tilde{\varepsilon}^2-\delta^2)(\tilde{\varepsilon}^2-k'^2\delta^2)}}\, \Theta(\tilde{\varepsilon}^2-\delta^2)(\tilde{\varepsilon}^2-k'^2\delta^2) \tag{4.24}$$

where we have introduced the parameter δ defined by $\delta^2 = e_1 - e_3$. Thus we obtain $e_1-e_2=k'^2\delta^2$ and $e_2-e_3=k^2\delta^2$. The density of states $N(\varepsilon)$ in (4.24) clearly indicates two energy gap structures in the spectrum: one corresponds to 2δ and the other $2(1-k')\delta$ as will be seen later in detail. The two independent parameters k_1 and Δ_1 introduced before are now related to the newly defined ones k and δ: $\Delta_1 = (1+k')\delta$, $k_1 = (1-k')/(1+k')$. The latter is known as the Landen transformation [22].

In order to determine the two unknown parameters k and δ for the problem let us first consider the self-consistent equation (4.14):

$$\begin{aligned} \Delta(\tilde{x}) &= g \sum_\nu \{f_{\nu+}(\tilde{x})f_{\nu-}(\tilde{x}) + f_{\nu-}(\tilde{x})f_{\nu+}(\tilde{x})\}f(\varepsilon_\nu) \\ &= -g \sum_\nu \frac{f(\varepsilon_\nu)}{\tilde{\varepsilon}_\nu} (\tilde{\partial}+2\Delta(\tilde{x}))f_{\nu+}(\tilde{x})f_{\nu+}(\tilde{x}) \end{aligned} \tag{4.25}$$

where we have used the first equation of (4.14). Substituting (4.21) into (4.25), we obtain after some manipulations

$$1/g = -1/L\cdot\sum_\nu f(\varepsilon_\nu)\tilde{\varepsilon}_\nu/(\wp-e_1+\tilde{\varepsilon}_\nu{}^2) \tag{4.26}$$

which yields the following form by utilizing (4.24)

$$\begin{aligned} \ln\frac{\delta_0}{\delta} &= \int_\delta^\infty \{f(\varepsilon+h) + f(\varepsilon-h)\}\frac{\varepsilon d\varepsilon}{\sqrt{(\varepsilon^2-\delta^2)(\varepsilon^2-k'^2\delta^2)}} \\ &+ \int_0^{k'\delta} \{f(h-\varepsilon) - f(h+\varepsilon)\}\frac{\varepsilon d\varepsilon}{\sqrt{(\varepsilon^2-\delta^2)(\varepsilon^2-k'^2\delta^2)}} \quad . \end{aligned} \tag{4.27}$$

The order parameter at $h=0$ and $T=0$ is introduced by $\delta_0=2\tilde{q}_s\cdot e^{-1/gN(0)}$ where $N(0)=1/(\pi\hbar v_F)$ and $\tilde{q}_s$ is the $v_F\hbar$ times momentum cutoff.

The free energy Ω per unit length is expressed as

$$\begin{aligned} \Omega &= -\frac{2}{\beta L}\sum_q \ln(1+e^{-\beta\varepsilon_q}) + \frac{1}{gL}\int_0^L \Delta^2(x)dx \\ &= -\frac{1}{\pi\hbar v_F}\{\tilde{q}_s(\tilde{q}_s+2h) + \frac{\pi h\delta}{K(k)} + \delta^2[(1+k'^2-2\frac{E(k)}{K(k)})(\ln\frac{\delta_0}{k\delta}+1) - \frac{1+k'^2}{2}]\} \\ &+ \frac{\delta}{\hbar v_F\beta K(k)}\ln\frac{(1+e^{-\beta(\delta+h)})(1+e^{-\beta(\delta-h)})}{(1+e^{-\beta(h-k'\delta)})(1+e^{-\beta(h+k'\delta)})} \\ &+ \frac{2}{\pi}\int_0^{k'\delta}\{f(h-\varepsilon) - f(h+\varepsilon)\}q(\varepsilon)d\varepsilon - \frac{2}{\pi}\int_0^\infty\{f(\varepsilon+h) + f(\varepsilon-h)\}q(\varepsilon)d\varepsilon \end{aligned} \tag{4.28}$$

where we have used (4.24) and (4.27). The minimum condition of Ω with respect to k or δ yields

$$\frac{\partial\Omega}{\partial\delta} = \frac{1}{\hbar v_F}\frac{1}{K(k)}F(T,\ h,\ \delta,\ k') = 0 \tag{4.29}$$

$$F(T,h,\delta,k') = -h + \frac{2}{\pi}\delta E(k) + \frac{1}{\beta}\ln \frac{(1+e^{-\beta(\delta+h)})(1+e^{-\beta(\delta-h)})}{(1+e^{-\beta(h-k'\delta)})(1+e^{-\beta(h+k'\delta)})}$$

$$+ \frac{2}{\pi}\int_0^{k'\delta}\{f(h-\delta) - f(h+\delta)\}[S(\sin^{-1}\frac{\varepsilon}{k'\delta}, k) + \varepsilon\delta \frac{E(k)-(1+k'^2-\varepsilon^2/\delta^2)K(k)}{\sqrt{(\delta^2-\varepsilon^2)(k'^2\delta^2-\varepsilon^2)}}]d\varepsilon$$

$$- \frac{2}{\pi}\int_\delta^\infty\{f(\varepsilon+h) - f(\varepsilon-h)\}[S(\sin^{-1}\frac{\delta}{\varepsilon}, k) - \varepsilon\delta \frac{E(k)-(k'^2\delta^2/\varepsilon^2)K(k)}{\sqrt{(\varepsilon^2-\delta^2)(\varepsilon^2-k'^2\delta^2)}}]d\varepsilon \quad (4.30)$$

where the Heumann function: $S(\varphi,k) = (E(k) - K(k))F(\varphi,k') + K(k)E(\varphi,k')$; $F(\varphi,k')$ and $E(\varphi,k')$ are the incomplete elliptic integrals of the first and second kind. Equations (4.27) and (4.29) constitute a complete set of the simultaneous equations for determining the two parameters k and δ.

It is easy to see that at T=0 the solution of (4.28) and (4.29) is given by

$$\delta = \frac{\delta_0}{k} \quad , \quad h = \frac{2}{\pi}\delta E(k) \quad . \quad (4.31)$$

Since $1 \leq E(k) \leq \pi/2$ for $0 \leq k \leq 1$, δ is always greater than or equal to δ_0. At finite temperatures we must solve these equations numerically.

5. Physical Properties of the Solution

5.1 Phase diagram

Let us consider the phase diagram in the T vs. h plane. From (4.15) or equivalently from $\Delta(\tilde{x}) = \delta(1-k')\mathrm{sn}((1+k')\delta\tilde{x}, k_1)$, it can be seen that the solution with k=1 corresponds to the ordinary BCS state characterized by the spatially homogeneous order parameter: $\Delta(\tilde{x})=\delta$, while the solution with $k\neq1$ (we call it the sn phase hereafter) corresponds to the spatially modulated superconducting state whose wave number is $\tilde{q} = \pi\delta/K(k)$ and the amplitude of the order parameter $\Delta(\tilde{x})$ is $\delta(1-k')$. Note that the roles of δ and k are mixed in the sn phase although δ can be identified as the usual BCS gap parameter in the homogeneous phase. When $k\to0$, $\Delta(\tilde{x})$ approaches the sinusoidal form. The order parameter δ in the BCS state is evaluated by putting k=1 in (4.26) [23]. The boundary line T_c between the normal and BCS state is determined by taking $\delta\to0$ and $T\to T_c$.

To determine the phase boundary between the normal and sn states we put k=0 in (4.26) and obtain,

$$\frac{\pi}{g} = - \frac{\pi}{L}\sum_q \frac{f(\varepsilon_q)(\varepsilon_q+h)}{(\varepsilon_q+h)^2 + \wp - e_1} = - \frac{1}{\hbar v_F}\int_0^{\tilde{q}_s} d\varepsilon\{f(\varepsilon-h) - f(-\varepsilon-h)\}\frac{\varepsilon}{\varepsilon^2-\delta^2} \quad .$$

After some manipulations we get

$$\ln\frac{T_{c0}}{T_c} + \mathrm{Re}[\psi(\tfrac{1}{2}) - \tfrac{1}{2}\psi(\tfrac{1}{2} + \frac{\delta+h}{2\pi T_c}i) - \tfrac{1}{2}\psi(\tfrac{1}{2} - \frac{\delta-h}{2\pi T_c}i)] = 0 \quad . \quad (5.1)$$

We note that $2\delta/\hbar v_F$ corresponds to the wave number of the sn phase when $k\to0$ and $\Delta(\tilde{x})\propto\sin2\delta\tilde{x}$ near the boundary. In order to get the other equation which determines δ and T_c along the boundary line, we impose the condition that the transition temperature T_c should be maximized with respect to δ, that is, $dT_c/d\delta=0$. It yields

$$\mathrm{Im}[\psi'(\frac{1}{2} + \frac{\delta+h}{2\pi T_c} i) - \psi'(\frac{1}{2} - \frac{\delta-h}{2\pi T_c} i)] = 0 \quad . \tag{5.2}$$

These (5.1) and (5.2) coincide with those in a previous work [24]. We can show that $T_c \propto h^{-1}$ when h becomes large. This asymptotic behavior of T_c is characteristic of the one-dimensional model [18,24] and different from that of the three-dimensional case [12-14]. In the latter case there is a finite critical field beyond which the superconductivity ceases to exist.

The phase boundary line between the BCS (k=1) and sn phase (k≠1) is determined by taking the limit $k\to 1$ and $\delta\to\delta_c$ in (4.27) and (4.30). From (4.31) this boundary is given by $h_{cr}=\frac{2}{\pi}\delta_0$ at T=0.

The phase diagram determined by solving those equations numerically is depicted in Fig.3.

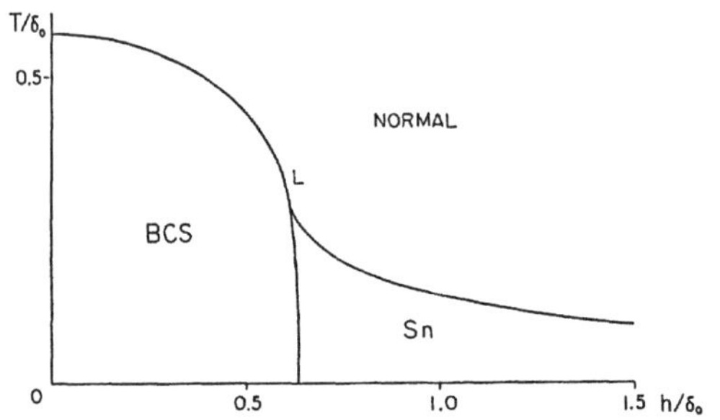

Fig.3 The phase diagram is the temperature (T) vs. field (h). All of the transition lines are second order and meet at the point L. T and h are normalized by δ_0

The three second-order phase transition lines meet at a point called the Leung point.

So far we considered only the real solution of the BdG equation. As mentioned before there also exists the complex solution described by $\Delta(x)=\Delta_0 e^{iqx}$, a helical solution. This has been investigated by FULDE and FERRELL [12] and TAKADA and IZUYAMA [14]. From the calculation for one dimension it can be seen that the phase transition from the BCS to the helical phase is of first order as coincides with the three-dimensional spherical Fermi surface model and that the critical field ($h_{cr}=0.68\delta_0$) at T=0 between the BCS and helical phase is greater than that of the sn phase case ($h_{cr}=\frac{2}{\pi}\delta_0$), which implies that the sn phase is more stable than the helical solution over the whole region. Physically this is because the kink lattice structure gives energy gaps right at the Fermi level to reduce the total energy. Hereafter we consider only the sn phase.

5.2 Energy spectrum

The dispersion relation for the sn phase obtained from (4.22) is shown in Fig. 4(a). The Fermi level stays in the midst of the gap at T=0. The number of the mid-gap states can be calculated from (4.23) and is $\delta/K(k)$ per unit length, which coincides with the number of kinks. As k approaches 1, the number of the mid-gap states decreases and the dispersion becomes

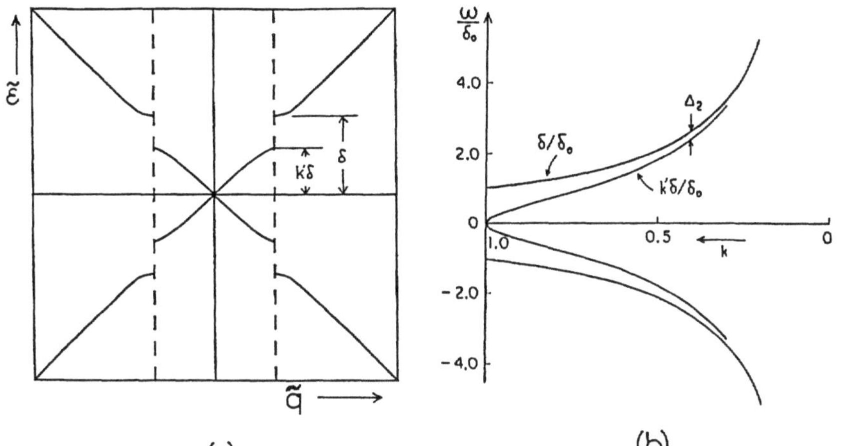

Fig.4 (a) The one-particle energy spectrum for the sn phase in $\tilde{\varepsilon}=\varepsilon-h$ vs. $\tilde{q}$. (b) The energy gap edges as a function of the modulus k, where k=1 corresponds to the BCS state and k≠1 to the sn state

that of the BCS superconductor when k=1 [Fig. 4(b)]. What is characteristic of the dispersion is the two-gap structure. Although the state has the periodic structure incommensurate with the underlying crystal, there are no more gaps associated with the higher harmonics. Physically this is because the energy of the electrons cannot be reduced farther by introducing more gaps.

5.3 Order parameter

The order parameter $\Delta(x)$ is determined by k and δ which can be calculated from (4.26) and (4.28). At T=0, we examine two limiting behaviors as follows: (1) As $h\to\infty$, we get that $k\to 1/h$ and $\delta\to h$. Then the amplitude of the order parameter tends to 1/2h and the wave number to 2h. (2) As $h/\delta_0\to 2/\pi+0$, the wave number tends to zero logarithmically: $\tilde{q}\cong\delta_0/\ell n(\delta_0/\Delta h)$, where $\Delta h = h - 2/\pi\cdot\delta_0$.

The amplitude and wave number calculated in the sn phase are shown in Fig. 5. Those quantities along T=0 line are also depicted in Fig. 6.

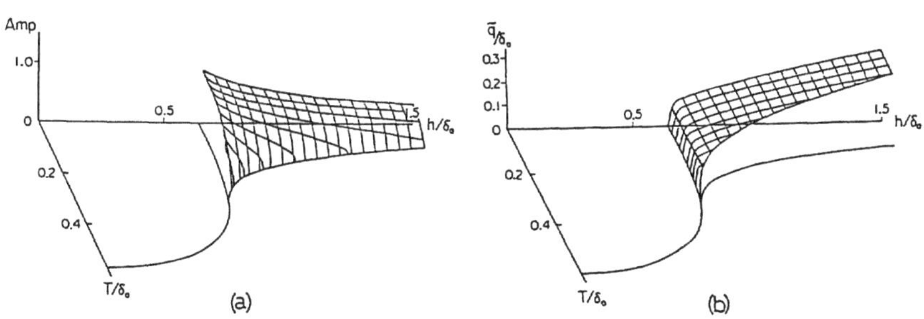

Fig.5 The stereo graph of (a) the amplitude $\delta(1-k')/\delta_0$ and (b) the wave number $\tilde{q}/\delta_0$ of the order parameter $\Delta(\tilde{x})$ in the T vs. h plane for the sn phase

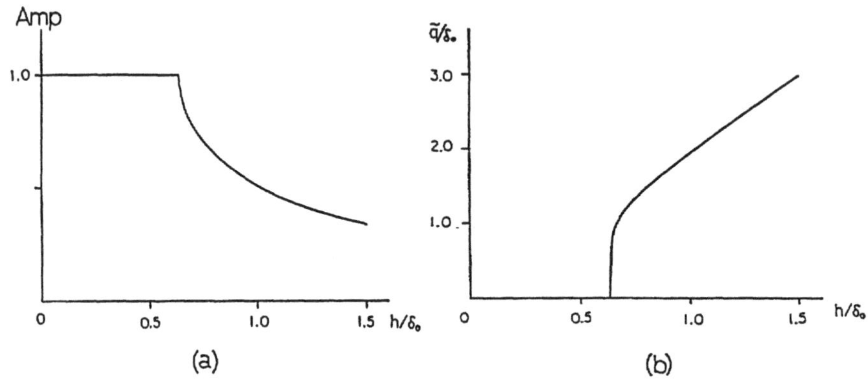

Fig.6 (a) The amplitude $\delta(1-k')/\delta_0$ and (b) the wave number $\tilde{q}/\delta_0$ of the order parameter $\Delta(\tilde{x})$ along the h axis at T=0

6. The Spin Density of the Conduction Electrons

The sn solution gives an inhomogeneous spin density while the helical solution gives a homogeneous one. To see this, we define the spin density by

$$S(x) = \langle\psi_\uparrow^+(x)\psi_\uparrow(x) - \psi_\downarrow^+(x)\psi_\downarrow(x) + \phi_\uparrow^+(x)\phi_\uparrow(x) - \phi_\downarrow^+(x)\phi_\downarrow(x)\rangle . \qquad (6.1)$$

Inserting (4.8) into this and using (4.13), we obtain

$$S(\tilde{x}) = 2\sum_\nu (f(\varepsilon_\nu) - \frac{1}{2})(|f_{\nu+}(\tilde{x})|^2 + |f_{\nu-}(\tilde{x})|^2) \quad . \qquad (6.2)$$

It can be proved that $|f_{\nu-}(\tilde{x})|^2 = |f_{\nu+}(\tilde{x}+\tilde{\omega}_1)|^2$. Then putting the solution of (4.21) into (6.2), we get

$$S(\tilde{x}) = 2S_0 + 2k^2\delta^2 S_1\{sn^2(\delta\tilde{x},k) + sn^2(\delta\tilde{x}+\delta\tilde{\omega}_1,k)\} \qquad (6.3)$$

$$\begin{cases} S_0 = \dfrac{1}{\pi v_F\hbar}[\displaystyle\int_0^{k'\delta} d\varepsilon + \int_\delta^\infty d\varepsilon](f(\varepsilon+h) - f(\varepsilon-h)) \dfrac{\sqrt{\varepsilon^2-\delta^2}}{\sqrt{\varepsilon^2-k'^2\delta^2}} & (6.4) \\ S_1 = \dfrac{1}{\pi v_F\hbar}[\displaystyle\int_0^{k'\delta} d\varepsilon + \int_\delta^\infty d\varepsilon](f(\varepsilon+h) - f(\varepsilon-h)) \dfrac{1}{\sqrt{(\varepsilon^2-\delta^2)(\varepsilon^2-k'^2\delta^2)}} . & (6.5) \end{cases}$$

Equation (6.3) can be expressed in a Fourier series,

$$S(\tilde{x}) = 2S_0 + 4S_1\delta^2\{1 - \frac{E}{K} - \frac{\pi^2}{K^2}\sum_{n=1}^\infty \frac{2n\cos(2\pi n\delta\tilde{x}/K)}{\sinh(2\pi K'/K)}\} \qquad (6.6)$$

where K=K(k), K'=K(k'), and E=E(k).

At T=0, (6.6) is reduced to

$$S(\tilde{x}) = -\frac{\delta_0}{\pi v_F\hbar}\cdot\frac{1}{kK}[1 + \sum_{n=1}^\infty \frac{2kK'\,\tilde{q}_n/\delta_0}{\sinh(kK'\,\tilde{q}_n/\delta_0)}\cos(\tilde{q}_n\cdot\tilde{x})] \qquad (6.7)$$

with $\tilde{q}_n = 2\pi n\delta_0/kK$. In the limit of $h\to\infty$, namely k=0 at T=0, we obtain

$$S(\tilde{x}) = -\chi_p \cdot h - \frac{8\delta_0}{\pi v_F \hbar}\left(\frac{\delta_0}{h}\right)^3 \ln\left(\frac{h}{\delta_0}\right)^4 \cdot \cos(4h\tilde{x}) \tag{6.8}$$

where χ_p is the Pauli susceptibility given by $\chi_p = 2N(0)$ and $N(0)$ is the density of states at the Fermi level given by $N(0) = 1/\pi v_F \hbar$. Note that near the boundary between the normal and sn phases, the uniform component becomes that of the normal electron described by the Pauli susceptibility.

On the other hand, in the limit of $h \to \frac{2}{\pi}\delta_0 + 0$, namely $k \to 1$, the asymptotic form of (6.7) becomes

$$S(\tilde{x}) = -\frac{\delta_0}{2\hbar v_F}\left[1 - \tanh^2(\delta_0 \tilde{x})\right] \quad . \tag{6.9}$$

This clearly indicates that the spin density is localized around the kinks of the order parameter $\Delta(\tilde{x})$ where $\Delta(\tilde{x})$ vanishes [Fig. 7(a)]. We can see also from (6.7) that the net density of spin at T=0 is proportional to the number of kinks, and therefore the number of mid-gap states for any k.

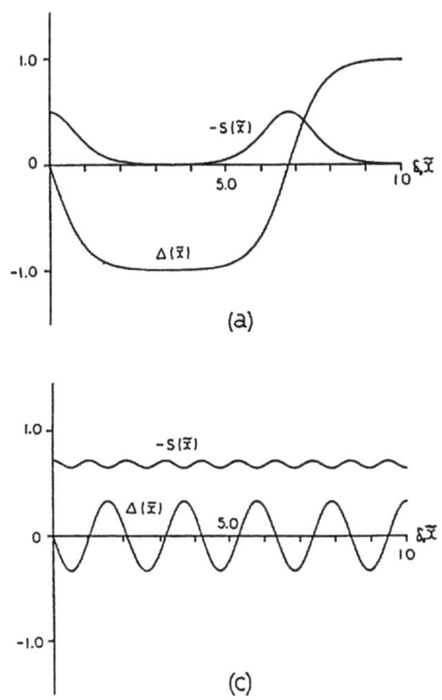

Fig.7 The spatial variation of the order parameter $\Delta(\tilde{x})/\delta_0$ and the spin density $S(\tilde{x}) v_F \hbar/\delta_0$ at T=0. (a) for k=0.99999 which corresponds to $h \sim 2/\pi$, (b) h=0.65, k=0.995, and (c) h=1.5, k=0.6

We show some examples of the spatial variation of the order parameter and the spin density in Fig. 7. Figure 8 illustrates the histogram of the Fourier components of $S(\tilde{q}_n)$ in the h vs. $\tilde{q}$ plane at a constant temperature $T/\delta_0 = 0.12$. This indicates that the position of the fundamental component (n=1) does not change appreciably as h increases except for a narrow region near $h_{cr} = \frac{2}{\pi}\delta_0$, and in that critical region many higher harmonics appear, rapidly diminishing in higher fields. Figure 9 exhibits the field dependence of the fundamental component (n=1) in (6.6).

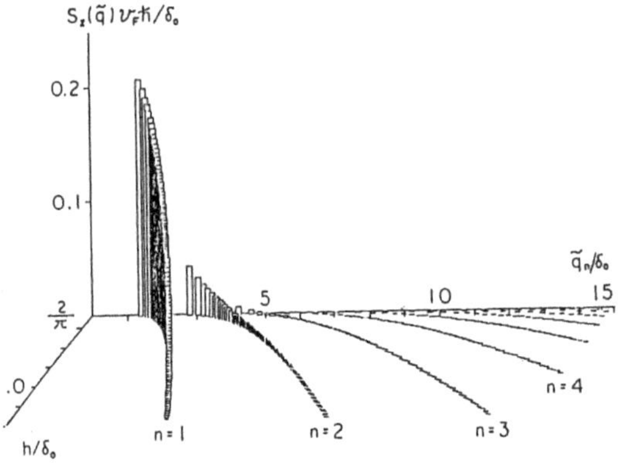

Fig.8 The histogram $S(\tilde{q}_n)v_F\hbar/\delta_0$ of the Fourier components ($n=1,2,3,\cdots$) of the spin polarization in the $\tilde{q}_n/\delta_0$ vs. h/δ_0 plane at $T/\delta_0=0.12$. Near the $h/\delta_0=2/\pi$ lines, $S(\tilde{q}_n)$'s tend to zero suddenly but it cannot be seen in this scale

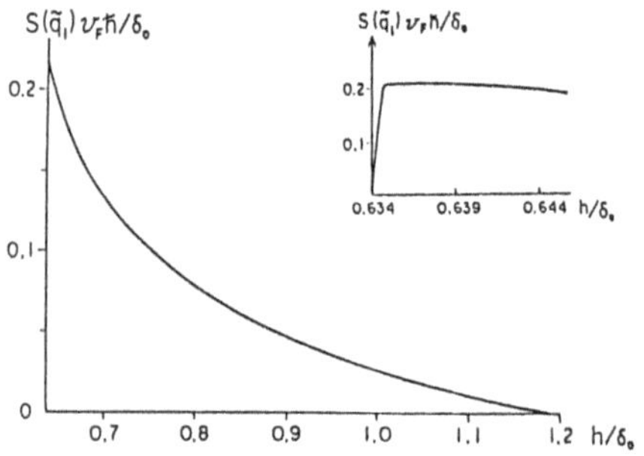

Fig.9 The field dependence of the fundamental Fourier component $S(\tilde{q}_1)v_F\hbar/\delta_0$ at $T/\delta_0=0.12$. The inset shows the low field region of the figure

7. Interpretation of Experiments on $ErRh_4B_4$

Let us apply the above theory to experiments on $ErRh_4B_4$(ERB), especially to the neutron experiment by SINHA et al.[5].

7.1 Strength of the ferromagnetic molecular field

The ferromagnetic molecular field H_{ex} arising from the sf exchange interaction (2.3) can be estimated from the g factor and angular momentum for a free Er ion ($g_J=6/5$, $J=7.5\mu_B$) and from the exchange integral which has been estimated by one of the authors ($I\cong 2.7\times 10^{-2}$ eV-atom)[25]. The estimated value H_{ex} would be about 400kG, which is temperature dependent through $\langle S\rangle$. On the other hand, the order of magnitude of the molecular field at which the sn phase is stabilized is determined by the critical field $h_{cr}=2/\pi\cdot\delta_0$. This is the same order of the Chandrasekhar-Clogston limiting field $H_p(=\delta_0/\sqrt{2}\mu_B)$ which is the order of $H_p\cong 160$kG for ERB when taking $T_{c_1}=8.7$K. These rough estimates indicate that there should exist a narrow temperature region where the sn phase is stable below the ferromagnetic transition temperature $T_M\cong 1.2$K.

7.2 Successive phase transitions

Upon decreasing temperature the BCS state appearing at $T=T_{c_1}$ continuously transforms into the sn state at $T=T_{sn}$ which corresponds to the critical field $h_{cr}=2/\pi\cdot\delta_0$ immediately after the 4f moment system begins to order ferromagnetically at T_M. The phase transition at $T=T_{sn}$ is of second order according to the present theory. This coincides with the neutron experiment, which reports that the satellite intensity grows continuously. As the temperature decreases further the ferromagnetic magnetization grows rapidly and exceeds the value of the upper critical molecular field at $T=T_c$ beyond which the sn state ceases to exist. This reentrant phase transition from the sn state to the normal ferromagnetic state should be of second order according to the present theory which considers only the conduction electron system and regards the 4f moment system as an external entity. However, if we take into account the reaction of the 4f system due to the formation of the sn state, this reentrant phase transition becomes first order as explained in a previous paper [16]. In fact the upper critical field H_{c2} measurement by CRABTREE et al.[26] on ERB who found the first-order transition when an external field is applied parallel to the easy axis of the magnetization is interpreted successfully by this idea[27].

7.3 Sinusoidally modulated magnetic phase

The sinusoidally modulated magnetic phase observed in the neutron experiment as satellite peaks is characterized by the wave length 100A which is independent of the temperature [5]. As already shown previously [15] the wave number q of the sn phase is estimated roughly by $q \cong k_F h\bar{q}/\varepsilon_F$, where $\bar{q}$ is the order of unity. Since the energy associated with the molecular field is $\sim\mu_B H_{ex}$ ($\cong 10^{-3}$eV) and the Fermi energy ε_F and momentum k_F are estimated as $\varepsilon_F^{-1} \sim 1$ eV^{-1} and $k_F \cong 1Å^{-1}$ respectively, we obtain $q \cong O(10^{-2} \sim 10^{-3} Å^{-1})$ which is within the range of the experimental value ($q \sim 0.06 Å^{-1}$). A similar estimate [15] of q gives a correct order of magnitude for HMS.

According to the present theory the periodicity of the conduction electron polarization S(x) associating with $\Delta(x)$ depends on the field h, in other words, the temperature in the context of the experimental situation. As seen from Fig. 8 the region where the wave number varies rapidly is limited very narrowly near the onset field of the sn state. Therefore except for the narrow temperature region near T_{sn} the temperature dependence of the periodicity is very weak, in agreement with the experiments [4,5]. In this connection an interesting fact is reported by SINHA and KJEMS [28] that at T=1.0K they observe a peak with the wave number much smaller than that of the satellite fully established at T≅0.7K. This finding might be related to our assertion that the periodicity of the spin modulation should change continuously. We expect that the satellites associated with many higher harmonics should appear in this temperature region near T_M=1.2K. The wave number becomes very small, however, as can be seen from Fig. 5 and the higher satellites merge to the center q=0 and thus are hard to observe because the ferromagnetic component located at q=0 might hinder these fine structures around the center and also the intensities of harmonics are weak. In spite of this difficulty we think that it is important to observe the higher harmonic satellites associating with the distorted sinusoidal modulation or the kink lattice structure, paying special attention to temperatures immediately below T_M=1.2K.

The temperature dependence of the satellite intensity in ERB is a decreasing function of temperature. On the other hand, the amplitude of the spin modulation in the sn state decreases upon increasing h as displayed in Figs. 8 and 9. This is opposite to the experiment. We emphasize here that the satellite scattering arises from the 4f moment system and the conduction electron spin modulation only triggers the modulated moment structure of the 4f system through the sf exchange interaction. Along the line developed in the present theory one of the authors gave an explanation [16] of the observed temperature dependence of the satellite intensity by using a phenomenological Ginzburg-Landau theory.

8. Summary and Discussion

We have shown that the superconducting state under the ferromagnetic molecular field can be solved exactly within the mean field approximation for one dimension. This state (sn state) gives the distorted sinusoidal modulation of the order parameter and the spin density. The energy spectrum of the state has two gaps, which is clearly different from that of the ordinary BCS state. We have also considered the helical solution, which has been studied by several authors, but its energy turns out to be higher than that of the sn state not only near the transition line but also throughout the region in the T vs. h plane.

Using the sn state, we interpret the experiments on $ErRh_4B_4$. The actual situation is three-dimensional, but the essential features are not different between the three-dimensional case and the one-dimensional case for the helical solution except that the inhomogeneous state is limited within the smaller region near the BCS phase. The direction of q should be determined by a nesting condition for the up spin and down spin bands, but once the single wave vector q is chosen, the system can be considered as one-dimensional.

The spin-modulated phase in ERB is interpreted as that triggered by the conduction electron spin modulation of the sn state. The temperature dependence of amplitude of the modulated phase can be interpreted through the reaction effects on the local spin system. The estimated order of the wave number falls on that obtained in the experiments.

The electro-magnetic interaction has not been taken into account. As for the effects on the conduction electron, it is negligible since the internal molecular field through the sf interaction reaches about several hundreds kG in spite of the smallness of the sf interaction. On the other hand,the dipole field $4\pi M$ coming from the electromagnetic interaction is of the order of a few kG which is the same order of H_{c1} for ERB. As for the stability of this state, however, it may be important because it carries the comparable order of energy with that of the condensation energy of the sn state. Therefore it requires further consideration.

We propose crucial experiments on ERB to observe the sn phase directly. An optical absorption and a tunneling experiment are able to reveal the two-gap structure which is rather common for an incommensurate phase [29]. The higher harmonics and their temperature dependences may be found by a neutron experiment near $T_M \cong 1.2K$.

Acknowledgements

The authors thank T. Ohmi, S.K. Sinha, T. Koyama, M. Tachiki, H. Yamamoto and M. Fujita for stimulating discussions.

References

1. V.L. Ginzburg: Sov. Phys. JETP 4 (1957) 153.
2. A.A. Abrikosov and L.P. Gor'kov: Sov. Phys. JETP 12 (1961) 1243.
3. See for example "Superconductivity in Ternary Compounds Vols.1 and 2", edited by M.B. Mapel and Ø.Fischer (Springer Verlag, New York, 1982).
4. D.E. Moncton, D.B. McWhan, P.H. Schmidt, G. Shirane, W. Thomlinson, M.B. Maple, H.B. MacKay, L.D. Woolf, Z. Fisk and D.C. Johnston: Phys. Rev. Lett. 45 (1980) 2060.
5. S.K. Sinha, G.W. Crabtree, D.G. Hinks and H.A. Mook: Phys. Rev. Lett. 48 (1982) 950.
   S.K. Sinha, G.W. Crabtree, D.G. Hinks, H.A. Mook and O.A. Pringle: J. Mag. Mag. Mat. 31-34 (1983) 489.
6. J.W. Lynn, D.E. Moncton, W. Thomlinson, G. Shirane and R.N. Shelton: Solid State Commun. 26 (1978) 493.
   J.W. Lynn, G. Shirane, W. Thomlinson and R.N. Shelton: Phys. Rev. Lett. 46 (1981) 368.
7. H. Matsumoto, H. Umezawa and M. Tachiki: Solid State Commun. 31 (1979) 157.
   M. Tachiki, A. Kotani, H. Matsumoto and H. Umezawa: ibid. 31 (1979) 927.
8. E.I. Blout and C.M. Varma: Phys. Rev. Lett. 42 (1979) 1079.
   H.S. Greenside, E.I. Blount and C.M. Varma: ibid. 46 (1981) 49.
9. R.A. Ferrell, J.K. Bhattacharjee and A. Bagchi: Phys. Rev. Lett. 43 (1979) 154.
10. M. Tachiki, H. Matsumoto, T. Koyama and H. Umezawa: Solid State Commun. 34 (1980) 19.
11. C.K. Kuper, M. Revzen and A. Ron: Phys. Rev. Lett. 44 (1980) 1545.
12. P. Fulde and R.A. Ferrell: Phys. Rev. 135 (1964) A550.
13. A.I. Larkin and Y.N. Ovchinnikov: Sov. Phys. JETP 20 (1965) 762.
14. S. Takada and T. Izuyama: Prog. Theor. Phys. (Kyoto) 41 (1969) 635.
15. K. Machida: J. Phys. Soc. Jpn. 51 (1982) 3462.
16. K. Machida: J. Phys. Soc. Jpn. 52 (1983) 2181.
17. S.A. Brazovskii, S.A. Gordynin and N.N. Kirova: JETP Lett. 31 (1980) 456.
18. J. Mertsching and H.J. Fischbeck: Phys. Stat. Sol.(b) 103 (1981) 783.
19. B. Horovitz: Phys. Rev. Lett. 46 (1981) 742.
20. P.G. de Gennes: Superconductivity of Metals and Alloys (Benjamin, New York, 1966).
21. H. Takayama, Y.R. Lin-Liu and K. Maki: Phys. Rev. B21 (1980) 2388. Also see H. Yamamoto and T. Ohmi: Prog. Theor. Phys. 66 (1981) 805.
22. E.T. Whittaker and G.N. Watson: A Course of Modern Analysis 4th ed. (Cambridge University Press, 1927).
23. K. Maki and T. Tsuneto: Prog. Theor. Phys. 31 (1964) 945.
24. M.C. Leung: Phys. Rev. B11 (1975) 4272.
25. K. Machida and D. Youngner: J. Low Temp. Phys. 35 (1979) 449 and 561.
    D. Youngner and K. Machida: ibid. 36 (1979) 617.
26. G.W. Crabtree, F. Behroozi, S.A. Campbell and D.G. Hinks: Phys. Rev. Lett. 49 (1982) 1342.
    F. Behroozi, G.W. Crabtree, S.A. Campbell and D.G. Hinks: Phys. Rev. B27 (1983) 6849.
27. K. Machida: J. Low Temp. Phys. 53 (1983) 403.
28. S.K. Sinha and J. Kjems: private communication.
29. Recently a tunneling experiment was done on a polycrystalline ERB. The data shows some unusual structures which may reflect the structure of the density of state for the intermediate phase. L.-J. Lin, A.M. Goldman, A.M. Kadin and C.P. Umbach: Phys. Rev. Lett. 51 (1983) 2151.

Part II

Antiferromagnetic Superconductors

Theory of Antiferromagnetic Superconductors

K. Levin, M.J. Nass, and Charles Ro

The James Franck Institute, The University of Chicago
Chicago, IL 60637, USA

Gary S. Grest

Exxon Research and Engineering Company, Annandale, NJ 08801, USA

I. General Review

It is useful to summarize the basic theoretical picture that we now have of antiferromagnetic (AF) superconductors. For definiteness we'll focus on the two types of ternary compounds $RERh_4B_4$ and $REMo_6S_8$ where RE denotes a rare earth element. There are a number of factors which aid the coexistence of antiferromagnetism and superconductivity. In the first place there are believed to be three distinct classes of electrons which play an important role in these systems. The RE f electrons undergo magnetic order. The d electrons associated with Rh or Mo are superconducting. Finally the electrons which mediate the RKKY interactions between the RE localized electrons have probably little to do with the superconducting electrons and constitute a distinct (third) class of electrons. It is important to recognize that in the ternary superconductors the electrons which mediate or undergo magnetic order are distinct from those which are superconducting. The second factor which aids coexistence has to do with the nature of the AF order parameter, called $\langle S_Q \rangle$. The fact that the wave vector Q of the AF order is considerably larger than ξ^{-1} means that on the size scale of a Cooper pair the average molecular field vanishes. By contrast,in ferromagnets the fact that $Q<\xi^{-1}$ is at the heart of the very strong incompatibility between ferromagnetism and superconductivity.

Now, let us consider the various ways in which AF order can undermine superconductivity. We start with the basic exchange Hamiltonian

$$H = J^{df} \int d^3r\, \vec{S}(r) \cdot \vec{\sigma}(r).$$

Here S and σ are the RE and superconducting spins respectively. This term is generally factored into two contributions: that deriving from mean field theory $J^{df}\langle S_Q \rangle \cdot \sigma$ and that deriving from spin disorder scattering $J^{df}(S-\langle S_Q \rangle)\cdot \sigma$. We define the molecular field $H_Q \equiv J^{df}\langle S_Q \rangle$.

A number of different groups [1-4] have studied the effect of this molecular field on superconductivity. It is generally believed that H_Q will weaken superconductivity whenever there are states $\vec{k}$ within the Debye energy ω_D of the Fermi surface such that $\varepsilon_k \sim \varepsilon_{k\pm Q}$. Along these $\vec{k}$ vectors the molecular field introduces a gap in the normal state density. This gap will reduce the effective number of electrons which can participate in the superconducting pairing, provided $H_Q \gtrsim \omega_D$.

The effect of the spin disorder scattering on superconductivity has also been discussed by a number of different authors [1-2],[5-6]. Spin disorder scattering, as in the usual Abrikosov-Gorkov theory, breaks Cooper pairs and hence weakens the superconducting state. Because the RE spins are on a lattice we have to consider the inelastic (magnon) as well as elastic spin fluctuation scattering. One interesting effect related to the spin disorder scattering was noted by RAMAKRISHNAN and VARMA [5]. It is known [1] that for a spherical Fermi surface when $H_Q = 0$, the effect of the freezing out of spin fluctuation scattering at the Néel temperature T_N causes decreased pairbreaking below T_N. Therefore to explain experimental observations [7] which suggest increased pairbreaking below T_N in some systems, RAMAKRISHNAN and VARMA studied the effects of non-spherical Fermi surfaces and calculated the function $\alpha^2F^{magnon}(\omega)$ which enters into the strong coupling superconducting gap equation.

Other authors [3-4] have pointed out that the effective $\alpha^2F^{phonon}(\omega)$ may be modified by the onset of AF order. In the particular case studied by ZWICKNAGL and FULDE [3] and ASHKENAZI, KUPER and RON [4], the phonon spectrum was assumed to be unchanged but it was argued that the paired states were not Cooper pairs but magnetic quasiparticles. As a consequence the effective coupling to the phonons was altered, leading to a reduction in $\alpha^2F^{phonon}(\omega)$.

Another interesting interaction between the AF and superconducting order comes from non-magnetic impurities which are presumably always present in the ternary superconductors. Because non-magnetic impurities are pairbreaking in the AF state [2], they will lead to a weakening of superconductivity below T_N. This fact may explain why the vast majority of AF superconductors have $T_N < T_c$. If the converse were true, T_c would be substantially depressed by any intrinsic or extrinsic disorder so that the superconducting state would have a hard time "turning on" for $T_N > T_c$.

Finally, TACHIKI, MATSUMOTO and UMEZAWA [8] have considered the effects of electromagnetic interactions, generated by the AF order, on superconductivity. These should be particularly important in compounds such as $DyMo_6S_8$, which are induced into a ferromagnetic state in moderate magnetic fields [9].

How big are the key parameters in the AF superconductors? From neutron data it is known that Q^{-1} is a few Å. Experiments [10] suggest that J^{df} $\sim$ 10meV so that $H_Q \sim 100$K. Note that the ratio H_Q/ω_D is not particularly small. The characteristic Néel temperature T_N for the AF superconductors is $\sim$ 1K. Finally it is believed that the fraction of the Fermi surface that is strongly affected by the gaps introduced by the molecular field is of the order of 10-20%. It should be emphasized that this number is not well known and that we need more experiments (combined with theoretical interpretation) to help pin it down.

The first major theoretical study of the AF superconductors is that of MAEKAWA and TACHIKI [6]. These authors set up an extremely general Green's function formalism. They studied spin-orbit and Pauli paramagnetic effects in some detail. However their detailed calculations were for the case $H_Q = 0$ so that $T > T_N$. The spin disorder scattering was assumed to be elastic so that they could lump all effects of AF order into an effective coupling constant and use the standard (weak coupling) framework to compute the upper critical field H_{c2} and other properties.

MACHIDA, NOKURA and MATSUBARA [1] extended the work of Ref. (6) to the case $H_Q \neq 0$. Here H_Q was assumed to be a "small" parameter so that it was treated perturbatively. Furthermore, only elastic spin disorder scattering processes were included. In this way the AF order again entered as a renormalization of the BCS coupling constant. With this renormalization one could then calculate $H_{c2}(T)$ and get rather good qualitative agreement with experiment.

Our own work [2], [11] treated the effects of the molecular field non-perturbatively. We also included inelastic spin fluctuation scattering, and a self-consistent treatment of impurity scattering. [12]. Consequently, one could no longer use a weak coupling formalism and our strong coupling calculations required detailed numerical studies. This work will be discussed in detail later on. Like a number of other theoretical approaches [1], [3-4] we also obtained good agreement between theory and experiment for $H_{c2}(T)$.

RAMAKRISHNAN and VARMA considered the case where the molecular field H_Q was unimportant but treated inelastic (as well as elastic) spin disorder scattering in detail. By considering nonspherical Fermi surfaces, they computed $\alpha^2F^{magnon}(\omega)$ and using a strong coupling formalism obtained the T dependence of the pairbreaking parameter. Their results also provided a qualitative explanation for the observed $H_{c2}(T)$ behavior.

Finally ZWICKNAGL and FULDE [3] and ASHKENAZI et al.[4] calculated $\alpha^2F^{phonon}(\omega)$, which contained modifications as a result of their assumption that the pairing is between magnetic quasiparticles. Both groups obtained good agreement with the measured $H_{c2}(T)$ curves. However (elastic and inelastic) spin fluctuation effects were ignored in these calculations.

2. SELF-CONSISTENT TREATMENT OF THE COUPLED GAP EQUATIONS

I will summarize here our own theory of AF superconductors and then compare it with that of ZWICKNAGL and FULDE [3] and ASHKENAZI et al.[4]. A complete calculation has to be done at several levels. First one has to treat the coupled mean field problem. Both the AF and BCS order are treated in mean field theory. Using this as a basis one can then include strong coupling effects (arising from inelastic spin fluctuations, impurities, etc.).

In the coupled mean field theory, the Hamiltonian is given by

$$H^{mf} = \sum_{k\sigma} \varepsilon_k C^+_{k\sigma} C_{k\sigma} \; \sum_{k\sigma} H_Q \sigma (C^+_{k\sigma} C_{k+Q\sigma} + C^+_{k+Q\sigma} C_{k\sigma})$$
$$+ \; - \sum_k \Delta C^+_{k\uparrow} C^+_{-k\downarrow} - \sum_k (\Delta_Q C^+_{k+Q\uparrow} C^+_{-k\downarrow} - \Delta_{-Q} C^+_{k\uparrow} C^+_{-k-Q\downarrow}) + C.C. \qquad (2.1)$$

where the self-consistent order parameters (with k' vectors near k_F) are

$$\Delta = \sum_{k'} V \langle C_{-k'\downarrow} C_{k'\uparrow} \rangle , \qquad \Delta_{-Q} = 2\sum_{k'} V \langle C_{-k'-Q\downarrow} C_{k'\uparrow} \rangle .$$

$$\Delta_Q = 2\sum_{k'} V \langle C_{k'\downarrow} C_{-k'+Q\uparrow} \rangle , \qquad H_Q = H^\circ_Q - 1/2 \sum_{k'\sigma} V\sigma \langle C^+_{k'\sigma} C_{k'+Q\sigma} \rangle \qquad (2.2)$$

and we define the bare molecular field H°_Q as

$$H^\circ_Q = J^{df} \langle S_Q \rangle \quad .$$

Here we have allowed for a general set of superconducting pairings, which, for isotropic V, are $\vec{k}$ independent. It can be shown that these order parameters satisfy the following equations:

$$\Delta = VT\sum_m \int \frac{d^3k'}{(2\pi)^3} K^{-1} [\Delta(\omega_m^2 + \varepsilon_s^2 + \varepsilon_a^2 - H_Q^2 + \Delta^2 - \Delta_Q^{-2} - \Delta_Q^{+2}) + 2H_Q(\varepsilon_a \Delta_Q^- + i\omega_m \Delta_Q^+)] \qquad (2.3a)$$

$$\Delta_Q^- = VT\sum_m \int \frac{d^3k'}{(2\pi)^3} K^{-1}[\Delta_Q^-(\omega_m^2+\varepsilon_s^2-\varepsilon_a^2+H_Q^2+\Delta^2+\Delta_Q^{-2}-\Delta_Q^{+2})-2\varepsilon_a(i\omega_m\Delta_Q^+-H_Q\Delta)], \qquad (2.3b)$$

$$\Delta_Q^+ = VT\sum_m \int \frac{d^3k'}{(2\pi)^3} K^{-1}[\Delta_Q^+(\omega_m^2+\varepsilon_s^2-\varepsilon_a^2-H_Q^2-\Delta^2-\Delta_Q^{-2}+\Delta_Q^{+2})-2i\omega_m(\varepsilon_a\Delta_Q^--H_Q\Delta)], \qquad (2.3c)$$

where $\Delta_Q^{\pm} \equiv 1/2(\Delta_{-Q}\pm\Delta_Q)$ and

$$K \equiv (\omega_m^2+\varepsilon_a^2+\varepsilon_s^2+H_Q^2+\Delta^2+\Delta_Q^{-2}-\Delta_Q^{+2})^2-4(\varepsilon_a\Delta_Q^-+i\omega_m\Delta_Q^+-H_Q\Delta)^2-4\varepsilon_s^2(H_Q^2-\Delta_Q^{+2})-4\varepsilon_s^2\varepsilon_a^2 \ . \qquad (2.3d)$$

Finally $\varepsilon_s \equiv (\varepsilon_k+\varepsilon_{k+Q})/2$ and $\varepsilon_a=(\varepsilon_k-\varepsilon_{k+Q})/2$ and V is the usual BCS coupling parameter, which is assumed to be unaffected by the AF order.

We have numerically solved Eqs. (2.3) for a spherical (three-dimensional) Fermi surface and find that the maximum value that the "anomalous" pairing parameter reaches is about (2%±2%) of the "Cooper" order parameter Δ. (Here the uncertainty is due to our estimated numerical error). This maximum value is achieved by varying the parameters H_Q/ω_D and ω_D/k_FQ. Thus it appears that there is a very small essentially negligible amount of "anomalous" pairing within our weak coupling calculation.

To go beyond mean field theory we make an ansatz for the renormalized Green's function and then use self-consistent Hartree-Fock theory. Our Green's function is written in the enlarged 8x8 basis:

$$\psi_k^+ = (C_{k\uparrow}^+ C_{-k\downarrow}^+ C_{k\uparrow} C_{-k\downarrow} C_{k+Q\uparrow}^+ C_{-k-Q\downarrow}^+ C_{k+Q\uparrow} C_{-k-Q\downarrow}) \qquad (2.4)$$

to treat both spin fluctuations and the coupling between states k and k+Q. In weak coupling (mean field theory)

$$G^{-1} = i\omega_n-\varepsilon_s\rho_3-\varepsilon_a\tau_3\rho_3-\Delta\rho_2\sigma_2+H_Q\tau_1\rho_3\sigma_3 \qquad (2.5)$$

where ρ,τ and σ are the usual Pauli (2x2) matrices. Using these as a basis, the renormalized Green's function is

$$\tilde{G}^{-1} = i\tilde{\omega}_n-\varepsilon_3\rho_3-\varepsilon_a\tau_3\rho_3-\tilde{\Delta}_n\rho_2\sigma_2+\tilde{H}_{Qn}\tau_1\rho_3\sigma_3+i\tilde{\Omega}_n\tau_1\rho_1\sigma_1 \ . \qquad (2.6)$$

Thus there are four renormalized parameters which must be solved for self-consistently. The resulting equations [2] are rather complicated so they will not be written down here. They are discussed in detail in Ref. (2).

Recently we [11] have extended our strong coupling theory to a calculation of the upper critical field H_{c2}. This quantity satisfies the equation

$$\tilde{\Delta}_n = 2\pi N(0)T\sum_m \frac{\lambda(n-m)\,\tilde{\Delta}_m}{\left[1-\frac{\tilde{H}_{Qm}^2}{\tilde{\omega}_m^2+\tilde{H}_{Qm}^2}\right]^{-1}\left[2\sqrt{\tilde{\omega}_m^2+\tilde{H}_{Qm}^2}+\frac{e\hbar v_F^2 H_{c2}}{6c\sqrt{\tilde{\omega}_m^2+\tilde{H}_{Qm}^2}}(\tau_1^{-1}-\tau_2^{-1})\right]} \ . \qquad (2.7)$$

Here $\lambda(n-m)$ includes both (inelastic and elastic) spin fluctuations and electron-phonon interactions; v_F is the Fermi velocity and τ_1 and τ_2 correspond to the lifetimes for non-magnetic and magnetic impurities, respectively.

In the limit $\tilde{H}_Q \to 0$, this equation is the same as that considered by RAMAKRISHNAN and VARMA. In the limit of very small $\tilde{H}_Q$ and elastic scattering it yields the usual weak coupling form for H_{c2} which is equivalent to that considered by MACHIDA *et al.* [1].

3. COMPARISON WITH OTHER PAIRING SCHEMES

Several groups [3-4] have proposed that the superconducting pairs in AF superconductors should consist of magnetic quasiparticles (of opposite spin but degenerate in energy). Our mean field Hamiltonian (Eq. (2.1)) can be expressed in terms of the magnetic quasiparticles defined by

$$a_{k\uparrow}=\alpha_k C_{k\uparrow}-\beta_k C_{k+Q\uparrow} \text{ and } b_{k\uparrow}=\beta_k C_{k\uparrow}+\alpha_k C_{k+Q\uparrow} . \tag{3.1a}$$

Also

$$a_{-k\downarrow}=\beta_k C_{-k\downarrow}-\alpha_k C_{-k-Q\downarrow} \text{ and } b_{-k\downarrow}=\alpha_k C_{-k\downarrow}+\beta_k C_{-k-Q\downarrow} . \tag{3.1b}$$

Here

$$\alpha_k{}^2-\beta_k{}^2 = |\varepsilon_k-\varepsilon_{k+Q}|/[(\varepsilon_k-\varepsilon_{k+Q})^2+4H_Q{}^2]^{1/2} \tag{3.2a}$$

and

$$\alpha_k\beta_k = -H_Q \operatorname{sgn}(\varepsilon_k-\varepsilon_{k+Q})/[(\varepsilon_k-\varepsilon_{k+Q})^2+4H_Q{}^2]^{1/2} . \tag{3.2b}$$

Since it is assumed that the dominant pairing is between the degenerate [3,4] quasiparticle states, of the form $\langle a_{k\uparrow} b_{-k\downarrow}\rangle$, this requires that $\Delta_Q = -\Delta_{-Q}$. Using this simplification the mean field Hamiltonian can be written in the quasiparticle (4x4) basis. In the new basis the two gap parameters are given by

$$\Delta^{ab} = (\beta_k^2-\alpha_k{}^2)\Delta + 2\alpha_k\beta_k\Delta_Q \tag{3.3a}$$

$$\Delta^{aa} = \Delta^{bb} = -2\alpha_k\beta_k\Delta-(\alpha_k{}^2-\beta_k{}^2)\Delta_Q . \tag{3.3b}$$

Imposing the constraint that the pairing corresponds to $\Delta^{aa}=\Delta^{bb}=0$ implies that

$$\frac{\Delta}{\Delta_Q} = \frac{\beta_k{}^2-\alpha_k{}^2}{2\alpha_k\beta_k} \tag{3.4}$$

so that there is a fixed relation between a k-dependent Δ_Q and Δ. Note that this last equation can be satisfied only if the "bare" electron-electron interaction V in Eqs. (2.2) is anisotropic. Substituting Eq. (3.4) into (3.3a) implies $\Delta^{ab}=\Delta/(\beta_k^2-\alpha_k^2)$ so that the magnitude of the quasiparticle gap is *enhanced* over Δ. By contrast, it follows from Eq. (3.3a), when the Δ_Q term is ignored, that $\Delta^{ab}=(\beta_k{}^2-\alpha_k{}^2)\,\Delta$. It appears that the *reduction* in the attractive interaction between quasiparticles found in Refs. (3) and (4) can be derived in this way. A transformation of the BCS interaction Hamiltonian $\sum_{kk'} V_{kk'} C^+_{k\uparrow} C^+_{-k\downarrow} C_{k'\uparrow} C_{-k'\downarrow}$ to the quasiparticle basis yields an effective

$V^{eff} = V_{kk'}(\alpha_k{}^2-\beta_k{}^2)(\alpha_{k'}^2-\beta_{k'}^2)$ when terms of the form a^+a^+ and b^+b^+, etc., are neglected [13]. It is assumed that V^{eff} is isotropic so that the k-dependent terms $(\alpha_k{}^2-\beta_k{}^2)^2$ are averaged over the Fermi surface. It should be stressed that the assumption that only magnetic quasiparticles participate in the pairing *cannot* be derived self-consistently when V is isotropic, since Eq. (3.4) cannot be satisfied. Furthermore for anisotropic V, if self-consistency is imposed (through Eq. (3.4)), then it would follow that the quasiparticle gap parameter is enhanced relative to the Cooper pairing order parameter Δ. The suppression of Δ^{ab} found in Refs. (3) and (4) is a reflection of the fact that terms of the type $\sum_{kk'} V_{kk'} C^+_{k\uparrow} C^+_{-k-Q\downarrow} C_{k'\uparrow} C_{-k'-Q\downarrow}$ are not assumed to contribute to the effective attraction between quasiparticles [13].

In summary, our self-consistent approach and the assumption of magnetic quasiparticle pairing are very different. In our case self-consistency is assumed to be of the utmost importance. References (3) and (4) are based on the premise that only energetically degenerate quasiparticles can participate in superconducting pairing.

4. RESULTS

In order to solve the four coupled integral equations for $\tilde{H}_{Qn}, \tilde{\omega}_n, \tilde{\Delta}_n$ and $\tilde{\Omega}_n$ we were forced to make an approximation, introduced earlier by BILBRO and MCMILLAN [14]. We assume that the Fermi surface is divided into two regions. Region I corresponds to states with $H_Q/|\varepsilon_k-\varepsilon_{k\pm Q}| \gtrsim 1$ and Region II corresponds to the remainder of the Fermi surface. In Region II we neglect the molecular field altogether since it is a weak perturbative effect, whereas in Region I we adopt a one-dimensional approximation so that $\varepsilon_k = -\varepsilon_{k+Q}$. Under these circumstances the integral equations simplify considerably and we can obtain their solution numerically. For simplicity, we assume the Fermi surface is spherical.

To treat spin fluctuation processes it is essential to impose the sum rule deriving from $\langle S_i^2 \rangle = S(S+1)$:

$$\langle S_Q \rangle^2 + \sum_{q,n} \chi^{zz}(q,\omega_n) + 1/2[\chi^{+-}(q,\omega_n) + \chi^{-+}(q,\omega_n)] = S(S+1) \,. \qquad (4.1)$$

Here χ represents the dynamical spin susceptibility. Because the term $\langle S_Q \rangle$ is subtracted out in our definition of χ, it must be added back in to satisfy the sum rule. Thus there is a correlation between the temperature dependence of the molecular field and that of the spin disorder scattering. Since the former enters into the superconducting gap equation as $J^{df}\langle S_Q \rangle$ and χ enters multiplied by $N(0)(J^{df})^2$, the relative importance of molecular field and spin fluctuation effects is characterized by the dimensionless parameter $N(0)J^{df}$.

Among the more interesting effects we have found [2] using our strong coupling formalism are the following. (1) Non-magnetic impurities in moderate concentrations are pairbreaking: they will introduce gaplessness until their concentration is sufficiently large so that the screened molecular field $\tilde{H}_Q$ is reduced to zero, at which point the density of states $N_s(\omega)$ and other properties become BCS-like. (2) Inelastic spin fluctuations (magnons) should lead to structure in $N_s(\omega)$ in the same way that phonons lead to structure in the density of states in strong coupling superconductors. (3) Finally we have calculated [11] the temperature dependence of the upper critical field using Eq. (2.7) and a simple Einstein model for the magnons and treating all impurity effects self-consistently. Unlike our density of states calculations, our results for $H_{c2}(T)$ can be compared with present experiments. Figs. 1 show $H_{c2}(T)$ for some AF chevrel compounds [7] (1a) and for $SmRh_4B_4$ [after Ref. (15)] (1b). In the first case increased pairbreaking is found. We believe that these two types of behavior are intimately connected to the size of the dimensionless parameter $N(0)J^{df}$. This can be seen from Fig. 2 which plots our theoretical results for a range of different H_Q, holding the spin fluctuation parameters constant. When

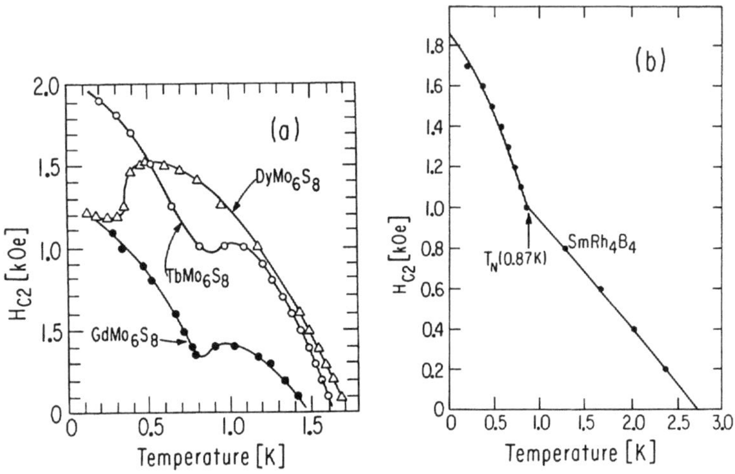

Fig. 1. Experimental results for H_{c2} in $SmRh_4B_4$ and in a series of chevrel compounds. Fig. 1a is from Ref. 7 and 1b is from Ref. 15

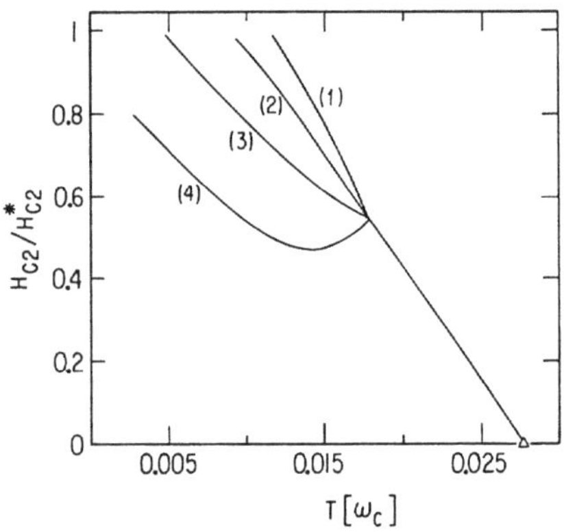

Fig. 2. Strong coupling calculation of the T dependence (in units of the Debye frequency $\omega_c \equiv \omega_D$) of H_{c2} for $H_Q°$ equal to 0.05 (1), 0.20 (2), 0.30 (3), and 0.40 (4) in units of ω_c. Here $H^*_{c2} = c\omega_c^2/\hbar v_F^2 e$

$N(0)J^{df} \gtrsim 10^{-3}$ spin disorder scattering dominates the behavior of H_{c2} and decreased pairbreaking is observed, whereas when $N(0)J^{df} \lesssim 10^{-3}$ increased pairbreaking is found and the molecular field dominates the behavior of H_{c2}. In the first case the freezing out of spin fluctuations causes H_{c2} to rise abruptly below T_N. In the second case the onset of a molecular field causes H_{c2} to drop abruptly at T_N as seen in the chevrel compounds. The physical values of the parameter $N(0)J^{df}$ are around 10^{-3} for the ternary compounds so that it seems possible for both types of behavior to be observed experimentally.

It should be stressed that there are presently three separate explanations which have been offered for the observed behavior of $H_{c2}(T)$. Our own calculation (and indirectly that of MACHIDA and co-workers [1]) suggests that it is the parameter $N(0)J^{df}$ which characterizes the shape of the upper critical field curves. RAMAKRISHNAN and VARMA argue that deviations from a spherical Fermi surface are important in explaining $H_{c2}(T)$ particularly in the chevrel compounds where increased pairbreaking is observed. Finally

ZWICKNAGL and FULDE [3] and ASHKENAZI and co-workers [4] have found that increased pairbreaking below T_N can be explained as a reduction in the effective electron-electron interaction,which reduction is due to the assumed pairing of magnetic quasiparticles. It seems likely that effects of a nonspherical Fermi surface [5] and, in some systems, electromagnetic interactions [8] may be important in addition to the competition between molecular field and spin fluctuation scattering that we have invoked here. However, the ansatz of magnetic quasiparticle pairing [3-4] which violates self-consistency at several levels appears to be highly controversial. In order to sort out the underlying physics of the AF superconductors, it is important to have detailed theoretical predictions which can be checked in the laboratory. In our calculation of $H_{c2}(T)$ we have noted [11] that the pairbreaking nature of non-magnetic impurities means that the *shape* of $H_{c2}(T)$ is sensitive to impurity effects. We have provided detailed numerical predictions which can be tested experimentally. Using these along with (hopefully different) predictions of other competing theories may help to sort out the nature of the interactions between antiferromagnetic and superconducting order.

We wish to thank Dana Browne for useful conversations. This work was supported in part by the NSF under Grant No. DMR-8115618 and by the NSF Materials Research Laboratory under Grant No. DMR-8216892.

REFERENCES

1. K. Machida, K. Nokora and T. Matsubara, Phys. Rev. B22, 2307 (1980).
2. M.J. Nass, K. Levin and G.S. Grest, Phys. Rev. B25, 4541 (1982); See also M.J. Nass, K. Levin and G.S. Grest, Phys. Rev. Lett. 46, 614 (1981).
3. G. Zwicknagl and P. Fulde, Z. Physik B43, 23 (1981).
4. J. Ashkenazi, C.G. Kuper and Amiram Ron, Phys. Rev. B28, 468 (1983).
5. T.V. Ramakrishnan and C.M. Varma, Phys. Rev. B24, 137 (1981).
6. S. Maekawa and M. Tachiki, Phys. Rev. B18, 4688 (1978).
7. C.F. Majkrzak, G. Shirane, W. Thomlinson, M. Ishikawa, Ø. Fischer and D.E. Moncton, Solid State Commun. 31, 773 (1979).
8. M. Tachiki, H. Matsumoto and H. Umezawa, Phys. Rev. B 20, 1915 (1979).
9. W. Thomlinson, G. Shirane, D.E. Moncton, M. Ishikawa and Ø. Fischer, Phys. Rev. B23, 4455 (1981).
10. M. Ishikawa and J. Muller, Solid State Comm. 27, 761 (1978); R. Odermatt, M. Hardiman and J. Van Neijel, ibid.32, 1227 (1979).
11. Charles Ro and K. Levin (preprint).
12. See also Y. Okabe and A.D.S. Nagi, Phys. Rev. B28, 6290 (1983).
13. J. Keller, Proceedings of the European Physical Society (to be published).
14. G. Bilbro and W.L. McMillan, Phys. Rev. B14, 1887 (1976).
15. H.C. Hamaker, L.D. Wolff, H.B. Mackay, Z. Fisk and M.B. Maple, Solid State Commun. 32, 289 (1979).

The Coexistence of Superconductivity and Antiferromagnetism

J. Ashkenazi, C.G. Kuper, and Amiram Ron

Physics Department, Technion: Israel Institute of Technology
Haifa, Israel

Abstract

We present a theory of the coexistence of superconductivity and antiferromagnetism. The diagonal part of the interaction between the magnetic ions and the conduction electrons is included with the electronic kinetic energies, to give new "magnetic Bloch states", in which electrons of opposite spin are also spatially displaced. This leads to a reduction in the superconducting pairing strength, linear in the exchange strength and the staggered magnetization. Agreement with experiment is good for $DyMo_6S_8$ and $TbMo_6S_8$.

A number of compounds have been reported [1, 2], in which magnetic order and superconductivity coexist. Typically, they are ternary compounds, containing a rare-earth element and a transition metal. The incomplete f shell contributes a localized moment, and, because of the smallness of their overlap with the conduction electrons, their interactions with these electrons are very weak. The weak residual interaction leads to an RKKY interaction between the localized moments themselves, which can be either ferromagnetic or antiferromagnetic. In the present paper, we shall restrict ourselves to the antiferromagnetic case, and more specifically to the case when the Néel temperature T_N lies below the superconducting transition temperature T_c. When this condition is satisfied, we can observe the influence of the onset of antiferromagnetism on the superconducting properties; the upper critical field H_{c2} is seen to behave anomalously near T_N, and $H_{c2}(T=0)$ is considerably smaller than the value expected from an extrapolation of the behavior [1, 3] for $T>T_N$.

In a nonmagnetic superconductor, a localized magnetic impurity has the effect of breaking pairs, as shown by ABRIKOSOV and GOR'KOV [4]. This arises because the BCS-paired states are pairs of one-electron states related by time reversal, and the scattering from a magnetic ion breaks the time-reversal invariance. When, however, there is an antiferromagnetic lattice of magnetic ions, BALTENSPERGER and STRÄSSLER [5] showed that a BCS condensed state can be formed by pairing states which are mutually time-reversed and displaced by a lattice vector.

Baltensperger-Strässler electrons interact both through the usual Fröhlich electron-phonon mechanism and through spin-spin scattering off the magnetic ions. Below T_N, the latter interaction may be described in terms of absorption and emission of magnons - a manifestly repulsive interaction. Well above T_N, this spin-spin scattering is just the Abrikosov-Gor'kov mechanism. RAMAKRISHNAN and VARMA [6] show that, unless there is strong Fermi surface nesting, there should be no drastic change in the superconducting pairing strength, due to spin-spin scattering, on passing through T_N.

The magnetic ion-electron interaction contains one term which will indeed differ above and below T_N: it is the diagonal part $JS_z s_z$ of the exchange interaction. Below T_N, a Fourier component $S_z(\vec{Q})$ becomes just the macroscopic staggered magnetization. By means of an analysis similar to that of ZWICKNAGL and FULDE [7], we show that the phonon-induced attraction between conduction electrons is weakened by the presence of staggered magnetization. The principal difference between our work and that of Zwicknagl and Fulde is that where they find numerical solutions for a particular model, we parametrize the problem, and find analytical solutions exhibiting the essential features.

We assume that the frequency of flipping of the magnetic ions is slow, on the time scale of typical electron or phonon frequencies. With this assumption, we can neglect the dynamics of the magnetic lattice, and replace $S_z(\vec{Q})$ by its expectation value $S_M = \langle S_z(\vec{Q})\rangle$; i.e.,we treat the staggered magnetization as an external field [8]. Thus our starting point is the Hamiltonian

$$H = H_e + H_p + H_{e-p} , \tag{1}$$

$$H_e = \sum_{\vec{k},\sigma} E(\vec{k})\, c^+_\sigma(\vec{k})\, c_\sigma(\vec{k}) + \frac{1}{2} J(\vec{Q}) S_z(\vec{Q}) \sum_{\vec{k},\sigma} \sigma\, c^+_\sigma(\vec{k}) c_\sigma(\vec{k}+\vec{Q}), \tag{2}$$

$$H_p = \sum_{\vec{q},j} \Omega_j(\vec{q})\, b^+_j(\vec{q}) b_j(\vec{q}) , \tag{3}$$

$$H_{e-p} = \sum_{j,\sigma,\vec{k},\vec{k}'} g_j(\vec{k},\vec{k}')\, c^+_\sigma(\vec{k}) c_\sigma(\vec{k}')[b_j(\vec{k}-\vec{k}') + b^+_j(\vec{k}'-\vec{k})]. \tag{4}$$

The notation is as follows: $c_\sigma(\vec{k})$, $b_j(\vec{q})$ are annihilation operators for conduction electrons and (j-th branch) phonons respectively, $\sigma = \pm 1$ is a spin index, $\Omega_j(\vec{q})$, $E(\vec{k})$ are the phonon and electron spectra [where, for simplicity, we consider only one electron band, containing the Fermi surface (FS)], and $g_j(\vec{k}, \vec{k}')$ are the matrix elements of the Fröhlich interaction.

Whether or not the magnetic lattice is ordered, $2\vec{Q}$ is always a reciprocal-lattice vector, and $\vec{k} + 2\vec{Q}$ is equivalent to $\vec{k}$. This fact enables us to diagonalize the electron Hamiltonian (2) exactly. The transformation which affects the diagonalization is:

$$c_\sigma(\vec{k}) = C_\sigma(\vec{k})\cos\alpha_{\vec{k}} - \sigma C_\sigma(\vec{k}+\vec{Q})\sin\alpha_{\vec{k}} , \tag{5}$$

where $\alpha_{\vec{k}} = -\alpha_{\vec{k}+\vec{Q}}$ and

$$\tan 2\alpha_{\vec{k}} = \frac{J(\vec{Q})S_M}{E(\vec{k}) - E(\vec{k}+\vec{Q})} . \tag{6}$$

Under this transformation, (4) becomes

$$H_{e-p} = \sum_{j\sigma\vec{k}\vec{k}'} C^+_\sigma(\vec{k}) C_\sigma(\vec{k}')\{\bar{g}_j(\vec{k},\vec{k}')[b_j(\vec{k}-\vec{k}') + b^+_j(\vec{k}'-\vec{k})]$$

$$+ \sigma\tilde{g}_j(\vec{k},\vec{k}')[b_j(\vec{k}-\vec{k}'+\vec{Q}) + b^+_j(\vec{k}'-\vec{k}-\vec{Q})]\} , \tag{7}$$

where

$$\bar{g}_j(\vec{k}, \vec{k}') = g_j(\vec{k}, \vec{k}')\cos\alpha_{\vec{k}} \cos\alpha_{\vec{k}'} + g_j(\vec{k} + \vec{Q}, \vec{k}' + \vec{Q})\sin\alpha_{\vec{k}} \sin\alpha_{\vec{k}'} , \tag{8a}$$

$$\tilde{g}(\vec{k}, \vec{k}') = g_j(\vec{k}, \vec{k}' + \vec{Q})\cos\alpha_{\vec{k}}\sin\alpha_{\vec{k}'} + g_j(\vec{k} + \vec{Q}, \vec{k}')\sin\alpha_{\vec{k}}\cos\alpha_{\vec{k}'} . \tag{8b}$$

The novel aspect of (7) is the appearance of the Umklapp term (8b) expressing the reduced translational symmetry of the magnetized lattice. The spin-dependent phase factor $\sigma = \pm 1$ expresses the fact that electrons of opposite spin are related by a spatial displacement of one lattice vector. Although (8b) is, explicitly, an electron-phonon interaction, it resembles an electron-magnon interaction insofar as it connects the phonon field to the spin density of the conduction electrons. Moreover, like a magnon interaction, it induces a weakening of the pairing interaction.

After eliminating the phonons [9], the effective electron-electron interaction parameter is altered, from its usual value λ, proportional to g^2, to $\tilde{\lambda}$, proportional to $g^2_{eff} = \tilde{g}^2 - \bar{g}^2$. If we make the simplifying assumptions that (a) the conduction electron spectrum is isotropic and (b) that there is no dispersion either in the phonon spectrum or in the electron-phonon interaction, we may then average over the Fermi surface. Thus the effective phonon-mediated electron-electron interaction parameter is renormalized to

$$\tilde{\lambda} = \lambda \langle \cos 2\alpha_{\vec{k}} \rangle^2_{\vec{k}\in FS} . \tag{9}$$

Of course, the transformation (5) also renormalizes the conduction electron spectrum when antiferromagnetic order is present. But in three dimensions it is usually reasonable to neglect this aspect, unless there is strong Fermi-surface nesting. In contrast to the general trend, $SmRh_4B_4$ shows enhancement of superconductivity below T_N; this anomaly can perhaps be understood in terms of such neglected effects.

The weakening of superconductivity is described by the factor $\langle \cos 2\alpha_{\vec{k}} \rangle^2$ in (9). In order to estimate its effect on H_{c2}, we make the approximation that for $\vec{k}$ on the Fermi surface $E(\vec{k}) - E(\vec{k} + \vec{Q})$ is linear in $\vec{k}$. [Note that for free electrons, this would be exact. It should still be good for more complicated band structures, since the main contributions to $\langle \cos 2\alpha_{\vec{k}} \rangle$ come from the small regions where $E(\vec{k}) - E(\vec{k} + \vec{Q})$ is small, and where a linear expansion in $\vec{k}$ is therefore valid.] From (2),

$$\langle \cos 2\alpha_{\vec{k}} \rangle_{\vec{k}\in FS} = (E_2 - E_1)^{-1} \int_{E_i}^{E_2} dE / \{1 + [J(\vec{Q})S_M/E]^2\}^{1/2} , \tag{10}$$

where E_1 and E_2 are the limiting values of $E(\vec{k}) - E(\vec{k} + \vec{Q})$, when $\vec{k}$ is on the Fermi surface. We assume that $E_1 < 0 < E_2$, and approximate $E_2 - E_1$ by the band width B. Evaluating the integral in (10), we get

$$\langle \cos 2\alpha_{\vec{k}} \rangle^2_{\vec{k}\in FS} = 1 - 4|J(\vec{Q})S_M/B| + O(J^2). \tag{11}$$

In order to discuss the weakening of superconductivity in real materials, it is convenient first to define an auxiliary hypothetical transition temperature T_S, the temperature at which the material would become superconducting if S_M where fixed at a value S. When S= 0, $T_S = T_c$. To first order in $J(\vec{Q})E$,

$$T_S/T_c = 1 - 4\gamma|J(\vec{Q})S_M/B| \ , \tag{12}$$

where the constant γ is approximately given by

$$\gamma \cong \lambda(1 + \lambda)/(\lambda - \mu^*)^2 \tag{13}$$

in McMillan's "two-square-well" model [10]. Here μ^* is the previously-neglected Coulomb repulsion parameter. From (12), the reduction is linear [11] in $J(\vec{Q})S_M/B$.

In real materials, the staggered magnetization S_M is not independent of temperature, and the temperature T_S is not observable. What is observed is a weakening of H_{c2} compared with the value H^0_{c2} [which is obtained by extrapolation from $T>T_N$ to the region $T<T_N$]. To predict the amount of this weakening, we assume that $H_{c2}(T)$ scales in a BCS-like way with $T_S(T)$: i.e.,

$$H_{c2}(T) = H^0_{c2}\ [TT_c/T_S(T)]T_S(T)/T_c \ . \tag{14}$$

Here $T_S(T)$ is known from (12), since $S_M(T)$ is measured; the latter is proportional to the square root of the intensity of the $\vec{Q}$ neutron diffraction peak (corrected for the background) - see the continuous curves in Fig. 1. Taking the ratio H_{c2}/H^0_{c2} at T= 0 as an adjustable parameter, we use the neutron scattering data to calculate $H_{c2}(T)$. For both $DyMo_6S_8$ and $TbMo_6S_8$, the agreement with experiment is good, except close to T_N. The dis-

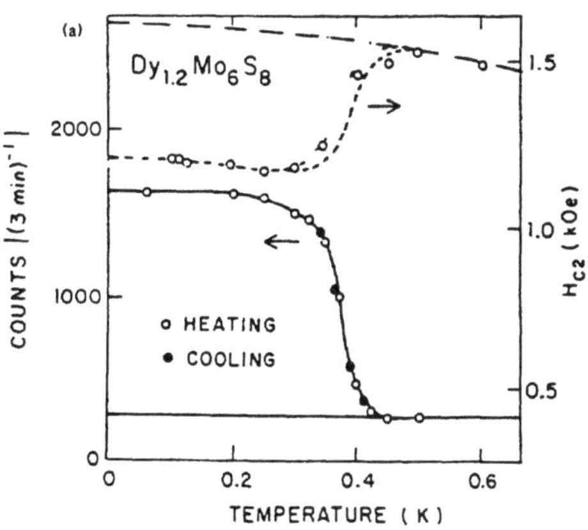

Fig. 1. Temperature dependence of the $\vec{Q}$ neutron-diffraction peak intensity [12] (solid curve), the experimental and extrapolated H^0_{c2} (open circles and coarse broken curve), and the present theoretical H_{c2} (fine broken curve) for (a) $Dy_{1.2}Mo_6S_8$, (b) $Tb_{1.2}Mo_6S_8$

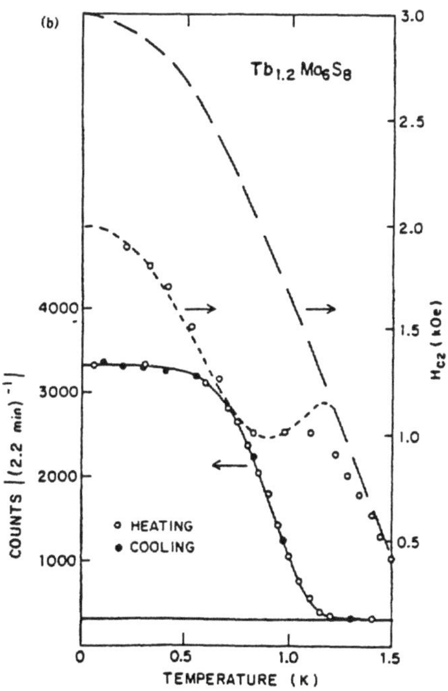

crepancies near T_N are not unexpected, since critical fluctuations should be important here. Hence the operator S_z should be replaced by a fluctuating moment, and not by the expectation value $<S_z(\vec{Q})>$.

For both the compounds discussed above, our choice of the ratio $H_{c2}(0)/H^0_{c2}(0)$ fixes the value $4\gamma|J(\vec{Q})S_M(T=0)/B| \simeq 0.3$. Now $S_M(T=0)$ is about 2.5, B is about 6×10^3 K [12], and γ lies in the range 2 to 4. Hence $J(\vec{Q})$ is estimated to be in the range 50 to 100 K, consistent with other estimates. [For example, we should except, from RKKY theory, that T_N is proportional to $J(\vec{Q})^2$].

A shorter version of this work has been published [13].

Acknowledgements

This work has been supported by the U. S. - Israel Binational Science Foundation, by the NCRD, Israel, by KFK, Germany, and by the Fund for the Promotion of Research at the Technion.

References

1 M. Ishikawa and Ø. Fischer: Solid State Commun. 24, 747 (1977); M. Ishikawa and J. Muller, ibid. 27, 761 (1978).
2 D. E. Moncton, G. Shirane, W. Tomlinson, M. Ishikawa, and Ø. Fischer: Phys. Rev. Lett. 41, 1133 (1978).
3 Ø. Fischer: Appl. Phys. 16, 1 (1978).
4 A. A. Abrikosov and L. P. Gor'kov: Zh. Eksp. i Teor. Fiz. 39, 1781 (1960).
5 W. Baltensperger and S. Strässler: Phys. Kondens. Materie 1, 20 (1963).
6 T. V. Ramakrishnan and C. M. Varma: Phys. Rev. B24, 137 (1981).
7 G. Zwicknagl and P. Fulde: Z. Phys. B43, 23 (1981).
8 A similar approximation is made by K. Machida, K. Nokura and T. Matsubara: Phys. Rev. B22, 2307 (1980), and by M. J. Nass, K. Levin and G. S. Grest: Phys. Rev. Lett. 46, 614 (1981); Phys. Rev. B25, 4541 (1982). However, these authors start from a BCS Hamiltonian, rather than from (1), and therefore fail to find the correct effect of antiferromagnetism on the electron-phonon interaction. Moreover, Nass et al. assume further that the Cooper pairing of an electron is affected only when both $\vec{k}$ and $\vec{k}+\vec{Q}$ are within a BCS cutoff from the Fermi surface.
9 cf. H. Fröhlich [Proc. Roy. Soc. A215, 291 (1952)]. Note that the electron mass renormalization is determined by $\bar{g}^2 + \tilde{g}^2 \simeq g^2$ for isotropic electrons.
10 W. L. McMillan: Phys. Rev. 167, 331 (1968).
11 Machida et al., loc.cit. also find a linear reduction, but only after making tne restrictive assumption that $Q = 2K_F$.
12 O. K. Anderson, W. Klose and H. Nohl: Phys. Rev. B17, 1209 (1978).
13 J. Ashkenazi, C. G. Kuper and A. Ron: Phys. Rev. B28, 418 (1983).

Effect of a Homogeneous Magnetic Field on One-Dimensional Antiferromagnetic Superconductors

Y. Suzumura

Department of Physics, Tohoku University, Sendai 980, Japan

Applying the mean-field theory, we investigate the effect of an external homogeneous magnetic field $\vec{H}_0$, and the staggered magnetic field $\vec{H}_Q$, arising from antiferromagnetism on superconductors with one-dimensional incommensurate bands. The superconducting order parameter is spatially homogeneous (BCS state) for the weak H_0 and oscillates in space (FF state) for the strong H_0. The superconducting domains on H_Q-H_0 plane as temperature decreases is shown by calculating the ground state and the second-order phase transition temperature. It is found that the interplay between $\vec{H}_0$ and $\vec{H}_Q$ leads to the crucial differences between the domain of $\vec{H}_0//\vec{H}_Q$ and that of $\vec{H}_0 \perp \vec{H}_Q$. Further, the effect of the one-dimensional fluctuation is examined by use of the bosonization method.

1. Introduction

Recently, there have been extensive studies on the interplay between superconducting and magnetic state. The coexistence of antiferromagnetism and superconductivity has been found experimentally in Chevrel compounds by ISHIKAWA and FISHER [1] and ternary compounds by HAMAKER et al. [2] and WOOLF et al. [3]. MACHIDA et al. have studied these materials theoretically and predicted a new pairing state [4] and further presented the theory of coexistence of spin density wave ordering and superconductivity [5]. In the mean-field theory, NASS et al. [6] have obtained the usual BCS-type pairing state especially for the antiferromagnetic superconductors with one-dimensional bands.

In the above materials, the superconductivity originates from the d electrons of Mo or Rh whereas the antiferromagnetism is ascribed to the f electrons of rare-earth atoms. Below the antiferromagnetic ordering temperature, the f-d interaction is treated as a staggered molecular field $\vec{H}_Q$, which acts on the d conduction electrons. The BCS pairing state is destroyed by $\vec{H}_Q$ which opens the insulating gap at the Fermi surface.

Besides the staggered field, it is also known that a homogeneous magnetic field $\vec{H}_0$ destroys the BCS pairing state because the down-spin electrons near the Fermi surface are replaced by up-spin electrons to gain the Zeeman energy [7,8]. The Fermi surface of the up(down)-spin electron moves so as to increase (decrease) the total number of the up(down)-spin electrons as H_0 increases. FULDE and FERRELL [9] first investigated the electron pairing around the new Fermi surface which produces the spatially oscillating order parameter (FF state). In the FF state, the effect of destruction by Ho weakens although the number of FF pairing electrons is smaller than that of the BCS pairing electrons. Consequently the transition from the BCS state into the FF state takes place as Ho increases. The spatially oscillating state with three-dimensional (3-D) band has been studied extensively in the presence of Ho [10,11].

In the present paper, we investigate the combined effect of $\vec{H}o$ and $\vec{H}_Q$ on the superconductors with one-dimensional (1-D) incommensurate band especially for $\vec{H}o//\vec{H}_Q$ and $\vec{H}o\perp\vec{H}_Q$ [12-16]. The superconducting region on H_Q-Ho plane is calculated by use of the mean-field theory. Several kinds of differences between $\vec{H}o//\vec{H}_Q$ and $\vec{H}o\perp\vec{H}_Q$ are shown for both the BCS state and the FF state. Further the effect of the 1-D fluctuation on the antiferromagnetic superconductors is studied by use of the bosonization method [17-19].

2. BCS state

Hamiltonian for $\vec{H}o//\vec{H}_Q$ and $\vec{H}o\perp\vec{H}_Q$ is given by [12-14]

$$\begin{aligned} H_{\substack{//\\ \perp}} &= \sum_{k,\alpha} \varepsilon_k a^+_{k\alpha} a_{k\alpha} \\ &- \Delta \sum_k (a^+_{k\uparrow} a^+_{-k\downarrow} + \text{h.c.}) \\ &- Ho \sum_{k,\alpha,\beta} (\sigma_z)_{\alpha\beta} a^+_{k\alpha} a_{k\beta} \\ &- H_Q \sum_{k,\alpha,\beta} (\sigma_{\substack{z\\x}})_{\alpha\beta} (a^+_{k\alpha} a_{k+Q\beta} + \text{h.c.}), \end{aligned} \tag{1}$$

where $a^+_{k\uparrow}$ is the creation operator of the conduction electron with momentum k and up spin. The first term is the kinetic energy with $\varepsilon_k = v_F(|k|-k_F)$ where v_F and k_F are Fermi velocity and momentum respectively. We take $\hbar = k_B = \mu_B = 1$. The second term is the BCS interaction in the mean-field approximation. The order parameter is defined by

$$\Delta = g \sum_k \langle a_{-k\downarrow} a_{k\uparrow} \rangle, \tag{2}$$

which is assumed to be real. g is the BCS coupling constant. The third term denotes the interaction between the conduction electron and the homogeneous magnetic field with $\vec{H}o=(0,0,-Ho)$. Matrices σ_z and σ_x are Pauli matrices. The fourth term shows the interaction between the conduction electron and the staggered magnetic field with $\vec{H}_Q=(0,0,2H_Q\cos Qz)$ and $\vec{H}_Q=(2H_Q \cos Qz,0,0)$ for $\vec{H}o//\vec{H}_Q$ and $\vec{H}o\perp\vec{H}_Q$ respectively. We assume that $Q = 2k_F$. By use of the thermal Green function, the self-consistency equation for Δ is written as

$$\frac{1}{\lambda} = \int_{-\omega_D}^{\omega_D} d\varepsilon_k \, \frac{1}{2} T \sum_{n=-\infty}^{\infty} \frac{\partial}{\partial \Delta^2} \ln(D_+ D_-) \ . \tag{3}$$

In (3), $D_\pm$ is defined by

$$D_\pm = \begin{cases} [(\omega_n - iHo)^2 + \varepsilon_k^2 + \Delta_k^2 + H_Q^2 \pm 2\Delta H_Q]^{1/2} & \text{for } \vec{H}o//\vec{H}_Q \\ [\omega_n^2 + \varepsilon_k^2 + \Delta^2 + Ho^2 + H_Q^2 \pm 2(\varepsilon_k^2 Ho^2 + \Delta^2 Ho^2 + \Delta^2 H_Q^2)^{1/2}]^{1/2} & \text{for } \vec{H}o\perp\vec{H}_Q, \end{cases} \tag{4}$$

where $\lambda = gN(0)$ and $N(0)=1/\pi v_F$. The Debye cutoff energy ω_D is assumed to be much smaller than $v_F k_F$. $\omega_n=(2n+1)\pi T$ where T is the temperature and n is the integer. From (3), the excess free energy to create Δ is obtained by

$$\delta F/N(0) = \int_0^{\Delta^2} d(\Delta')^2 \, (\Delta')^2 d(1/\lambda)/d(\Delta')^2. \tag{5}$$

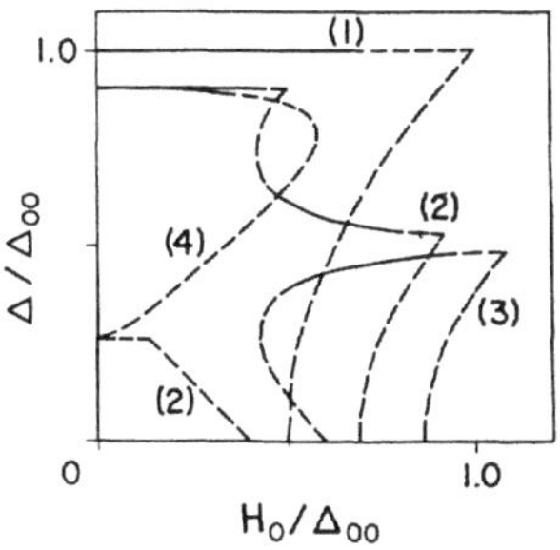

Fig. 1 Normalized order parameter $\Delta/\Delta oo$ at T=0 as a function of normalized external field $Ho/\Delta oo$ with $H_Q/\Delta oo$ fixed at 0(1), 0.4(2), 0.6(3) for $\vec{H}o//\vec{H}_Q$ and 0.4(4) for $\vec{H}o\perp\vec{H}_Q$

First we investigate the case T=0. In Fig.1, the normalized order parameter $\Delta/\Delta oo$ is shown as a function of $Ho/\Delta oo$ with H_Q fixed at several values where $\Delta oo(=2\omega_D \exp[-1/\lambda])$ corresponds to the order parameter in the absence of Ho and H_Q. The solution of (3) which gives the negative and lowest δF is shown by the solid curve and other solutions are shown by broken curves. For $H_Q/\Delta oo=0$ (curve(1)), Δ is independent of Ho and exists when $Ho/\Delta oo < 1/\sqrt{2}$[7]. For $\vec{H}o//\vec{H}_Q$ (curves (2) and (3)), the Ho-dependent order parameter appears when $H_Q/\Delta oo > 0.22$. The corresponding superconducting state is gapless,i.e., the electronic density of states at the Fermi surface remains finite. The gapless superconducting state exists even for the large value of $H_Q/\Delta oo$ if $Ho/\Delta oo \simeq H_Q/\Delta oo$. On the other hand, the order parameter in the case $\vec{H}o\perp\vec{H}_Q$ (curve(4)) simply decreases when both $Ho/\Delta oo$ and $H_Q/\Delta oo$ increase. For example, $\Delta/\Delta oo$ is expanded as $1-(H_Q/\Delta oo)^2/2-(H_Q Ho/\Delta oo)^2/3+\cdots$, for $Ho/\Delta oo$, $H_Q/\Delta oo \ll 1$.

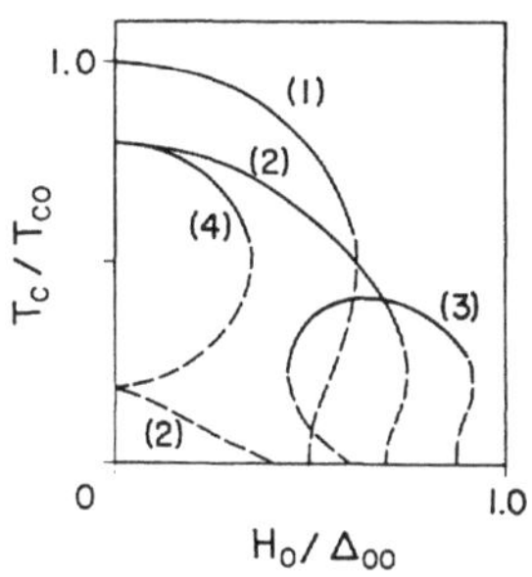

Fig. 2 Normalized transition temperature T_c/T_{co} as a function of $H_o/\Delta oo$ with H_Q fixed at the same value as Fig.1

Next, the second-order phase transition temperature T_c is examined by expanding (5) in terms of Δ^2. The temperature T_c, which is calculated by equating the coefficient of Δ^2 in (5) to zero, is called the stable (unstable) T_c when the corresponding coefficient of Δ^4 is positive (negative). In Fig.2, the stable (unstable) T_c is shown by the solid (broken) curve, where $T_{co}(=1.13\omega_D\exp[-1/\lambda])$ is T_c in the absence of Ho and H_Q. Hereafter, we use T_c in place of the stable T_c. For $H_Q=0$ (curve(1)), T_c decreases monotonically and vanishes at $(Ho/\Delta oo,H_Q/\Delta oo)=(0.61,o.57)$ as Ho increases[7,8]. For $\vec{H}o//\vec{H}_Q$, the Ho-dependence of T_c is similar to curve (2) (curve(3)) when $H_Q/\Delta oo<0.49$ ($H_Q/\Delta oo>0.49$). For $\vec{H}o\perp\vec{H}_Q$, T_c decreases additively as both $Ho/\Delta oo$ and $H_Q/\Delta oo$ increase .

Based on Fig.1 and 2, the superconducting regions on H_Q-Ho plane are shown in Fig.3. For $\vec{H}o//\vec{H}_Q$, the region of the BCS state obtained from the ground state (the second-order phase transition temperature) is shown by the domain enclosed by Ho axis, H_Q axis and curves (1) and (3) (curves (4)

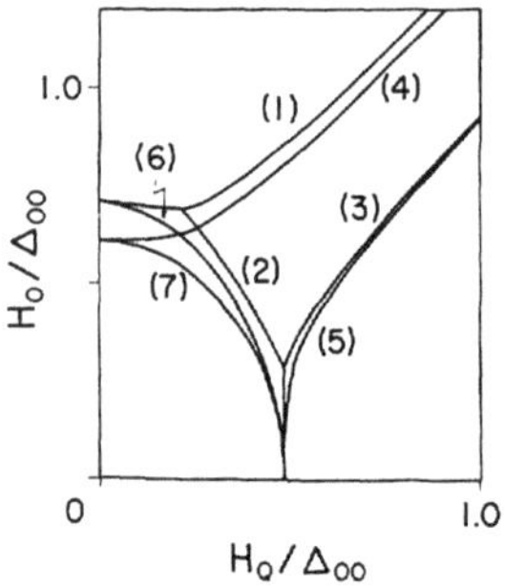

Fig. 3 Superconducting region on H_Q-Ho plane for the BCS state where curves (1)-(5) (curves(6) and (7)) denote boundaries corresponding to the case $\vec{H}o//\vec{H}_Q$ ($\vec{H}o\perp\vec{H}_Q$)

and (5)), where the curve (2) separates the domain of the usual BCS state from that of the gapless superconducting state at T=0. The superconducting region of $\vec{H}o\perp\vec{H}_Q$ obtained from the ground state (the second-order phase transition temperature) corresponds to the domain enclosed by Ho axis, H_Q axis and curve (6) (curve (7)). From the difference between the domain of T=0 and that of $T=T_c$, it is found that the first-order phase transition as a function of T/T_{co} takes place in the domain enclosed by curves (1) and (4) for $\vec{H}o//\vec{H}_Q$ (curves (6) and (7) for $\vec{H}o\perp\vec{H}_Q$).

In order to understand the significant difference between the superconducting region of $\vec{H}o//\vec{H}_Q$ and that of $\vec{H}o\perp\vec{H}_Q$, as is shown in the domain enclosed by curves (1),(3) and (6) of Fig.3, we investigate the band spectrum $E_k(=i\omega_n)$, which is calculated from (4). In the case $\vec{H}o//\vec{H}_Q$, the gap at $k=k_F$ for $\Delta=0$ is given by $|Ho-H_Q|$. When Ho increases and becomes nearly equal to H_Q, the gap decreases to zero so that the BCS pairing state is realized even for the large $H_Q/\Delta oo$. In the case $\vec{H}o\perp\vec{H}_Q$, the corresponding spectrum at $k=k_F$ is given by $(Ho^2+H_Q^2)^{1/2}$. Therefore the BCS pairing state is destroyed additively when both $Ho/\Delta oo$ and $H_Q/\Delta oo$ increase .

Several differences between $\vec{H}o//\vec{H}_Q$ and $\vec{H}o\perp\vec{H}_Q$ also exist in the magnetic susceptibility $\chi(=(N\uparrow-N\downarrow)/2HoN(0))$ [15] where $N\uparrow(N\downarrow)$ is the total number of electrons with up (down) spin. In the case $\vec{H}o//\vec{H}_Q$, χ at T=0 is zero (finites) for Ho < $|H_Q-\Delta|$ (H_Q > $|H_Q-\Delta|$) which corresponds to the usual (gapless)BCS state. In the case $\vec{H}o\perp\vec{H}_Q$, χ is finite even for the small $Ho/\Delta oo$ due to the spin-flip interaction.

Here we comment on the existence of another order parameter originating from the staggered magnetic field $\vec{H}_Q$ [13,14]. In addition to (1), we consider the interaction deduced from $\vec{H}_Q$. The corresponding order parameter is written as

$$\Delta_Q = g_1 \sum_k \langle a_{-k\downarrow} a_{k+Q\uparrow} \rangle, \tag{6}$$

where g_1 is the coupling constant for the interaction. In the case $\Delta_Q=\Delta_{-Q}$, which gives the lowest free energy, the excess free energy corresponding to δF is expanded in terms of Δ and Δ_Q as

$$\delta F/N(0) = A_2\Delta^2 + A_1\Delta\Delta_Q + Ao\Delta_Q^2 + \cdots. \tag{7}$$

In (7), we obtain Ao > 0 and $A_1 \propto HoH_Q\cos\theta$, where θ is the angle between $\vec{H}o$ and $\vec{H}_Q$. Therefore, it is concluded that Δ and Δ_Q coexist at the second-order phase transition temperature, when both $\vec{H}o$ and $\vec{H}_Q$ exist and $\theta \neq \pi/2$.

3. Spatially oscillating state

FULDE and FERRELL (FF) have found that the spatially oscillating order parameter in 3-D is realized for the large $Ho/\Delta oo$ [9]. We study the order parameter of the FF state which is written as $\langle\Psi\downarrow(z)\Psi\uparrow(z)\rangle = \Delta_q \exp[iqz]$ for 1-D band. The 1-D band is favorable for the FF pairing state because of the large nesting of the Fermi surface. In order to investigate the FF state, the second term of (1) is replaced by

$$H_s = -\sum_p (\Delta_q a^+_{p+q/2\uparrow} a^+_{-p+q/2\downarrow} + h.c.), \tag{8}$$

where

$$\Delta_q = g \sum_p \langle a_{-p+q/2\downarrow} a_{p+q/2\uparrow} \rangle . \tag{9}$$

The wave number q is determined so as to minimize the free energy and the BCS (FF) state corresponds to $q = 0$ ($q \neq 0$). Assuming $|q| << k_F$, the self-consistency equation for Δ_q is given by (3) with $D_\pm$ in (4) replaced by

$$D_\pm = \begin{cases} [\bar{\omega}_n^2 + \varepsilon_k^2 + \Delta_q^2 + H_Q^2 - \delta^2 \pm 2(\Delta_q^2 H_Q^2 - \delta^2 \bar{\omega}_n^2 - \delta^2 H_Q^2)^{1/2}]^{1/2} & \text{for } \vec{H}o // \vec{H}_Q \\ \prod_{s=\pm} [\omega_n^2 + \varepsilon_k^2 + \Delta_q^2 + H_s^2 + H_Q^2 \pm 2(\varepsilon_k^2 H_s^2 + \Delta_q^2 H_s^2 + \Delta_q^2 H_Q^2)^{1/2}]^{1/4} & \text{for } \vec{H}o \perp \vec{H}_Q, \end{cases} \tag{10}$$

where $\bar{\omega}_n = \omega_n - iHo$, $\delta = V_F q/2$ and $H_+ = Ho + \delta$ ($H_- = Ho - \delta$).

Substituting Δ_q into Δ of (5), the second-order phase transition is investigated by expanding (5) in terms of Δ_q^2 as

$$\delta F/N(0) = A(T,\bar{q})\, \Delta_q^{\,2} + B(T,\bar{q})\, \Delta_q^{\,4} + \cdots, \tag{11}$$

where $\bar{q} = v_F q/(2Ho)$. The transition temperature T_c, as a function of $\bar{q}$, is obtained from $A(T_c,\bar{q}) = 0$. The maximum of T_c is realized at $\bar{q} = \bar{q}o$ which corresponds to the lowest free energy. We write T_c as T_c^{BCS} (T_c^{FF}) when $\bar{q}o = 0$ ($\bar{q}o \neq 0$). In Fig.4, T_c as a function of $Ho/\Delta oo$ is shown with fixed $H_Q/\Delta oo$ where the solid curve and the broken curve denote the stable T_c ($B(Tc,\bar{q}o)>0$) and the unstable T_c ($B(T_c,\bar{q}o)<0$) respectively. The cross indicates the point at which T_c^{FF} (curves (1),(2) and (3)) is separated from T_c^{BCS} (curves (1)',(2)' and (3)'). First we investigate the case $H_Q/\Delta oo=0$ (curves (1) and (1)').The cross is situated at $Ho/\Delta oo=0.61$ where the corresponding $B(T_c,0)$ is zero. As $Ho/\Delta oo$ increases, T_c^{FF} decreases ($\bar{q}o$ increases from zero) monotonically with the asymptotic form $T_c^{FF}/T_{co} \simeq \Delta oo/(4Ho)$ ($\bar{q}o \simeq 1-0.024(\Delta oo/Ho)^4$) for $Ho/\Delta oo >> 1$. The fact that $B(T_c,qo) >0$ in the curve (1) is understood from the equality, $B(T,\bar{q}) = 2\partial^2 A(T,q)/\partial q^2$, which is positive except for the cross. The first-order phase transition T_1^{BCS} of the BCS state exists in the region of $0.61 < Ho/\Delta oo < 1/\sqrt{2}$. However, it turns out that T_c^{FF} is larger than T_1^{BCS}. For $\vec{H}o // \vec{H}_Q$ (curves (2) and (2)'), the FF state is more favorable. As the cross is located in the curve of the stable T_c^{BCS}, $B(T_c^{FF},qo)$ is always positive and T_1^{BCS} is smaller than T_c^{FF}. As $Ho/\Delta oo$ increases, the curve (2) approaches curve (1) so that there is no upper bound of $Ho/\Delta oo$ for T_c^{FF}. In the case $\vec{H}o \perp \vec{H}_Q$, T_c^{FF} does not exist for the large $Ho/\Delta oo$ (curve(3)). From the fact that $B(T,0) < 0$ at the point of the cross, it is found that the unstable T_c^{BCS}

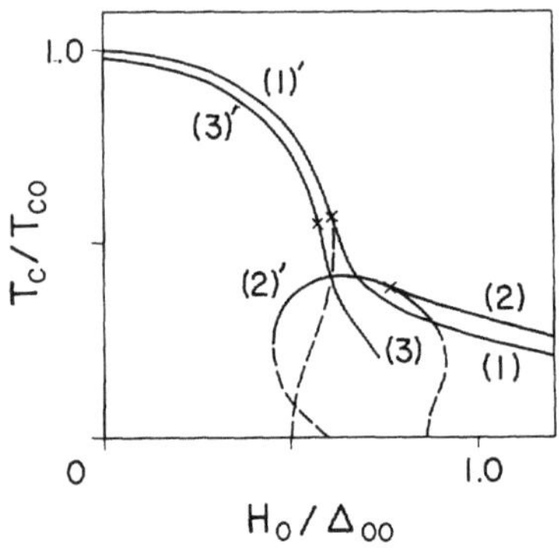

Fig. 4 Normalized transition temperature T_c^{FF}/T_{co} and T_c^{BCS}/T_{co} with H_Q fixed at 0(curves (1) and (1)'), 0.6 (curves (2) and (2)' for $\vec{H}o//\vec{H}_Q$) and 0.16 (curves(3) and (3)' for $\vec{H}o\perp\vec{H}_Q$)

moves into the unstable T_c^{FF} across the point and that T_1^{BCS} is larger than T_c^{BCS} and T_c^{FF} near the point. Therefore the first-order phase transition takes place around the cross as T/T_{co} decreases.

The ground state is obtained as follows. As Ho/Δoo increases, Δ^{BCS} with H_Q/Δoo=0 jumps into the Ho-dependent Δ^{FF} at Ho/Δoo = 0.68 where Δ^{BCS} (Δ^{FF}) denotes the order parameter of the BCS (FF) state. For the large Ho/Δoo, we obtain the asymptotic form given by $\Delta^{FF} \simeq \Delta oo^2/(4Ho)$ and $\bar{q}o \simeq 1 - (\Delta oo/Ho)^4/64$. When $\vec{H}o//\vec{H}_Q$ and H_Q/Δoo > 0.26, the order parameter and $\bar{q}o$, as a function of Ho/Δoo, change continuously from the gapless BCS state into the FF state. As Ho/Δoo increases, the value of Δ^{FF} approaches that of H_Q/Δoo=0. In the case $\vec{H}o\perp\vec{H}_Q$, there is the upper bound of Ho/Δoo for Δ^{FF} as is found in the curve of T_c^{FF}.

Based on the results of the second-order phase transition and the ground state, the superconducting region as temperature decreases is summarized in Fig.5 where R1, R2,···,R6 correspond to the different kinds of phase transition. The region Ro denotes Δq = 0 at any temperature. In the case $\vec{H}o$//

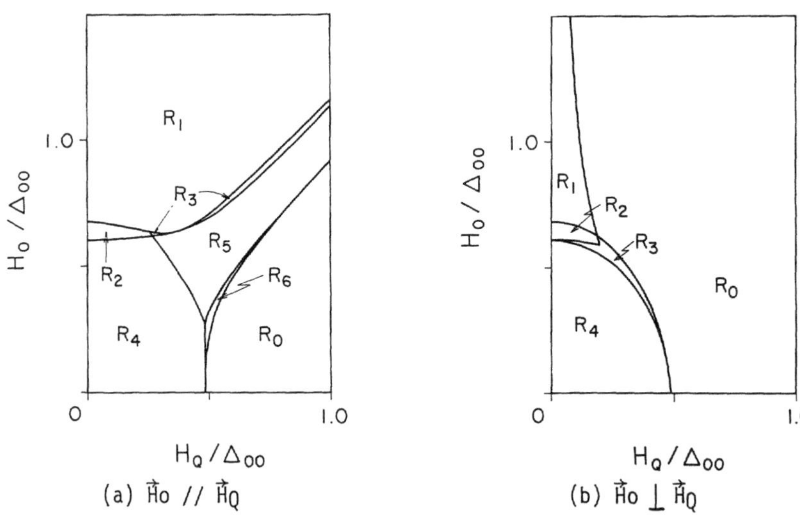

Fig. 5 Phase diagram of the superconducting state on the H_Q-Ho plane for $\vec{H}o//\vec{H}_Q$ (a) and $\vec{H}o\perp\vec{H}_Q$ (b) [16]

$\vec{H}_Q$, the second-order phase transition into the FF state takes place in the regions R1, R2 and R3 whereas the first (second)-order phase transition into the usual (the gapless) BCS state follows in the region R2 (R3). In the region R4 (R5), the second-order phase transition into the usual (the gapless) BCS state occurs. The second-order phase transition into the BCS state in the region R6 is followed by the first-order phase transition into the normal state. In the case $\vec{H}o\perp\vec{H}_Q$, the second-order phase transition into the FF state occurs in the region R1 and R2 whereas the first-order phase transition into the BCS state follows in the region R2. The first (second)-order phase transition into the BCS state takes place in the region R3 (R4).

The remarkable difference between the superconducting region of $\vec{H}o//\vec{H}_Q$ and that of $\vec{H}o\perp\vec{H}_Q$ is the existence of the region with $Ho/\Delta oo \gtrsim H_Q/\Delta oo \gtrsim T$ in the case of $\vec{H}o//\vec{H}_Q$. It is understood by use of the band spectrum (10) as follows. In the case $\vec{H}o//\vec{H}_Q$ ($\vec{H}o\perp\vec{H}_Q$), the insulating gap due to $\vec{H}_Q$ is produced by a pair of electrons with parallel (antiparallel) spins so that the gap remains at $|k| = k_F$ (moves to the new Fermi point) in the presence of $\vec{H}o$. Therefore, the pairing electrons of the FF state with the wave number around the new Fermi point is not destroyed (is destroyed) for $\vec{H}o//\vec{H}_Q$ ($\vec{H}o\perp\vec{H}_Q$).

Here we comment on the order parameter proposed by LARKIN and OVCHINNIKOV (LO) [11], who investigated the case of 3-D band. The excess free energy ((11) is the even function of q so that the order parameter with $\Delta q\exp[iqz]$ + h.c., (the LO state) is examined in comparison with the FF state. Near the second-order phase transition temperature, the excess free energy of the LO state is written as

$$\delta F/N(0) = A(|\Delta_q|^2+|\Delta_{-q}|^2) + B(|\Delta_q|^4+|\Delta_{-q}|^4) + 2B'|\Delta_q|^2|\Delta_{-q}|^2 + \cdots . \quad (12)$$

Coefficients A and B in (12) are the same as $A(T,\bar{q})$ and $B(T,\bar{q})$ in (11) whereas B' is different from B except for q = 0. The coefficient B_{eff} corresponding to B(T,q) in (11) is written as (B + B')/2 when (12) takes a minimum, i.e., $|\Delta_q| = |\Delta_{-q}|$. We found that $0 < B_{eff} \leq B$, especially $B_{eff} = B/2$ for $H_Q = 0$. Therefore it is concluded $T_c^{FF} = T_c^{LO}$ and that the second-order phase transition into the LO state is realized.

4. On the one-dimensional fluctuation

In the preceding section, superconductors with 1-D band have been studied by treating the BCS interaction in the mean-field approximation because we have considered that the electron hopping perpendicular to the 1-D axis (chain) is large enough to suppress the 1-D fluctuation. When such an electron hopping decreases to zero, the 1-D fluctuation becomes large and the transition temperature T_c goes down to zero. In this section, we investigate the case where the electron moves only along the 1-D axis (z axis).

It is known that the low-lying excitation of the 1-D electron system is described by collective modes of the phase variables $\theta_+(z)$ and $\phi_+(z)$ [17-19] defined by

$$\begin{Bmatrix} \theta_+(z) \\ \phi_+(z) \end{Bmatrix} = \frac{i}{2} \sum_{k\neq 0} \frac{2\pi}{Lk} \exp[-\frac{\alpha}{2}|k| - ikz] \sum_p a^+_{p+k,\alpha} a_{p,\beta} \begin{Bmatrix} (1)_{\alpha\beta} \\ (\sigma_z)_{\alpha\beta} \end{Bmatrix}, \quad (13)$$

where the summation is taken only for $p(p+k) > 0$. The quantity α is the cutoff parameter of the order of the lattice constant. L is the length of the system. The phase $\theta_+(z)$ ($\phi_+(z)$) is related with the local charge density wave (spin density wave) by $\nabla_z\theta_+(z)/\pi$ ($\nabla_z\phi_+(z)/(2\pi)$). We define $P_+(z)$ ($M_+(z)$) as the variable canonical conjugate to $\theta_+(z)$ ($\phi_+(z)$). By use of the boson representation, the Hamiltonian in the presence of $\vec{H}_Q$ (parallel to the z axis) is written as

$$H = \int_0^L dz\, [A_\sigma(\nabla_z\phi_+(z))^2 + C_\sigma M_+^2(z) + \frac{v_F\lambda}{2\pi\alpha^2}\cos 2\phi_+(z)$$

$$+ A_\rho(\nabla_z\theta_+(z))^2 + C_\rho P_+^2(z) + \frac{2}{\pi\alpha} H_Q \sin\theta_+(z)\sin\phi_+(z)], \tag{14}$$

where

$$A_{\substack{\sigma\\ \rho}} = \frac{1}{4\pi^2} \quad C_{\substack{\rho\\ \sigma}} = \frac{v_F}{4\pi}(1 \pm \frac{\lambda}{2}), \tag{15}$$

and $\lambda = g/(\pi v_F)$. The case $T = 0$ is investigated. The phase $\phi_+(z)$ ($\theta_+(z)$) is rewritten as $\phi_S + \hat{\phi}$ ($\theta_S + \hat{\theta}$) where $\hat{\phi}(\hat{\theta})$ denotes the quantum fluctuation around the classical value ϕ_S (θ_S). Using the self-consistent harmonic approximation and minimizing (14) with respect to ϕ_S and θ_S, the third and the sixth terms of (14) are rewritten as

$$\cos 2\phi_+(z) = \cos 2\phi_S\,(1 + 2\langle\hat{\phi}^2\rangle - 2\phi^2), \tag{16}$$

$$\sin\theta_+(z)\sin\phi_+(z) = \sin\theta_S \sin\phi_S\,[1 + (\langle\hat{\phi}^2\rangle + \langle\hat{\theta}^2\rangle - \hat{\phi}^2 - \hat{\theta}^2)/2], \tag{17}$$

$$\phi_S = \begin{cases} \sin^{-1}(\exp[-\,3\langle\hat{\phi}^2\rangle/2 + \langle\hat{\theta}^2\rangle/2]\,\tilde{H}_Q/\lambda) \\ \pi/2 \end{cases} \tag{18}$$

where $\theta_S = 3\pi/2$ and $\tilde{H}_Q = H_Q/(\pi v_F)$. Quantities $\langle\hat{\phi}^2\rangle$ and $\langle\hat{\theta}^2\rangle$ are calculated self-consistently from (14)-(18) as

$$\begin{Bmatrix} \langle\hat{\phi}^2\rangle \\ \langle\hat{\theta}^2\rangle \end{Bmatrix} = \eta_{\substack{\sigma\\ \rho}}\, \ln[\frac{1 + (1 + 4X_{\sigma\rho})^{1/2}}{2X_{\substack{\sigma\\ \rho}}}], \tag{19}$$

where

$$X^2_{\substack{\sigma\\ \rho}}(1 \pm \lambda/2) = \lambda X_\sigma^{2\eta_\sigma} \begin{Bmatrix} (\cos\phi_S)^2 \\ (\sin\phi_S)^2 \end{Bmatrix}, \tag{20}$$

$$\eta_\sigma = \eta_\rho^{-1} = [(1 - \lambda/2)(1 + \lambda/2)]^{1/2}. \tag{21}$$

In (21), $\lambda > 0$ so that $\eta_\sigma < 1$. It is noted that $4X_\nu(C_\nu A_\nu)^{1/2}/\alpha$, ($\nu = \sigma$ and ρ) denote the gap of the excitation spectrum for the phase $\phi_+(z)$ and $\theta_+(z)$ respectively. One of the solutions of (18)-(20) is given by $X_\rho = \phi_S = 0$. Especially for $\lambda \ll 1$, we obtain

$$X_\sigma = (\frac{\lambda}{1 + \lambda/2})^{\frac{1/2}{1-\eta_\sigma}} \equiv X_0, \tag{22}$$

and the corresponding energy of (14) is given by $E = -E_0$ where

$$E_0 = \frac{1}{2\pi\alpha^2} L v_F\,(1 - \eta_\sigma)(\frac{\lambda}{1 + \lambda/2})^{\frac{\eta_\sigma}{1-\eta_\sigma}}\ . \tag{23}$$

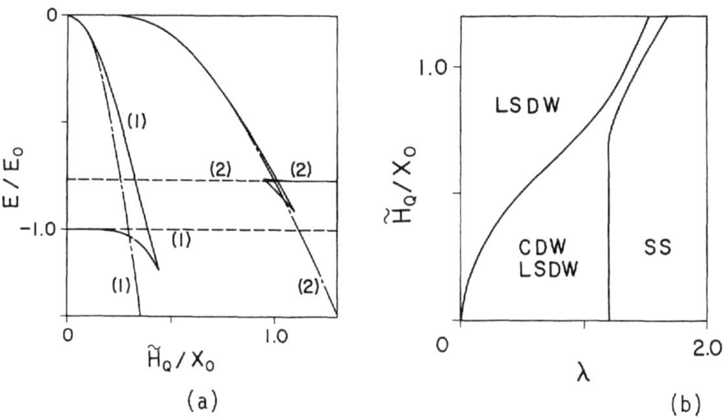

Fig. 6 Normalized energy E/E_0 as a function of $\tilde{H}_Q/X_0$ with λ fixed at 0.2 (1) and 1.4 (2) (a), and the corresponding phase diagram on λ-H_Q plane (b) where E_0 and X_0 are defined by (23) and (22) respectively

Another solution of (18)-(19) (i.e., $0 < \phi_s < \pi/2$) is calculated numerically. The corresponding energy E as a function of $\tilde{H}_Q$ is shown in Fig.6(a) where the broken line (dot-dashed curve) denotes $\phi_s = 0$ ($\phi_s = \pi/2$) and ϕ_s changes from zero to $\pi/2$ on the solid curve. When $\lambda < 1.2$ (> 1.2), the $\tilde{H}_Q$ dependence of E is similar to curves (1) (curves (2)). Several kinds of phases are related with ϕ_s as follows ($0 < \phi_s < \pi/2$). Space-correlation functions of the order parameter $\chi(z)$ are calculated for the longitudinal spin density wave (LSDW), the charge density wave (CDW) and the singlet superconductivity (SS) respectively as ($|z| \gg \alpha$)

$$\chi_{\substack{LSDW\\CDW}}(z) \propto \left(\left(\begin{smallmatrix}\sin\\ \cos\end{smallmatrix}\phi_s\right)^2 + O(\exp[-|z|2X_\sigma/\alpha]\right) X_\sigma^{\eta_\sigma} X_\rho^{\eta_\rho}, \tag{24}$$

$$\chi_{SS}(z) \propto \left((\cos\phi_s)^2 + O(\exp[-|z|2X_\sigma/\alpha]\right) X_\sigma^{\eta_\sigma} X_\rho^{1/\eta_\rho} \exp[-|z|/\xi], \tag{25}$$

where $\xi = \alpha\eta_\rho/\pi X_\rho$. Equation (25) leads to the short-range order for SS, as far as $X_\rho \neq 0$, i.e., $\phi_s \neq 0$. The response function for SS is dominant when $X_\rho = 0$. Then it turns out that $\phi_s = 0$ corresponds to SS and $0 < \phi_s < \pi/2$ indicates the coexistence of CDW and LSDW. Based on Fig.6(a), in which the phase with the lowest E is realized, phase diagram of λ-$\tilde{H}_Q$ is shown in Fig.6(b). When $\tilde{H}_Q = 0$, the superconducting state exists even for the small λ. In the presence of $\tilde{H}_Q$, the superconducting state exists only for $1.2 < \lambda$ where there is the upper bound of $\tilde{H}_Q/X_0$ for SS.

We comment on the effect of a homogeneous magnetic field. By use of the phase representation, the last term of (1) is rewritten as $-H_0\nabla_z\phi_+(z)/\pi$ which is added to (14). In the absence of $\tilde{H}_Q$, it is found as follows [20]. The classical part ϕ_s is homogeneous in space for the weak H_0 whereas ϕ_s becomes spatially dependent, i.e., $\chi_{SS}(z)$ varies in space when

$$H_0 > (1 - \eta_\sigma)\, X_0^{\eta_\sigma} [(1 + \lambda/2)\lambda]^{1/2} v_F/\pi\alpha \;. \tag{26}$$

Then, for pure 1-D antiferromagnetic superconductors, it is predicted that the spatially dependent order parameter will appear for the strong H_0 even in the presence of $\vec{H}_Q$ ($//\vec{H}_0$).

Acknowledgements

I am thankful to Prof. T. Tsuzuki, Dr. K. Ishino and Dr. H. Yamamoto for useful comments.

References

1. M. Ishikawa and Ø. Fisher, Solid State Commun. 24, 747 (1977) ; M. Ishikawa, Ø. Fisher and J. Muller, J. Phys. (Paris) 39, C6-1379 (1978).
2. H.C. Hamaker, L.D. Woolf, H.B. MacKay, Z. Fisk and M.B. Maple, Solid State Commun. 32, 289 (1979).
3. L.D. Woolf, S.E. Lambert, M.B. Maple, H.C. Ku, W. Odoni and H.R. Ott, Physica 108B, 761 (1981).
4. K. Machida, K. Nokura and T. Matsubara, Phys. Rev. Lett. 44, 821 (1980).
5. K. Machida, K. Nokura and T. Matsubara, Phys. Rev. B22, 2307 (1980).
6. M.J. Nass, K. Levin and G.S. Grest, Phys. Rev. Lett. 46, 614 (1981).
7. G. Sarma, J. Phys. Chem. Solids 24, 1029 (1963).
8. K. Maki and T. Tsuneto, Prog. Theor. Phys. 31, 945 (1964).
9. P. Fulde and R.A. Ferrell, Phys. Rev. 135, A550 (1964).
10. S. Takada and T. Izuyama, Prog. Theor. Phys. 41, 635 (1969) ; S. Takada, Prog. Theor. Phys. 43, 27 (1970).
11. A.I. Larkin and Y.N. Ovchinnikov, Zh. Eksp. Teor. Fiz. 47,1136 (1964) [Sov. Phys. JETP 20, 762 (1965)].
12. Y. Suzumura and A.D.S. Naqi, Solid State Commun. 40, 651 (1981) ; Phys. Rev. B24, 5103 (1981).
13. Y. Suzumura and A.D.S. Nagi, J. Low Temp. Phys. 46, 397 (1982).
14. K. Ishino and Y. Suzumura, Prog. Theor. Phys. 68, 1009 (1982) ; Y. Suzumura and K. Ishino, Solid State Commun. 45, 637 (1983).
15. K. Ishino and Y. Suzumura, Prog. Theor. Phys. 68, 1776 (1982).
16. Y. Suzumura and K. Ishino, Prog. Theor. Phys. 70, 654 (1983).
17. A. Luther and V.J. Emery, Phys. Rev. Lett. 33, 589 (1974).
18. Y. Suzumura, Prog. Theor. Phys. 61, 1 (1979).
19. For review, H. Fukuyama and H. Takayama, in Electronic Properties of Inorganic Quasi-One-Dimensional Compounds, ed. by P. Monceau (1984, D. Reidel Publishing Company)
20. K. Maki and H. Takayama, Phys. Rev. 20, 3223 (1979).

Effect of Impurities on Antiferromagnetic Superconductors

Y. Okabe

Department of Physics, Tohoku University, Sendai 980, Japan

1. Introduction

The coexistence of antiferromagnetism and superconductivity has been the subject of recent studies [1,2]. The mean-field theory of antiferromagnetic superconductors (AFS) has been developed by introducing the antiferromagnetic (AF) molecular field into the BCS model [3-6].

NASS, LEVIN and GREST (NLG) [6] have studied the effect of impurities on AFS using the lowest-order Born approximation theory of Abrikosov and Gorkov [7]. They have pointed out some similarities in the behavior of nonmagnetic impurities in AFS with that of magnetic impurities in usual superconductors. It is well known that the thermodynamic properties of isotropic BCS superconductors remain unchanged by nonmagnetic impurities (the Anderson theorem [8]). In the case of AFS, the nonmagnetic impurities have a pair-breaking effect because of the presence of AF order.

In the present paper, we investigate the effect of nonmagnetic impurities on AFS in detail. Starting from the exact treatment of the single impurity problem, we consider the finite impurity concentration by the t-matrix approximation. Our method is on the same line as the Shiba-Rusinov theory [9-11] of magnetic impurities in usual superconductors. We give an analytical investigation of the interesting behavior of the impurity concentration dependence of the density of states [12].

2. Model and Single-Impurity Problem

Our mean-field Hamiltonian for an AFS is given by

$$H_0 = \sum_{k,\alpha} \varepsilon_k C^+_{k,\alpha} C_{k,\alpha} - \Delta \sum_k (C^+_{k\uparrow} C^+_{-k\downarrow} + \text{h.c.})$$

$$- H_Q \sum_{k_{\parallel}<0,\alpha,\beta} (\sigma_z)_{\alpha\beta} (C^+_{k+Q,\alpha} C_{k,\beta} + \text{h.c.}) \quad , \qquad (2.1)$$

where ε_k is single-particle energy measured from the Fermi level, H_Q is the AF molecular field, Q is the wave vector characterizing the AF order and $k_{\parallel}$ is the component of the wave vector k parallel to Q. The superconducting order parameter Δ is determined self-consistently by

$$\Delta = g \sum_k \langle C_{-k\uparrow} C_{k\downarrow} \rangle \quad , \qquad (2.2)$$

where g is the BCS coupling constant. We have taken Δ as real and positive.

By using the eight-dimensional basis

$$\psi_k^+ = (C_{k\uparrow}^+, C_{-k\downarrow}^+, C_{k\uparrow}, C_{-k\downarrow}, C_{k+Q\uparrow}^+, C_{-k-Q\downarrow}^+, C_{k+Q\uparrow}, C_{-k-Q\downarrow}) \ , \tag{2.3}$$

the unperturbed Matsubara Green's function is given by

$$G_k^0(i\omega_n) = [i\omega_n - \varepsilon_s\rho_3 - \varepsilon_a\tau_3\rho_3 - \Delta\rho_2\sigma_2 + H_Q\tau_1\rho_3\sigma_3]^{-1} \tag{2.4}$$

with

$$\varepsilon_s = \frac{1}{2}(\varepsilon_k + \varepsilon_{k+Q}) \tag{2.5}$$

$$\varepsilon_a = \frac{1}{2}(\varepsilon_k - \varepsilon_{k+Q}) \ , \tag{2.6}$$

where σ_i, ρ_i and τ_i are the Pauli matrices operating on the ordinary spin states, the electron-hole states, and the positive- and negative-momentum states, respectively.

The interaction of a conduction electron with a nonmagnetic impurity is described by the Hamiltonian

$$H_{imp} = \frac{1}{N}\sum_{kk'} \psi_k^+ U \psi_k \ , \quad \text{with} \tag{2.7}$$

$$U = (V_1 + V_2\tau_1)\rho_3 \ , \tag{2.8}$$

where V_1 (V_2) is the scattering potential with small (large) momentum transfer. The important role played by V_2 in the pinning mechanism in charge density wave is well known [13,14].

MACHIDA [15] has considered the single-impurity problem in connection with the coexistence of spin density wave and superconductivity in organic conductors, so that we only sketch it here for later use. The Green's function for the single impurity is easily obtained as

$$G_{kk'}(i\omega_n) = G_k^0(i\omega_n)\delta_{kk'} + G^0(i\omega_n)t(i\omega_n)G_{k'}^0(i\omega_n) \ , \tag{2.9}$$

where the t matrix $t(i\omega_n)$ is given by

$$t(i\omega_n) = [1 - UF^0(i\omega_n)]^{-1}\,U \tag{2.10}$$

with

$$F^0(i\omega_n) = \frac{1}{N}\sum_k G_k^0(i\omega_n) \ . \tag{2.11}$$

In the above we include all terms with respect to V_1 and the even-order terms with respect to V_2. We assume a short-range interaction and we set $V_1 = V_2 = V$ at the end of the calculation. In this paper we restrict ourselves to the case of a one-dimensional (1D) electron band which satisfies the nesting condition $\varepsilon_{k+Q} = -\varepsilon_k$ for k near $-k_F$, that is, $\varepsilon_s = 0$, because the essential new results of the present problem lie in the 1D character.

Investigating the poles of the t matrix for the real frequency ω, we find that the bound state within the BCS gap due to the impurity exists if and only if $H_Q > \Delta$. The energy of bound states, $\pm\omega_B$, is

$$\omega_B^2 = (H_Q - \Delta)^2 - \left\{ \frac{2-\zeta}{2(1-\zeta)} [4(1-\zeta)\Delta^2 + \zeta^2 H_Q^2]^{1/2} - 2\Delta - \frac{\zeta^2}{2(1-\zeta)} H_Q \right\} H_Q \tag{2.12}$$

with

$$\zeta = 4v^2/(1 + v^2)^2 \quad . \tag{2.13}$$

Here

$$v = \pi N_0 V \tag{2.14}$$

and N_0 is the density of single-particle states at the Fermi level in the normal metal. The variable ζ characterizes the strength of the electron-impurity scattering and takes the values between 0 and 1. When $\zeta \to 0$, $\omega_B \to (H_Q - \Delta)$. It corresponds to the case of the weak scattering limit. Further when $\zeta \to 1$, $\omega_B \to 0$. In Fig. 1, the position of bound states is shown for $\zeta = 0.5$ [curve (a)], 0.8 [curve (b)] and 0.95 [curve (c)]. The hatched area in Fig. 1 represents the continuum of states with peaks at $\omega = \Delta + H_Q$ and $\omega = |\Delta - H_Q|$.

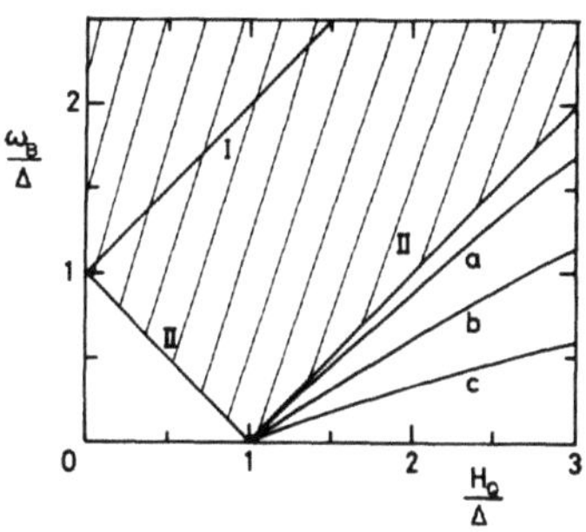

Fig. 1 Position of bound state due to single nonmagnetic impurity for $\zeta = 0.5$ [curve (a)], 0.8 [curve (b)] and 0.95 [curve (c)]. The hatched area shows the continuum of states, which has two peaks at $\omega = \Delta + H_Q$ [line (I)] and $|\Delta - H_Q|$ [line (II)]

3. Finite Concentration Problem

Let us consider the finite impurity concentration problem. We assume that the impurities are randomly distributed and their concentration n_i is low enough so that the impurity-impurity interaction is negligible. The Green's function averaged over the impurity distribution is connected with the self-energy, $\Sigma_k(i\omega_n)$, by

$$G_k(i\omega_n) = \overline{G}_{kk'}(i\omega_n)$$

$$= [G_k^0(i\omega_n)^{-1} - \Sigma_k(i\omega_n)]^{-1} \quad . \tag{3.1}$$

The self-energy is taken as the t matrix, Eq.(2.10), times n_i with G_k^0 replaced by G_k. This procedure is the same as that used in the problem of magnetic impurities in usual superconductors [9-11].

Making the ansatz that the Green's function is given by

$$G_k(i\omega_n) = [i\tilde{\omega}_n - \epsilon_k\tau_3\rho_3 - \tilde{\Delta}_n\rho_2\sigma_2 + i\tilde{\Omega}_n\tau_1\rho_1\sigma_1 + \tilde{H}_{Qn}\tau_1\rho_3\sigma_3]^{-1}, \qquad (3.2)$$

we can set up the self-consistency equation for $\tilde{\omega}_{n\pm}$ and $\tilde{\Delta}_{n\pm}$:

$$\tilde{\omega}_{n\pm} - \omega_n = \frac{n_i}{\pi N_0}\frac{v^2}{D}\frac{1}{\lambda_\mp}(x\,\tilde{\omega}_{n\mp} \mp y\,\tilde{\Delta}_{n\mp}) \qquad (3.3)$$

$$\tilde{\Delta}_{n\pm} - (\Delta \pm H_Q) = \frac{n_i}{\pi N_0}\frac{v^2}{D}\frac{1}{\lambda_\mp}(x\,\tilde{\Delta}_{n\mp} \pm y\,\tilde{\omega}_{n\mp}) \qquad (3.4)$$

with

$$\tilde{\omega}_{n\pm} = \tilde{\omega}_n \pm \tilde{\Omega}_n \qquad (3.5)$$

$$\tilde{\Delta}_{n\pm} = \tilde{\Delta}_n \pm \tilde{H}_{Qn} \quad . \qquad (3.6)$$

Here we have used the notations

$$\lambda_\pm = (\tilde{\omega}^2_{n\pm} + \tilde{\Delta}^2_{n\pm})^{1/2} \qquad (3.7)$$

$$x = 1 + v^2\,\frac{\tilde{\omega}_{n+}\tilde{\omega}_{n-} + \tilde{\Delta}_{n+}\tilde{\Delta}_{n-}}{\lambda_+\lambda_-} \qquad (3.8)$$

$$y = v^2\,\frac{\tilde{\omega}_{n-}\tilde{\Delta}_{n+} - \tilde{\omega}_{n+}\tilde{\Delta}_{n-}}{\lambda_+\lambda_-} \qquad (3.9)$$

$$D = x^2 + y^2 \quad . \qquad (3.10)$$

The branch of the square root of $\lambda_\pm$ is chosen such that the real part of this function is positive. In deriving Eqs. (3.3) and (3.4) we have discarded the self-energy part which contains odd-order terms with respect to v. This part, which is related to the shift of the chemical potential or the deviation from the complete nesting, has an interesting effect on the property of AFS [16]. However, this problem will be left for a seperate investigation.

Defining $u_{n\pm} = \tilde{\omega}_{n\pm}/\tilde{\Delta}_{n\pm}$, Eqs. (3.3) and (3.4) can be combined to give

$$\omega_n = u_{n\pm}(\Delta \pm H_Q) \pm \frac{1}{\tau_1}\,\frac{(u_{n+} - u_{n-})(1 + u^2_{n\pm})^{1/2}}{(2-\zeta)(1+u^2_{n+})^{1/2}(1+u^2_{n-})^{1/2} + \zeta(1+u_{n+}u_{n-})} \qquad (3.11)$$

where

$$\frac{1}{\tau_1} = \frac{1}{\tau_1^0}\,\frac{1}{(1+v^2)^2} = \frac{2n_i v^2}{\pi N_0}\,\frac{1}{(1+v^2)^2} \quad . \qquad (3.12)$$

Eq. (3.11) is our generalized equation for $u_{n\pm}$ and Eq. (3.1) of Ref. 6 is retrieved from it by putting $\zeta = 0$ and replacing τ_1 by τ_1^0. The lowest-order Born approximation of NLG is included in our theory as the weak scattering limit.

We make a comment on the "dirty" limit behavior of the system. We follow a method similar to the one used by Fulde and Maki [17] while considering the spin-orbit scattering in usual superconductors in a magnetic field. In the short mean-free-path limit, the equation has the same structure as the one obtained in the usual superconductors [7]. For the present case of the AFS, the effective pair-breaking parameter α_s is given by

$$\alpha_s = \tau_1 H_Q^2/\Delta \quad . \tag{3.13}$$

In the extremely "dirty" limit $\alpha_s \to 0$, we obtain the BCS result. That is, the effect of AF order is screened by nonmagnetic impurities.

4. Density of States

Several thermodynamic and transport properties can be calculated by using the u equation (3.11). Here we consider the superconducting density of state $N(\omega)$, which is given by

$$N(\omega) = \frac{N_0}{2} \, \mathrm{Im} \left[\frac{u_+}{(1-u_+^2)^{1/2}} + \frac{u_-}{(1-u_-^2)^{1/2}} \right] \quad , \tag{4.1}$$

where Im stands for the imaginary part of the function. Here $u_\pm$ is the analytically continued value of $u_{n\pm}$ by replacing $iu_{n\pm} \to {}^{\pm}u_+$. We solve the self-consistency equation by the complementary Broyden method [18-21] and calculate the density of states. In the present paper, we consider the order parameter Δ as given and scale various quantities with respect to Δ.

The density of states is qualitatively different for the cases $H_Q < \Delta$ and $H_Q > \Delta$. As an example of the former case, the $N(\omega)$ for $H_Q = 0.4\Delta$, $\zeta = 0$, 0.5, 1.0 and $\alpha = 0.2$, 1.0 and 5.0, respectively, are shown in Figs. 2(A), 2(B) and 2(C), where

$$\alpha = 1/(\tau_1 \Delta) \quad . \tag{4.2}$$

The ζ-dependence in various figures should be noted. The large concentration behavior, shown in Fig. 2(C), is easily understood by the "dirty" limit analysis given at the end of the last section.

Now we consider the case $H_Q > \Delta$ where the impurity bound states arise. In Fig. 3 we show the density of states $N(\omega)$ for $H_Q = 2.5\Delta$. The arrow indicates the position of the single-impurity bound state. One may note that the "impurity band" grows [curve (a)-(b)] and that the gaplessness appears as the impurity concentration increases [curve (c)]. For a much larger impurity concentration the gap appears again [curve (d)] and the overall behavior of the density of state becomes Abrikosov-Gorkov-like [curve (e)].

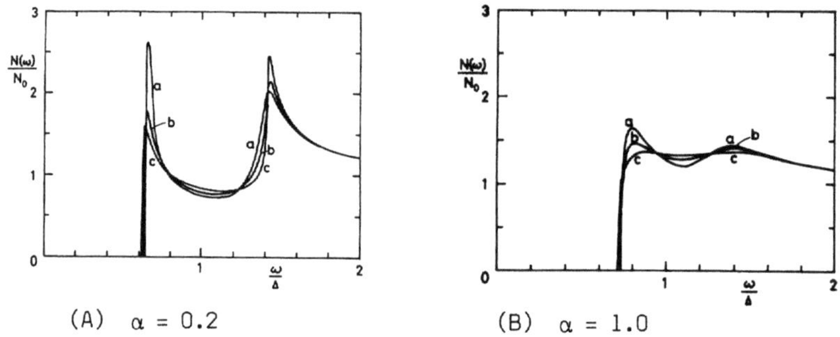

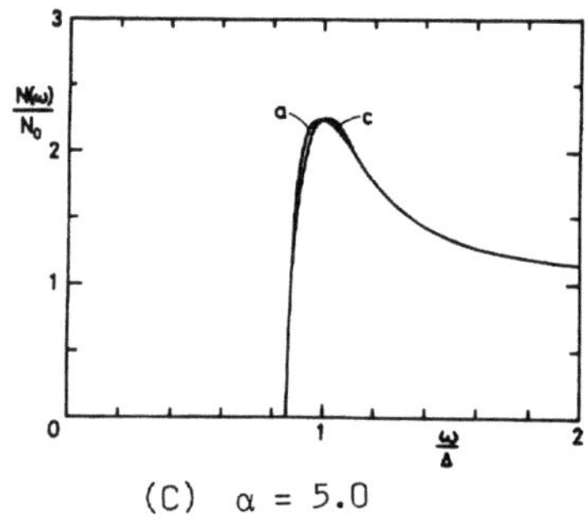

Fig. 2 Density of states $N(\omega)$ for $H_Q = 0.4\Delta$ and $\zeta = 0$ [curve (a)], 0.5 [curve (b)] and 1.0 [curve (c)]

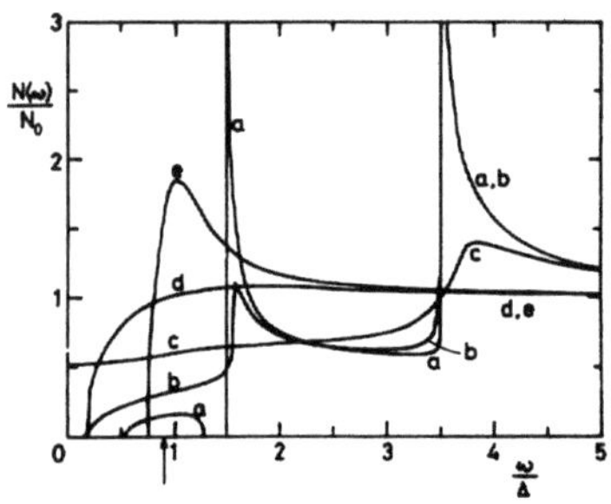

Fig. 3 Impurity concentration dependence of $N(\omega)$ for $H_Q = 2.5\Delta$ and $\zeta = 0.8$. The value of α is taken as 0.05 [a], 0.2 [b], 1.0 [c], 10.0 [d] and 100.0 [e]. The arrow indicates the position of single-impurity bound state

Next let us consider the condition for the appearance of the gaplessness in more detail. We note that from the analytical properties of the Green's function, the real part of $u_\pm$ is zero at $\omega = 0$. Then the density of states at $\omega = 0$ becomes finite only if the imaginary part of $u_\pm$ is nonzero. Of course, there is no possibility of the gaplessness for $H_Q < \Delta$, which was already pointed out by NLG [6]. For $H_Q > \Delta$, we find that the condition for a finite density of states at $\omega = 0$ is

$$\alpha_1 < \alpha < \alpha_2 \quad \text{where} \tag{4.3}$$

$$\alpha_1 = \frac{H_Q^2 - \Delta^2}{H_Q \Delta}(1 - \zeta) \tag{4.4}$$

$$\alpha_2 = \frac{H_Q^2 - \Delta^2}{\Delta^2} \tag{4.5}$$

We should note that the upper condition for the gaplessness can be written as $\alpha_s > 1$ for $H_Q \gg \Delta$. The imaginary part of u_+ is always positive. On the other hand, the imaginary part of u_- is positive for $\alpha_1 < \alpha < \alpha_3$ and negative for $\alpha_3 < \alpha < \alpha_2$, where α_3 is given by

$$\alpha_3 = \frac{(H_Q - \Delta)[H_Q + (1-\zeta)\Delta]}{\Delta(H_Q\Delta)^{1/2}} . \qquad (4.6)$$

In AFS the energy band splits off because of the AF order. As the nonmagnetic impurity increases, the effect of H_Q is weakened. The role of the superconduncting order parameter Δ remains the same. Thus, even if $H_Q > \Delta$, the system behaves like usual superconductors. We may regard α_3 as the critical value of α whether the system is H_Q-dominant or Δ-dominant. The condition for the gaplessness is graphically shown in Fig. 4. In Fig. 5 we show the imaginary part of $u_\pm$ at $\omega = 0$ as a function of impurity concentration for the parameters used in Fig. 3. In this case $\alpha_1 = 0.42$, $\alpha_2 = 5.25$ and $\alpha_3 = 2.56$.

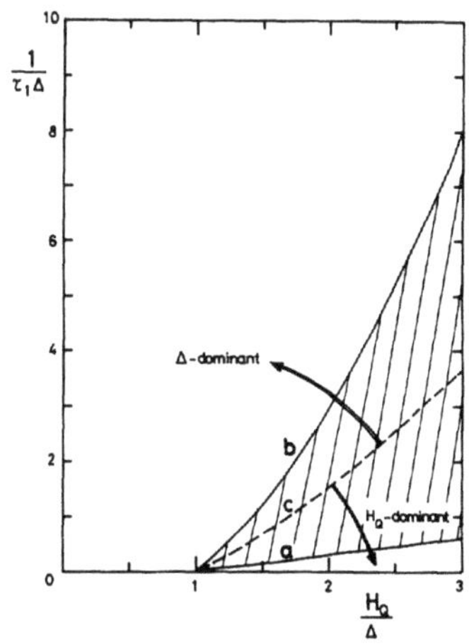

Fig. 4 Condition for the gaplessness for $\zeta = 0.8$. The hatched area shows the gapless region. The lower bound α_1, the upper bound α_2 and the critical value α_3 are denoted by curves (a), (b) and (c)

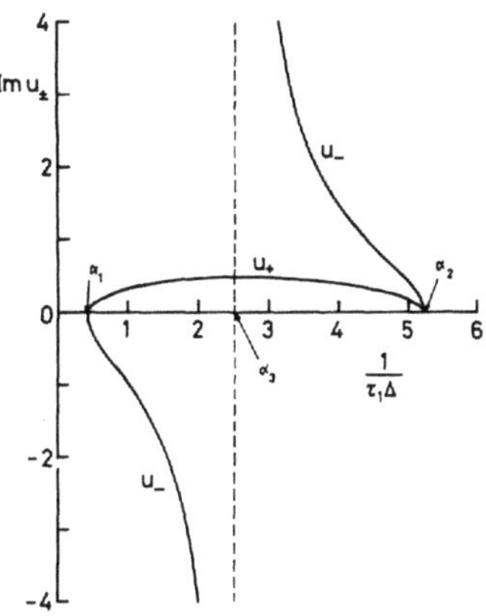

Fig. 5 Imaginary part of $u_\pm$ at $\omega = 0$ as a function of α for $H_Q = 2.5\Delta$ and $\zeta = 0.8$

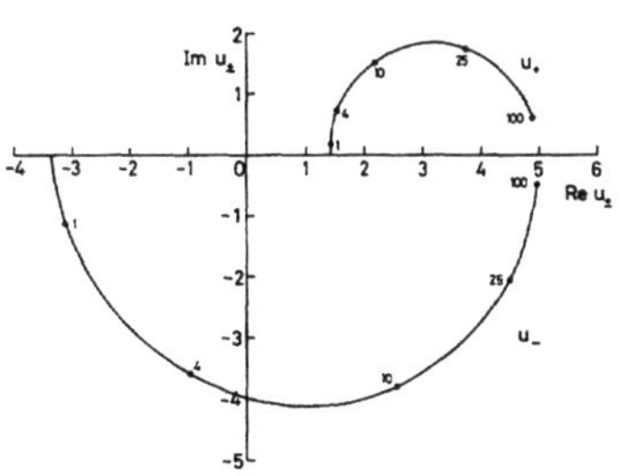

Fig. 6 Variation of $u_\pm$ at $\omega = 5\Delta$ with the impurity concentration for $H_Q = 2.5\Delta$ and $\zeta = 0.8$. A circle denotes the value of α

We also show the impurity concentration dependence of $u_{\pm}$ at $\omega = 5\Delta$ in Fig. 6. The value of α is represented by a circle. From this figure we can see how the structure of $u_{\pm}$ varies as the impurity concentration increases.

5. Summary

We have studied the effect of nonmagnetic impurities on AFS in detail. The bound states due to nonmagnetic impurities appear when $H_Q < \Delta$. The growth of the "impurity band" and the appearance and the disappearance of the gaplessness are shown. The "dirty" limit behavior of the system and the condition for the gaplessness are analytically investigated. The direct experimental research on the density of states of AFS, for example, a tunneling experiment, is highly desirable.

Acknowledgement

A part of this work was done during my stay at University of Waterloo in Canada in collaboration with Professor A.D.S. Nagi. Financial support provided by the Natural Sciences and Engineering Research Council of Canada is gratefully acknowledged.

References

1. For review articles, see "Superconductivity in Ternary Compounds II", edited by M.B. Maple and Ø. Fischer (Springer-Verlag, Berlin, 1982).
2. K. Levin : in this symposium.
3. K. Machida, K. Nokura and T. Matsubara : Phys. Rev. Lett. 44, 821 (1980); Phys. Rev. B22, 2307 (1980).
4. M.J. Nass, K. Levin and G.S. Grest : Phys. Rev. Lett. 46, 614 (1981).
5. Y. Suzumura and A.D.S. Nagi, Solid State Commun. 40, 651 (1981); Phys. Rev. B24, 5103 (1981); J. Low. Temp. Phys. 46, 397 (1982).
6. M.J. Nass, K. Levin and G.S. Grest : Phys. Rev. B25, 4541 (1982).
7. A.A. Abrikosov and L.P. Gorkov : Zh. Eskp. Teor. Fiz. 39, 1781 (1960) [Sov. Phys. JETP 12, 1243 (1961)].
8. P.W. Anderson : J. Phys. Chem. Solids 11, 26 (1959).
9. H. Shiba : Prog. Theor. Phys. 40, 435 (1968).
10. A.I. Rusinov : Zh. Eskp. Teor. Fiz. 56, 2047 (1969) [Sov. Phys. JETP 29, 1101 (1969)].
11. Y.Okabe and A.D.S. Nagi : Phys. Rev. B28, 1320 (1983); B28, 1323 (1983).
12. Y. Okabe and A.D.S. Nagi : Phys. Rev. B28, 6290 (1983).
13. P.A. Lee, T.M. Rice and P.W. Anderson, Solid State Commun. 14, 703 (1974).
14. S. Kurihara : J. Phys. Soc. Jpn. 41, 1488 (1976).
15. K. Machida : J. Phys. Soc. Jpn. 51, 1420 (1982).
16. Y. Suzumura and K. Ishino : Prog. Theor. Phys. 67, 1668 (1982).
17. P. Fulde and K. Maki : Phys. Rev. 141, 275 (1966).
18. C.G. Broyden : Math. Comput. 19, 577 (1976).
19. J.E. Dennis, Jr. and J.J. More : SIAM Rev. 19, 46 (1977).
20. P. Bendt and A. Zunger : Phys. Rev. B26, 3114 (1982).
21. V.A. Singh and P. Bendt : Phys. Rev. B27, 6464 (1983).

Part III

Organic Superconductors

Organic Superconductors

E.W. Fenton
National Research Council of Canada, Ottawa, Canada K1A 0R6

1. Brief Review of Experiments and Electron Band Structure

The first observation of superconductivity in an organic material, in the selenium-organic compound $(TMTSF)_2PF_6$ (ditetramethyltetraselenafulvalene-hexafluorophosphate), was reported by JEROME, MAZAUD, RIBAULT, and BECHGAARD in 1980 [1]. Following this, superconductivity in $(TMTSF)_2X$ with X = PF_6, ClO_4, AsF_6, SbF_6, and TaF_6 was observed by the Orsay-Copenhagen group mentioned above, the Bell Labs group [2], the IBM (San Jose)-U.C.L.A. group [3], the IBM (Yorktown Heights) group [4], Japanese groups at the Institute for Molecular Science at Okazaki and at the Electrotechnical Laboratory at Ibaraki [5], and by others. Superconductivity is observed below roughly 1.5K, at zero pressure for X = ClO_4 and above threshold pressures ranging between 5 and 15 kilobars for other members of the $(TMTSF)_2X$ family. It should be mentioned that the application of pressure on organic conductors to obtain superconductivity had been strongly advocated earlier by Meyer Weger. Spin density wave (SDW) states or at least antiferromagnetically ordered states are observed below roughly 15K [2,4,6,7,8] at pressures lower than the threshold pressure for superconductivity (excepting for X = ClO_4). For rapidly cooled specimens of X = ClO_4 at zero or near-zero pressure, a SDW state is observed in a temperature range $1K \lesssim T \lesssim 5.5K$ and superconductivity occurs below roughly 1K [9]. Evidence for the SDW consists of antiferromagnetic resonance, spin-flop transitions, transport properties, and NMR. Coexistence of SDW and superconducting states has not been proven. Neutron scattering has been done neither for the SDW state nor for any suspected SDW-superconductivity coexistence state, and some evidence which might be taken as favoring the coexistence state can equally well be due to an inhomogeneous specimen containing superconductivity-only regions and SDW-only regions.

The existence of superconductivity is well established and not in dispute. Of particular interest is an experiment by ANDRES et al.[2] at Bell Labs. which shows the Meissner effect, full shielding currents, 50% flux exclusion, and highly anisotropic critical fields H_{c1} and H_{c2}, done in such a way that it is hard to doubt their results and conclusions.

In 1982, PARKIN, ENGLER, SCHUMAKER, LAGIER, LEE, SCOTT and GREENE [3] observed superconductivity in the sulfur-organic compound $(BEDT\text{-}TTF)_4(ReO_4)_2$ [10]. To my knowledge, nothing more has been reported on superconductivity in the new BEDT-TTF family of organic conductors, and for the rest of this discussion we will concern ourselves only with the $(TMTSF)_2X$ family.

We show in Fig. 1 the molecular donors and acceptors of presently known superconducting organic compounds. The triclinic $P\bar{1}$ crystal structure of, e.g., $(TMTSF)_2PF_6$ is illustrated schematically in Fig. 2, looking down along the b^*axis (b^* is not the b axis but is the axis perpendicular to a and c

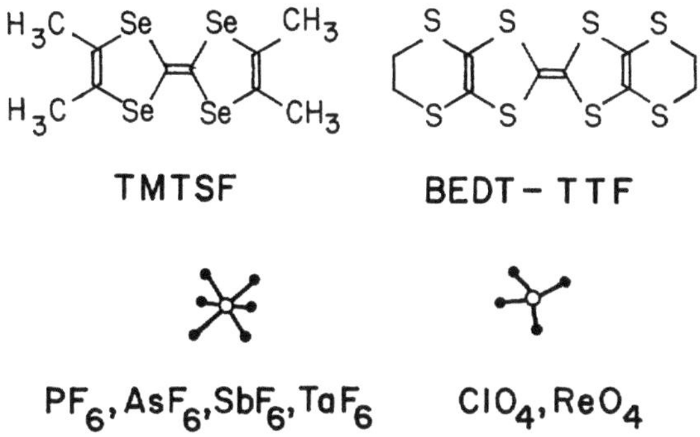

Fig. 1 Electron-donor and acceptor molecules of presently known organic superconductors

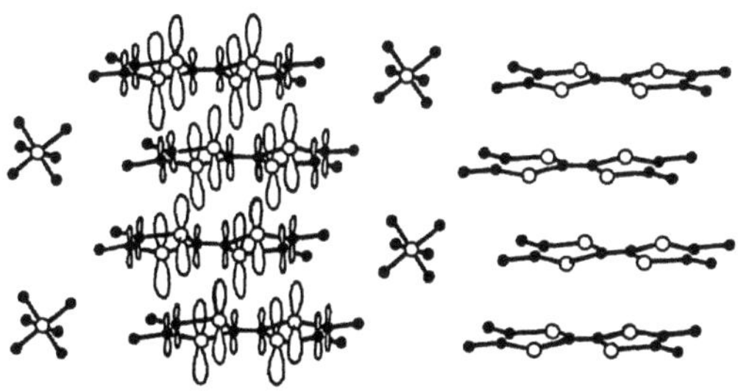

Fig. 2 Structure of $(TMTSF)_2PF_6$ looking down along b* axis

axes). Perpendicular to the plane of the figure, the stacks of TMTSF cations repeat edge to edge with each other. Transport and plasma frequency measurements by a number of groups [1,3,11,12] and highly simplified (necessarily) band structure calculations by GRANT [13], shown schematically in Fig. 3, provide a consistent picture where the stacking a-axis electron band dispersion width is comparable to 1.5 eV, the stack-to-stack b-axis dispersion width is comparable to 0.1 eV, the hole Fermi energy is roughly 0.3 eV, and the c-axis dispersion width (not shown) is comparable to 10^{-3} eV. Anisotropy of the electron dispersion both near to and far from E_f is about 15 to 1 comparing a and b axes and between 300 to 1 and 1000 to 1 comparing a and c axes. One-half of the b-axis dispersion width, ~ 500 k_B, or even one-quarter of it, ~ 250 k_B, is much greater than the electron lifetime broadening below 50K, determined by transport and plasma-frequency measurements combined with gross features of the band structure [13], and is much greater than energy scales relevant for superconductivity with $T_c \lesssim 1.5$K or a SDW state with $T_{SDW} \lesssim 15$K. Activation energies, e.g. the specific heat at temperatures well below T_c or T_{SDW}, show 2Δ is near the BCS mean field value 3.5 k_BT_c or 3.5 k_BT_{SDW}. According to various criteria for one-dimensionality derived by Meyer WEGER and his colleagues [14,15] and according to common sense, the electron system at low temperatures should be regarded as anisotropic quasi-two-dimensional (the Fermi surface is almost certainly not closed in 2-D), not quasi-one-dimensional. Grant uses a criterion by Weger to show that the occurrence of a density wave with transition temperature $\lesssim$ 15K would probably not annihilate the entire Fermi surface, although the electrical conductivity indicates that much more of it is annihilated than the roughly 50% Fermi surface annihilation by the SDW in chromium. The estimates of the electron energy band dispersions and anisotropy mentioned above are not in dispute to my knowledge, however the Orsay group do not agree with the conclusion that the electron band structure is not quasi-one-dimensional at $T \lesssim$ 50K.

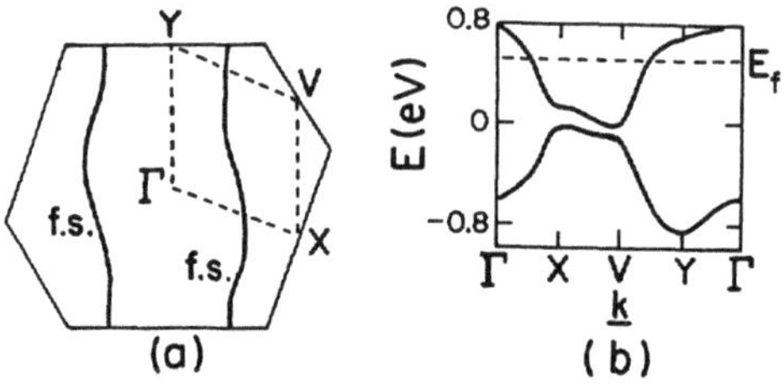

Fig. 3 Schematic drawing of band-structure results by GRANT [13]: (a) Brillouin zone and Fermi surface (f.s.); (b) energy E versus wavevector k

The phase diagram for $(TMTSF)_2X$ is a function of temperature T, pressure P, magnetic field H [16] and sometimes the rapidity with which the specimen has been cooled to 4.2K through temperatures below roughly 50K [16]. The phase diagram in the T-P plane with H = 0 and with the specimen slowly cooled is shown schematically in Fig. 4 for all X except ClO_4. For ClO_4, the threshold pressure for the superconductivity region of the phase diagram is zero rather than in the 5 to 15 kilobar range for the other members of the family. In Fig. 5 we show a T-H phase diagram for X = ClO_4 and PF_6. At magnetic fields above approximately 6 Tesla, Shubnikov-deHaas oscillations observed by, e.g., KWAK et al. [17] show small closed electron or hole k-space orbits and 2-D electron gas behavior (orbits depend only on $H\cos\theta_{\underline{H},\underline{c}^*}$). A

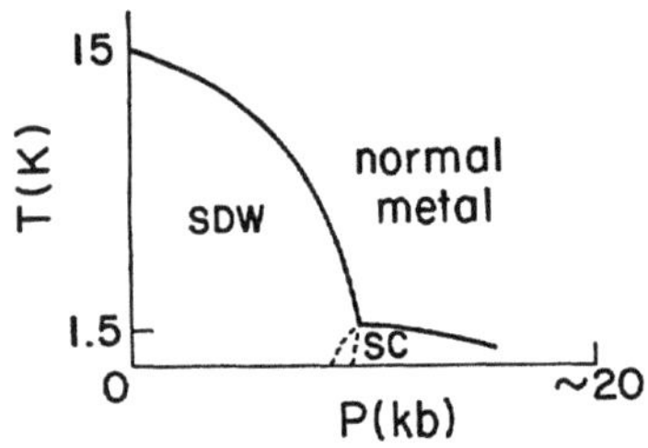

Fig. 4 Schematic phase diagram in temperature - pressure plane with H = 0, for X = PF_6, AsF_6, SbF_6, TaF_6. SC - superconductor

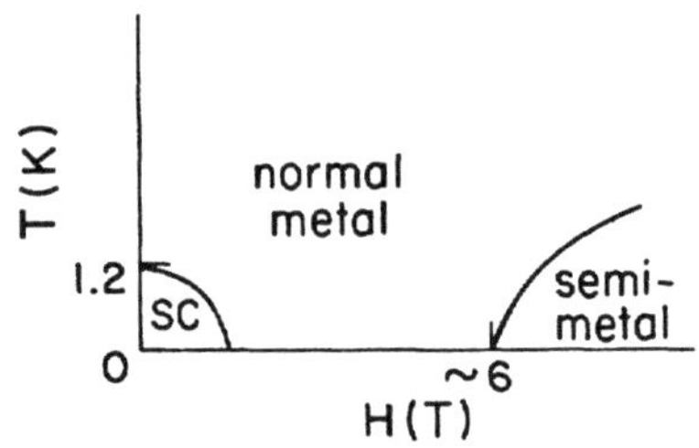

Fig. 5 Schematic phase diagram in temperature - magnetic field plane at P = 0 for X = ClO_4 and at P = 12 kbar for X = PF_6

phase transition line was established by specific heat measurements [18]. The nature of the ordering responsible for what is apparently new reciprocal lattice vectors giving rise to the Fermi-surface translations and the closed orbits is not known at present. We show in Fig. 6 two T-P phase diagrams for X = ClO_4 at H = 0 for specimens slowly cooled to liquid helium temperature, and for specimens rapidly cooled. The ClO_4 anion is not centrosymmetric in contrast to PF_6 and other X members, and has two low-energy orientations in the crystal. X-ray experiments have shown that as T decreases through 22K, and decreases slowly, the anions order with a wavevector for the ordering thought to be the commensurate vector $(0,\pi/b,0)$ [19]. Superconductivity still occurs below roughly 1.2K, however there is no evidence for a SDW phase for X = ClO_4 when the specimen is slowly cooled [16].

The superconductivity that occurs in $(TMTSF)_2X$ at high pressures (above threshold) has very anisotropic critical magnetic field and the transition temperature is very sensitive to nonmagnetic impurities, i.e., T_c decreases

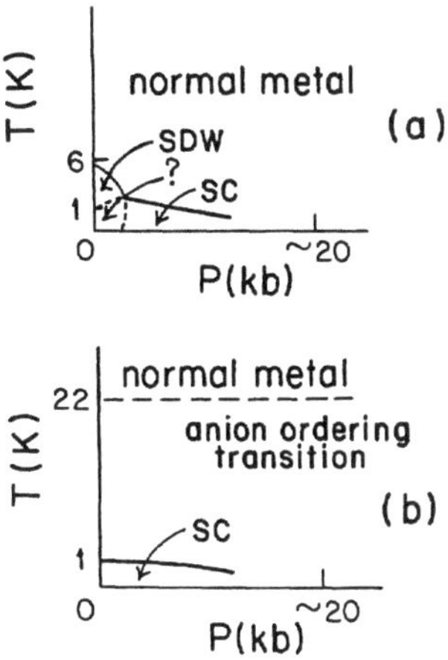

Fig. 6 Schematic phase diagram in temperature - pressure plane with H = 0: (a) quench-cooled X = ClO_4; (b) slowly cooled X = ClO_4

very rapidly for X = ClO_4 specimens which have been irradiated in a strong x-ray beam or which have IO_4 impurities and for X = PF_6 specimens irradiated with protons [20,21]. The anisotropy of the critical magnetic field in the (a,b) plane is however too small (only a factor of two) to relate to the band structure anisotropy by Ginzburg-Landau theory. The a-axis value is thought to be limited by the Pauli pair-breaking effect when the electron-spin Zeeman energy is larger than the pair-binding energy [20]. This interpretation is however disputed by JEROME [22].

This completes my summary of the experimental situation for the organic superconductors. I emphasize that only some of the major points have been touched. The phenomena observed besides a more or less normal metal include superconductivity, a SDW state, a magnetic-field-induced metal to semi-metal transition, and a transition from anion-disorder to anion-order. An intriguing question is just why do five different phases have nearly the same free energy? Two phenomena not proven to occur as yet are coexistent with SDW and superconductivity states, and quantized Hall conductance (which may or may not be expected to occur). Experimentalists are so busy trying to find new organic superconductors, and trying to bring some general order to the knowledge for those already discovered, that the coexistence state has apparently not been of much interest to any experimentalist except Mickie Walsh of Bell Labs. The possibility of the quantum Hall effect occurring in $(TMTSF)_2X$ has been investigated by an eminent experimentalist who prefers not to be identified. Plateaux in ρ_{xy} occur but do not occur at h/ne^2 with n an integer.

A great deal of work has been done by Denis Jerome and colleagues regarding possible superconducting fluctuations at T up to 30 or 40K and a possible large superconductivity pseudo-gap $2\Delta \approx 4$ meV. This work includes tunnelling measurements in Europe, and infrared experiments by a Canadian group (T. Timusk and H.K. Ng of McMaster University). A controversy rages at present over the existence or non-existence of large fluctuation amplitudes at higher T. I personally see no experimental evidence for the fluctuations or for a large superconductivity pseudo-gap, and see evidence that the electron system is not quasi-one-dimensional. This is not to say that Jerome's claims regarding fluctuations and a pseudo-gap are not correct, but only that I see no evidence as yet in support of them. Because that subject has been extensively discussed in publications or conference proceedings, and because of time and space limitations, I will not review the troops involved in that war, and refer you to publications of the Orsay group, of the IBM (San Jose)-U.C.L.A. group, and various conference proceedings (e.g., Boulder in 1981 and Les Arcs in 1982).

2. Theory

The largest effort has been to attempt to relate the theory of quasi-one-dimensional systems to physical phenomena in the $(TMTSF)_2X$ compounds, which appear to me not to be quasi-one-dimensional systems. I will give only two examples of this. The first is a modification of the famous g-ology phase diagram by Hanoch GUTFREUND and collaborators [23] to include effects of retarded (i.e. phonon-mediated) electron-electron interactions. The non-modified diagram does not have a singlet-superconductor to SDW phase transition line, and of course would not be descriptive of the $(TMTSF)_2X$ superconductors. The modified diagram is shown in Fig. 7, where g_2 is the electron-electron coupling constant for forward scattering, g_1 represents backward scattering with momentum transfer $\pm 2k_f$, and λ_2 and λ_1 are components due to the phonon-mediated interaction, all for a quasi-one-dimensional system. The modified diagram does have a SDW-singlet superconductivity transition line. The second example of quasi-one-dimensional theory is a theory of electron-electron $4k_f$ Umklapp scattering by EMERY, BRUINSMA, and BARISIC [24]. They introduce a new Umklapp electron-electron coupling g_3 to the g-ology diagram to account for absence of the Peierls-CDW lattice distortion for $(TMTSF)_2X$ superconductors. The relevance of all such quasi-one-dimensional theories seems highly questionable to me in view of the apparent two-dimensional electron band structure for $(TMTSF)_2X$. Features of the g-ology diagram arising because of interference of logarithmic singularities in electron-hole and electron-electron channels should not occur when the a-axis to b-axis electron band dispersion widths are in the relatively small ratio 15 to 1.

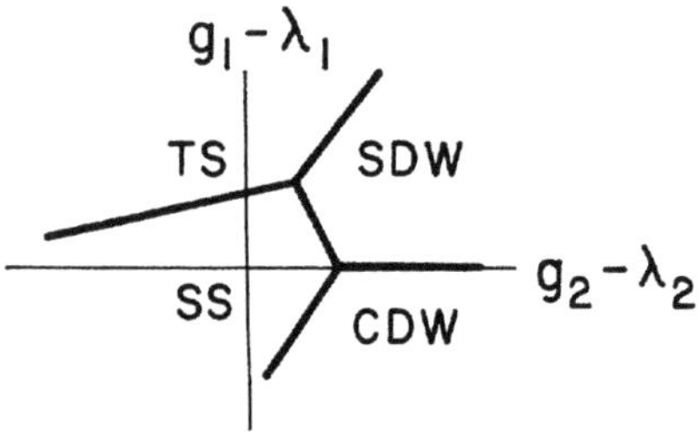

Fig.7 Phase diagram by GUTFREUND [23] in $(g_1-\lambda_1)$ - $(g_2-\lambda_2)$ plane of electron-electron interactions for a quasi-one-dimensional electron assembly. TS - triplet superconductivity; SS - singlet superconductivity; SDW - spin density wave; CDW - charge density wave

MACHIDA and MATSUBARA [25] have taken a model similar to one used for coexistence of superconductivity with a charge density wave (CDW) by Bilbro and McMillan. The SDW annihilates part but not all of the Fermi surface (f.s.) and Cooper pairing over the non-annihilated part drives superconducting correlations over the entire f.s. The Hamiltonian is:

$$H = \sum_{\underset{\sim}{k}\alpha} \varepsilon_{\underset{\sim}{k}} c^{+}_{\underset{\sim}{k}\alpha} c_{\underset{\sim}{k}\alpha} - \Delta \sum_{\underset{\sim}{k}} c^{+}_{\underset{\sim}{k}\uparrow} c^{+}_{-\underset{\sim}{k}\downarrow} + \text{h.c.} - \sum_{\substack{\underset{\sim}{k}\varepsilon\,\text{region 1} \\ \alpha}} m\,\alpha\, c^{+}_{\underset{\sim}{k}+\underset{\sim}{Q}\alpha} c_{\underset{\sim}{k}\alpha} + \text{h.c.} \quad (1)$$

They obtain a coexistence state only when the f.s. density of states $N_2(0)$ for the non-nested f.s. region 2 where the SDW does not occur is not zero. $N_1(0) = N(0)-N_2(0)$ is the f.s. density of states for region 1 where the SDW does occur. Given that an attractive electron-electron interaction for superconductivity occurs when averaged over the entire f.s., and given that a repulsive electron-electron interaction for the SDW occurs when averaged over region 1, then the superconductivity order parameter for singlet Cooper pairing, Δ, and the SDW order parameter m behave as shown in Fig. 8 for a particular example with $N_1(0)/N(0) = 0.2$ and $T_{co}/T_{SDWo} = 0.5$. T_{co} for example means T_c if the SDW did not occur. YAMAJI [26] has taken a model band structure with cosine band dispersions for the electrons in both a and b directions and has found that for this case no SDW-superconductivity coexistence state occurs. It is probable however that no general proof of non-occurrence is possible, particularly in view of the fact that in Chevrel phases coexistence does occur, and for that case the mean field Hamiltonian is the same as (1) but with m determined by an effectively external system of rare earth ions rather than by the self-consistency equations of Machida and Matsubara (for electron states within the conduction electron system). Physically the difference is that the SDW would not occur without annihilating a large fraction of the f.s., e.g.,~ 50% in chromium, whereas the rare-earth magnetism with wave-vector $\underline{Q}$ may nest very little (or none) of the f.s. In a solid as complex as $(TMTSF)_2X$, and with the energy cut-off for the electron-electron pairing interaction for superconductivity very much smaller than the energy cut-off for the electron-hole pairing interaction of the SDW, it is not improbable that the interactions assumed to be present by Machida and Matsubara do actually exist.

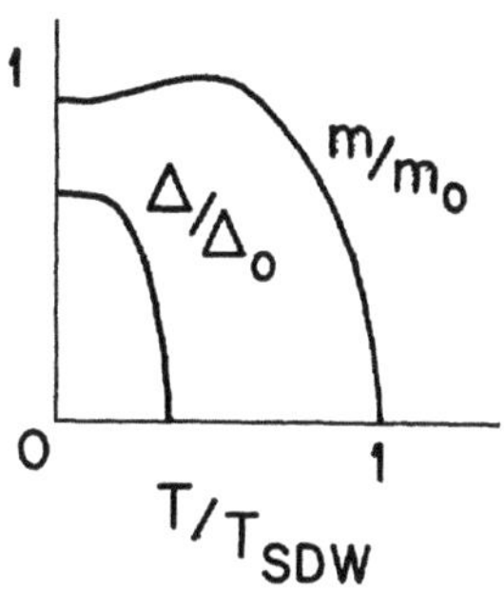

Fig. 8 Example of SDW magnetization m and SS order parameter Δ versus temperature for SDW - superconductivity coexistence phase. M_0, Δ_0, T_{co}, and T_{SDWo} refer to values if either the SDW or SS phase occurs alone. For this case, $T_{co}/T_{SDWo} = 0.5$ and $N_1(0)/N(0) =$ $= 0.2$

The theory described to this point has been concerned principally with whether or not SDW, superconductivity, or coexistence phases occur. For inorganic metals, just where among solids the superconductivity or SDW phases occur has been the last thing about these phases to be understood, e.g.,it is still not understood why lithium with $\lambda \approx 0.4$ is not a superconductor, or why chromium has a SDW but molybdenum and tungsten with very similar electron band structure do not. (For the SDW, spin-dependent density functional results agree with experiment but do not explain why.) In collaboration with G.C. Psaltakis, my own efforts have tended towards attempting to understand the superconductivity, SDW, and possible SDW-superconductivity coexistence states for $(TMTSF)_2X$, given that these states do occur [27,28,29,30,31]. Although the matter is still in dispute, it seemed to us that the critical field H_{c2} with $\underline{H}\|\hat{a}$ is spin-limited as claimed by Rick GREENE and collaborators [21], suggesting singlet Cooper pairs for superconductivity at high P where no SDW occurs. At the same time T_c is reduced quickly to zero by nonmagnetic impurities or defects. Whether for singlet or triplet Cooper pairing, strong depression of T_c by nonmagnetic impurities suggests exactly the same thing, a highly anisotropic pair-orbital function. In this sense, the distinction between these pairings is only that orbital anisotropy is required in all cases for triplet pairing, due to quantum symmetry, but it may or may not occur for singlet pairing. In the highly anisotropic $(TMTSF)_2X$ compounds, and with the compressibility along the a axis ten times greater than for perpendicular directions [32], it seems entirely reasonable that an orbital function for spin-singlet Cooper pairing would be highly anisotropic. It also seems reasonable that with the b-axis dispersion of the electron band structure two or more orders of magnitude larger than $k_B T_{SDW}$ or $k_B T_c$ (0.1 eV compared to less than 2×10^{-3} eV), any SDW would fail to annihilate the entire f.s. as in the model of Machida and Matsubara. However, differing from Machida and Matsubara (MM), we think it essential when discussing possible coexistence of superconductivity with an inhomogeneous magnetic state that the proper singlet spinor for the Cooper pair $(i\sigma_y)_{\alpha\beta}$ be included, and that the entire problem be expressed with the arbitrary spin-coordinate axes not taken as a special choice. The Hamiltonian incorporating these ideas is

$$H = H_0 + H_{SS} + H_{SDW} + H_{TS} \quad \text{with} \tag{2}$$

$$H_0 = \sum_{\underset{\sim}{k},\alpha} \varepsilon_{\underset{\sim}{k}} c^+_{\underset{\sim}{k}\alpha} c_{\underset{\sim}{k}\alpha}$$

$$H_{SS} = -\frac{1}{2} \sum_{\substack{\underset{\sim}{k} \\ \alpha,\beta}} (i\sigma_y)_{\alpha\beta} \{\Delta(\underset{\sim}{k}) c^+_{\underset{\sim}{k}\alpha} c^+_{-\underset{\sim}{k}\beta}\} + \text{h.c.}$$

$$H_{SDW} = - \sum_{\substack{k\varepsilon \text{region } 1 \\ \alpha,\beta}} (\underset{\sim}{\sigma}\cdot\underset{\sim}{n})_{\alpha\beta} \{m(k^-,k^+) c^+_{k^-\alpha} c_{k^+\beta}\} + \text{h.c.}$$

$$H_{TS} = -\frac{1}{2} \sum_{\substack{k\varepsilon \text{region } 1 \\ \alpha,\beta}} (\underset{\sim}{\sigma}\cdot\underset{\sim}{n} i\sigma_y)_{\alpha\beta}$$

$$\{\Delta_{-Q}(k) c^+_{-k^-\alpha} c^+_{k^-\beta} + \Delta_Q(k) c^+_{-k^+\alpha} c^+_{k^+\beta}\} + \text{h.c.}$$

$k^\pm$ for region 1 denotes $k \pm k_f$ and $-k^\pm$ denotes $-k \pm k_f$, where the k direction is parallel to $\hat{a}$. The SDW wavevector is $\underset{\sim}{Q} = 2k_f\hat{a}$ and the SDW polarization is $\underset{\sim}{n} \perp \hat{a}$. The complete set of electron Green functions for the above model has been obtained by the equations of motion method. The results for region 1 and for $\Delta_{\pm Q}(-k) = -\Delta_{\pm Q}(k)$ can be expressed as

$$\langle\langle c_{k^-\alpha}; c^+_{k^-\beta}\rangle\rangle_\omega = \delta_{\alpha\beta} G_1$$

$$\langle\langle c^+_{-k^+\alpha}; c^+_{k^-\beta}\rangle\rangle_\omega = (i\sigma_y)_{\alpha\beta} G_2$$

$$\langle\langle c_{k^+\alpha}; c^+_{k^-\beta}\rangle\rangle_\omega = (\underset{\sim}{\sigma}\cdot\underset{\sim}{n})_{\alpha\beta} G_3$$

$$\langle\langle c^+_{-k^-\alpha}; c^+_{k^-\beta}\rangle\rangle_\omega = (i\sigma_y \underset{\sim}{\sigma}\cdot\underset{\sim}{n})_{\alpha\beta} G_4$$

where the functions G_i $(i = 1,..,4)$ are determined by the following matrix equation:

$$\begin{bmatrix} \omega-\varepsilon_{k^-} & -\Delta & m(k^-,k^+) & \Delta_{-Q}(k) \\ -\Delta^* & \omega+\varepsilon_{k^-} & \Delta^*_Q(k) & m(-k^-,-k^+) \\ m^*(k^-,k^+) & \Delta_Q(k) & \omega+\varepsilon_{k^-} & -\Delta \\ \Delta^*_{-Q}(k) & m^*(-k^-,-k^+) & -\Delta^* & \omega-\varepsilon_{k^-} \end{bmatrix} \cdot \begin{bmatrix} G_1 \\ G_2 \\ G_3 \\ G_4 \end{bmatrix} = \begin{bmatrix} 1 \\ 0 \\ 0 \\ 0 \end{bmatrix} . \tag{3}$$

A similar equation occurs for $k^- \rightarrow k^+$. The 4×4 matrix A on the left hand side of the above can be easily inverted and when $H_{TS} = 0$,

$$<<c^+_{-k+\frac{Q}{2},\alpha};c^+_{k-\frac{Q}{2},\beta}>>^{SS}_\omega = (\det.A)^{-1}(i\sigma_y)_{\alpha\beta}\Delta^*[\omega^2-\varepsilon^2_{k^-}-|\Delta|^2 + m(-k^-,-k^+)m^*(k^-,k^+)] \tag{4}$$

$$<<c_{k+\frac{Q}{2},\alpha};c^+_{k-\frac{Q}{2},\beta}>>^{SDW}_\omega = -(\det.A)^{-1}(\underset{\sim}{\sigma}\cdot\underset{\sim}{n})_{\alpha\beta}[|\Delta|^2 m^*(-k^-,-k^+) + m^*(k^-,k^+)\{\omega^2-\varepsilon^2_{k^-}-|m(-k^-,-k^+)|^2\}] \tag{5}$$

$$<<c^+_{-k-\frac{Q}{2},\alpha};c^+_{k-\frac{Q}{2},\beta}>>^{TS}_\omega = -(\det.A)^{-1}(i\sigma_y\underset{\sim}{\sigma}\cdot\underset{\sim}{n})_{\alpha\beta}(\omega+\varepsilon_{k^-})\Delta^*[m^*(k^-,k^+) + m^*(-k^-,-k^+)] \tag{6}$$

with

$$\det.A = [\varepsilon^2_{k^-} - \omega^2 + (m+\Delta)^2][\varepsilon^2_{k^-} - \omega^2 + (m-\Delta)^2]$$

(written for brevity with m and Δ real and m independent of k). Three anomalous electron correlations unavoidably occur when only two correlation energies of H_{SDW} and H_{SS} appear in H, and so H_{TS} is necessary for self-consistency. With or without $\Delta_{\pm Q}(k) = 0$, the anomalous quasiaverage of (6) changes sign as it must for Fermion operators, $<<c^+_{k^-\beta};c^+_{-k^-,\alpha}>>_{-\omega} = -<<c^+_{-k^-\alpha};c^+_{k^-\beta}>>_\omega$. This quasiaverage describes a triplet electron pair with center-of-mass momentum $-\hbar Q$ and with odd-parity orbital function within the c.m. system, requiring that $\Delta_{\pm Q}(k) = -\Delta_{\pm Q}(-k)$ in H_{TS}. For this symmetry pairing the interaction implicit in H_{TS} is not identically zero but is likely to be small, so $\Delta_{\pm Q}$ and H_{TS} are probably negligible. However the triplet pair amplitude is not small.

If the SDW occurs alone, the electron field operator (e.f.o.) for occupied states is $\alpha^{SDW}_{k\alpha} = u_{k\alpha}c_{k\alpha} + v_{k\alpha}\sum_\beta(\underset{\sim}{\sigma}\cdot\underset{\sim}{n})_{\alpha\beta}c_{k\pm 2k_f,\beta}$ for $k \lessgtr 0$ and k smaller than k_f. If singlet superconductivity occurs alone, the e.f.o. is $\alpha^{SS}_{k\alpha} = w_{k\alpha}c_{k\alpha} + x_{k\alpha}(i\sigma_y)_{\alpha,-\alpha}c^+_{-k,-\alpha}$. When both occur, $c_{k\pm 2k_f,\beta}$ and $c^+_{-k,-\alpha}$ are related to the same operator $c_{k\alpha}$ and therefore related to each other, the third relation corresponding to the e.f.o. $\alpha^{TS}_{k\alpha} = y_{k\alpha}c_{k\alpha} + z_{k\alpha}\sum_\beta(i\sigma_y\underset{\sim}{\sigma}\cdot\underset{\sim}{n})_{\alpha\beta}c^+_{-k\pm 2k_f,\beta}$ for triplet Cooper pairing with pair momentum $\pm\hbar Q$. The same α^{TS} correlation occurs if instead we first diagonalize $H_0 + H_{SDW}$ and then include H_{SS} expressed in terms of α^{SDW}, or if we first diagonalize $H_0 + H_{SS}$, etc. The mixed spin-triplet and spin-singlet Cooper pairing occurs for a fundamental reason - the SDW breaks time reversal symmetry ($TH_{SDW}+H_{SDW}T = 0$) and mixes electron spin states.

For $\Delta_{\pm Q}(k) \neq 0$, the electron Green functions and gap equations simplify for the case $\Delta_{-Q}(k) = -\Delta_Q(k) = \Delta_{-Q}\mathrm{sign}k$, where $\sin Q\cdot r$ superconductivity maximally avoids $\cos Q\cdot r$ magnetic polarization. Taking matrix elements of V_{ee} as $V_{TS}(k^\pm,k'^\pm) = V_{TS}\,\mathrm{sign}k\,\mathrm{sign}k'$, V_{SDW} and V_{SS} constant, and for $m(k,k') = m = m^*$, gap equations written for brevity as if region 1 existed alone are

$$\Delta = -\tfrac{1}{2}V_{SS}\sum_{k,\pm}\left(\frac{\Delta\pm m}{2E_\pm}\right)\tanh\frac{E_\pm}{2T} \tag{7}$$

$$m = \tfrac{1}{2}V_{SDW}\sum_{k,\pm}\left(\frac{m\pm\Delta}{2E_\pm}\right)\tanh\frac{E_\pm}{2T} \tag{8}$$

$$\Delta_{-Q} = -\tfrac{1}{2}V_{TS}\sum_{k,\pm}\left(\frac{\Delta_{-Q}\pm|\epsilon_{k^-}|}{2E_\pm}\right)\tanh\frac{E_\pm}{2T} \tag{9}$$

where $E_\pm = [(\Delta_{-Q}(k) \pm \epsilon_{k^-})^2 + (\Delta\pm m)^2]^{\frac{1}{2}}$. Including additional terms of standard form for $\Delta(\underset{\sim}{k})$ involving region 2, solutions of the full gap equations have one T_c for superconductivity. With $T_{SDW} \gg T_c$, $\Delta_{-Q} = \frac{V_{TS}N_1(0)\Delta_1}{2 + V_{TS}N_1(0)}$ at T_c^-, where i = 1,2 denotes region i and $N(0) = N_1(0) + N_2(0)$ is the f.s. one-spin density of electron states. Because of large pair momentum, $\Delta_{-Q} = 0$ except when $m,\Delta \neq 0$. Similar to results obtained by MM,[18] T_c does not go to zero but is only reduced for a range of $T_{SDW} > T_c$, however $T_{SDW} \to 0$ when $T_c > T_{SDW}$. With $\Delta_{\pm Q}$ negligible or zero, changes from the MM solutions still occur because V_{SS} is allowed to vary over the f.s. We consider an example at high pressures where $T_{SDW} = 0$ and take $V_{SS}(\underset{\sim}{k}_i,-\underset{\sim}{k}_i;\underset{\sim}{k}_j',-\underset{\sim}{k}_j')N_j(0) = \lambda_{ij}$ (phonon-mediated and coulomb parts) with only λ_{22} attractive:

$$T_c = \frac{2\gamma}{\pi}\,\omega_{cutoff}e^{-1/\sqrt{2}\lambda}$$

$$\text{for } \lambda_{11} = \lambda_{12} = \lambda_{21} = -\lambda_{22} = \lambda > 0 \tag{10}$$

$$\Delta_1 = \frac{-\Delta_2}{1+\sqrt{2}}\ .$$

This compares with $T_c = \frac{2\gamma}{\pi}\,\omega_{cutoff}e^{-1/2\lambda}$ and $\Delta_1 = \Delta_2$ for $\lambda_{ij} = -\lambda$. With scattering of electrons by nonmagnetic impurities expressed by τ_{ij}^{-1} and $\tau_{12} = \tau_{21}$, $\tau_{11} = \tau_{22}$,

$$T_c = T_{co}\left[1 - \frac{\pi\left(1+\frac{1}{\sqrt{2}}\right)}{8\tau_{12}T_{co}}\right] \ ;\quad \tau_{12}T_{co} \gg 1 \tag{11}$$

T_{co} is T_c in the absence of impurities, given by (10). For this case, with τ_{12} decreasing Δ_1 approaches $-\Delta_2$, not Δ_2, and so T_c approaches zero at $h/\tau_{12} \approx k_BT_c$, not some $T_c \neq 0$ limit for a $\Delta_1 = \Delta_2$ state. T_c is negligibly sensitive to nonmagnetic impurities for the coexistence state with $T_{SDW} \gg T_c$ in comparison to sensitivity for the high-P superconductivity-only regime:

$$T_c = T_{co}\left[1 - 0\left(\frac{1}{\tau_{12}T_{SDWo}}\right)\right] \ ;\quad \tau_{12}T_{co} \gg 1\ . \tag{12}$$

T_{SDW} is also reduced by nonmagnetic impurities:

$$T_{SDW} = T_{SDWo}\left[1 - \frac{\pi}{4\tau_{11}T_{SDWo}} - \frac{\pi}{8\tau_{12}T_{SDWo}}\right] \ ;\quad \tau_{ij}T_{SDWo} \gg 1\ . \tag{13}$$

Qualitatively similar results occur for a large number of λ_{ij} parameter choices.

We have calculated the spin susceptibility for the coexistence state using an eight-component field

$$\left(c^{+}_{k^-\uparrow}, c^{+}_{k^-\downarrow}, c_{-k^+\uparrow}, c_{-k^+\downarrow}, c^{+}_{k^+\uparrow}, c^{+}_{k^+\downarrow}, c_{-k^-\uparrow}, c_{-k^-\downarrow}\right),$$

with the same result for arbitrary choice of spin coordinate axes. With the magnetic field parallel to the SDW polarization ($\underset{\sim}{H} \parallel \underset{\sim}{n} \parallel \hat{z}$spin), and so perpendicular to the stacking axis $\hat{a}$, the parallel spin susceptibility is given by

$$\chi_{\parallel} = -\frac{1}{4}\mu_B^2 T \sum_{j=1,2}\ \sum_{k\varepsilon \text{region} j}\ \sum_{\omega_n} \mathrm{Tr}\{\rho_3\sigma_3 G^0(\underset{\sim}{k}, i\omega_n)\rho_3\sigma_3 G^0(\underset{\sim}{k}, i\omega_n)\} \tag{14a}$$

whilst for $\underset{\sim}{H} \perp \underset{\sim}{n}$ the transverse spin susceptibility is given by

$$\chi_{\perp} = -\frac{1}{4}\mu_B^2 T \sum_{j=1,2}\ \sum_{k\varepsilon \text{region} j}\ \sum_{\omega_n} \mathrm{Tr}\{\rho_3\sigma_1 G^0(k, i\omega_n)\rho_3\sigma_1 G^0(k, i\omega_n)\}\ . \tag{14b}$$

Here $G^0(k,i\omega_n)$ denotes the Fourier transform of the 8×8 matrix Green function

$$G^0(k,\tau-\tau') = -\langle\tau\{c(k,\tau)c^+(k,\tau')\}\rangle$$

and σ_i, ρ_i denote the Pauli matrices on two- and four-dimensional vector spaces respectively. At T = 0, our results simplify to

$$\chi_{\parallel}(T=0) = 0 \tag{15a}$$

$$\chi_{\perp}(T=0) = \mu_B^2 N_1(0)\left\{1 + \frac{m^2-\Delta_1{}^2}{2m\Delta_1}\ \ln\left|\ \frac{m+\Delta_1}{m-\Delta_1}\ \right|\right\}\ . \tag{15b}$$

From (15b) we conclude that for $T^2_{SDW} >> T^2_c$ and for $\underset{\sim}{H} \perp \underset{\sim}{n}$ the spin susceptibility of the condensate approaches its normal-metal value in region 1,

$$\chi_{\perp}(T=0) = 2\mu_B^2 N_1(0)\quad ,\quad \text{for } T^2_{SDW} >> T^2_c\ . \tag{16}$$

In agreement with the result of (16), we have also shown using (4) and (6) that for $T^2_{SDW} >> T^2_c$, the spin-triplet Cooper pair amplitude at E_f in region 1 is much larger than the spin-singlet amplitude.

$$\left|\frac{\langle\text{triplet}\rangle}{\langle\text{singlet}\rangle}\right|_{\substack{\text{f.s.}\\ \text{region 1}}} = \left|\frac{\Delta_{-Q}(E_- + E_+)}{(\Delta_1+m)E_- + (\Delta_1-m)E_+}\right|_{\varepsilon_{k^-}=0}\ .$$

The denominator as well as the numerator is proportional to Δ_{-Q} times some function of Δ_{-Q}. With $\Delta_{-Q}\to 0$, or near zero,

$$\left|\frac{\langle\text{triplet}\rangle}{\langle\text{singlet}\rangle}\right|_{\substack{\text{f.s.}\\ \text{region 1}}} \sim \begin{cases} >>1 \text{ for } m>>\Delta_1 \\ 1 \text{ for } m=\Delta_1 \\ <<1 \text{ for } m<<\Delta_1 \end{cases}$$

Larger values of Δ_{-Q} increase the triplet amplitude. This causes the non-zero spin susceptibility of (15b), and is not special to SDW-superconductivity coexistence (with forced triplet superconductivity amplitude) but should occur also but to perhaps a much smaller extent for Chevrel phases. (For the SDW to occur at all, a significant fraction of the f.s. must be region 1, whereas for an effectively external source for $m(\underline{k})$ such as rare earth ions, the wavevector of the antiferromagnetic order may nest a negligible or zero part of the f.s.) The a-axis critical field for $T \ll T_c$ (and no orbital shielding currents) in the coexistence state becomes

$$H_c = \frac{1}{\mu_B\sqrt{2}}\left[\frac{N_1(0)\Delta_1{}^2 + N_2(0)\Delta_2{}^2}{N_2(0)}\right]^{\frac{1}{2}} \quad \text{for} \quad T^2_{SDW} \gg T^2_c \tag{17}$$

$$\underset{\sim}{H} \parallel \hat{a},\ \underset{\sim}{H} \perp \underset{\sim}{n}$$

which is larger than

$$H_c = \frac{1}{\mu_B\sqrt{2}}\left[\frac{N_1(0)\Delta_1{}^2 + N_2(0)\Delta_2{}^2}{N_1(0) + N_2(0)}\right]^{\frac{1}{2}}, \quad \text{for } T_{SDW} = 0 \tag{18}$$

$$\underset{\sim}{H} \parallel \hat{a},\ \underset{\sim}{H} \perp \underset{\sim}{n}$$

for the high-P superconductivity-only regime, for the same Δ_1 and Δ_2 values. For $T_{SDW} \gg T_c$, indications of increased spin susceptibility below T_c (X = PF_6 and AsF_6) exist but are not yet strong evidence.

Why do both triplet and singlet Cooper pair amplitudes for superconductivity necessarily occur in the presence of antiferromagnetic order and with only the singlet Cooper pair interaction significant? First of all, as we argued above, the SDW and singlet-superconductivity correlations link two objects to a third and therefore to each other, with this link the correlation of triplet superconductivity. Secondly, the electron spin for the eigenstates of the Hamiltonian including only H_0 and H_{SDW} (and with band structure cutoffs so that all harmonics of the SDW are zero amplitude) is much different than for the normal metal:

$$\frac{1}{2}\left[\phi^{SDW^*}_{\underline{k}\alpha}(\underline{r})\underline{\sigma}\phi^{SDW}_{\underline{k}\alpha}(\underline{r}) + (\underline{k}\rightarrow-\underline{k})\right] = \left(|u_{\underline{k}}|^2 - |v_{\underline{k}}|^2\right)\alpha\hat{z} + 2|v_{\underline{k}}|^2\alpha(\underline{n}\cdot\hat{z})\underline{n}$$
$$+\, 2|u_{\underline{k}}||v_{\underline{k}}|(\cos \underline{Q}\cdot\underline{r})\underline{n}\ . \tag{19}$$

We have taken plane wave basis states for the normal metal, and $u_{\underline{k}}$ and $v_{\underline{k}}$ are parameters of the canonical transformation to the operators for electrons for the SDW field:

$$\alpha^{SDW}_{\underline{k}\alpha} = u_{\underline{k}}\, c_{\underline{k}\alpha} + v_{\underline{k}} \sum_{\beta}(\underline{\sigma}\cdot\underline{n})_{\alpha\beta}c_{\underline{k}\pm\underline{Q}\beta}$$

with $|u_{\underline{k}}|^2 + |v_{\underline{k}}|^2 = 1$.

At the SDW gap edge, $|u_{\underline{k}}|^2 = |v_{\underline{k}}|^2 = \frac{1}{2}$ and this in (19) means that at that edge $\binom{1}{0}$ and $\binom{0}{1}$ spinors of the normal-metal state are totally lost. (This would not be seen from (19) with only the special choice $\underline{n} = \hat{z}$.) This modification of the electron spinors is stronger than any spin-orbit coupling effect. Spin-orbit coupling in fact, no matter how strong, does not force a mixed singlet and triplet Cooper pairing state for a homogeneous superconductor [33]. This lack of mixing no doubt would also occur for the SDW-superconductivity state in the $Q \to 0$ homogeneous limit, i.e., ferromagnetism, however the $c_{\underline{k}\alpha}$ to $\alpha_{\underline{k}\alpha}$ canonical transformation cannot be used for $Q \to 0$ because in this case the energy-band cutoff required to eliminate SDW harmonics (which make the transformation non-canonical) tends to zero.

It is probably true, but we have certainly not been able to prove it, that any inhomogeneous magnetism mixes spin-triplet and spin-singlet Cooper pairing. Gorkov and Rusinov said a very long time ago, "... the spin of a pair is not conserved in the process of scattering by the spin of an impurity" [34].

References

1. D. Jerome, A. Mazaud, M. Ribault, and K. Bechgaard: J. Physique Lettres 41, L95 (1980)
2. K. Andres, F. Wudl, D.B. McWhan, G.A. Thomas, D. Nalewajek, and A.L. Stevens: Phys. Rev. Lett. 45, 1449 (1980)
3. R.L. Greene and E.M. Engler: Phys. Rev. Lett. 45, 1587 (1980)
4. K. Mortensen, Y. Tomkiewicz, T.C. Schultz, and E.M. Engler: Phys. Rev. Lett 46, 1234 (1981)
5. K. Murata, T. Ukachi, H. Anzai, K. Kajimura, G. Saito, T. Ishiguro: J. Phys. Soc. Japan 51, 695 (1982)
6. K. Bechgaard, C.S. Jacobsen, K. Mortensen, H.J. Pedersen, and N. Thorup: Solid State Comm. 33, 1119 (1980)
7. W.M. Walsh, F. Wudl, G.A. Thomas, D. Nalewajek, J.J. Hauser, P.A. Lee, and T. Peohler: Phys. Rev. Lett. 45, 829 (1980)
8. A. Andrieux, D. Jerome, and K. Bechgaard: J. Physique Lett. 42, L87 (1981)
9. W.M. Walsh, F. Wudl, L.W. Rupp, R. Aharon-Shalom, and G.A. Thomas: Mol. Cryst. Liq. Cryst. 86, 259 (1982)
10. S.S.P Parkin, E.M. Engler, R.R. Schumaker, R. Lagier, V.Y. Lee, J.C. Scott, and R.L. Greene: Phys. Rev. Lett. 50, 270 (1982)
11. S.S.P. Parkin, M. Ribault, D. Jerome, and K. Bechgaard: J. Phys. C 14, 5305 (1981)
12. C.S. Jacobsen, D.B. Tanner, and K. Bechgaard: Phys. Rev. Lett. 46 1142 (1981)
13. P.M. Grant: Phys. Rev. B 26, 6888 (1982), and several papers in proceedings of "Synthetic Low Dimensional Conductors and Superconductors", Les Arcs, France, 11-18 Dec. 1982, to appear in J. Physique
14. B. Horowitz, H. Gutfreund, and M. Weger: Phys. Rev. B 12, 3174 (1975)
15. M. Weger: Bull. Europ. Phys. Soc. 9, 7 (1978)
16. T. Takahashi, D. Jerome, and K. Bechgaard: J. Physique Lett. 43, L565 (1982)
17. J.F. Kwak, J.R. Schirber, R.L. Greene, and E.M. Engler: Phys. Rev. Lett. 46, 1296 (1981)
18. P. Garoche, R. Brusetti, D. Jerome, and K. Bechgaard: J. Physique Lett. 43, L147 (1982)
19. K. Bechgaard: proceedings of "Synthetic Low Dimensional Conductors and Superconductors", Les Arcs, France, 11-18 Dec., 1982, to appear in J. Physique
20. R.L. Greene, P. Haen, S.Z. Huang, E.M. Engler, M.Y. Choi, and P.M. Chaikin: Mol. Cryst. Liq. Cryst. 79, 183 (1982); M.Y. Choi, P.M. Chaikin, P. Haen and R.L. Greene: Solid State Comm. 41, 225 (1982)
21. M.Y. Choi, P.M. Chaikin, S.Z. Huang, P. Haen, E.M. Engler, and R.L. Greene: Phys. Rev. B 25, 6208 (1982)

22. D. Jerome: Proc. d- and f-Band Superconductors V, Karlsruhe, 28-30 June, 1982
23. H. Gutfreund: Physica 109 and 110B, 1866 (1982).
24. V.J. Emery, R. Bruinsma, and S. Barisic: Phys. Rev. Lett. 48, 1039 (1982)
25. K. Machida: J. Phys. Soc. Japan 50, 2195 (1981); K. Machida and T. Matsubara: J. Phys. Soc. Japan 50, 3231 (1981)
26. K. Yamaji: J. Phys. Soc. Japan, 1361 (1983)
27. E.W. Fenton and G.C. Psaltakis: Solid State Comm. 45, 5 (1982)
28. G.C. Psaltakis and E.W. Fenton: J. Phys. C 16, 3913 (1983)
29. E.W. Fenton and G.C. Psaltakis: Solid State Comm. 45, 421 (1982)
30. E.W. Fenton and G.C. Psaltakis: Solid State Comm. 47 767 (1983)
31. E.W. Fenton and G.C. Psaltakis: J. Physique (to be published)
32. B. Morosin, J.E. Schirber, R.L. Greene, and E.M. Engler: Phys. Rev. 26, 2660 (1982)
33. E.W. Fenton: Solid State Comm. 34, 917 (1980)
34. L.P. Gorkov and A.I. Rusinov: Soviet Physics JETP 19, 922 (1964)

Theory of the SDW and Superconducting Phases in the Bechgaard Salts

Kunihiko Yamaji

Electrotechnical Laboratory, Sakura-mura, Ibaraki 305, Japan

1. Introduction

The discovery [1] of superconductivity in $(TMTSF)_2X$, $X=PF_6$, under pressure opened a very active field of organic superconductors. Subsequently Bechgaard salts with other electron acceptors $X=AsF_6$, ClO_4,etc.,were found superconducting [2]. They have proved to be extraordinarily rich systems showing not only superconductivity but also spin density wave (SDW) and anion orderings [3,4,5]. Among hot interests have been the dimensionality of the electronic band, the pressure dependence of the SDW states, galvanomagnetic properties, the possibility of the coexistence of the superconducting and SDW orders and the possible triplet nature of the superconductivity. In this review of the author's work we try to get an overall consistent understanding on them on the basis of a simplified model.

The dimensionality of their electronic band was controversial in the first step to establish the theoretical model to clarify their rich properties; whether it is essentially one-dimensional as is assumed in the g-ology model or whether it is rather two-dimensional as was suggested by experimental observations on the actual systems. The divergent opinions have recently converged to a two-dimensional (2d) tight-binding band picture with consented values of transfer energies in the two crystallographic directions. On the basis of this band model with addition of the on-site Coulomb interaction we can make a mean field theory of SDW. It clearly imposes a condition, i.e., an upper bound to the smaller transfer energy for SDW to exist stably at moderate temperatures [6]. Remarkable closeness of this value to that estimated for the Bechgaard salts strongly supports a viewpoint that the disappearance of SDW under pressure or depending on the kind of electron acceptor X is due to violation of the above condition. Then we can easily understand the pressure dependence. This viewpoint leads to conclude that a purely superconducting phase intrudes into the SDW phase, denying the coexistence of both orders [7]. In a narrow boundary parameter region in the midst of which this condition is violated the SDW wave vector proves to shift slightly [8], which allows to interpret galvanomagnetic results nicely.

The next section gives account of the 2d band model. The third section is devoted to the SDW phase and the fourth to the superconducting phase. Summary is given in the last section.

2. 2d Band Model

Planar TMTSF molecules are stacked, forming columns along the a-axis direction. The columns are tightly arranged in the b direction, forming a sheet parallel to the ab plane [9]. Therefore, electron transfer in each sheet is very much favored. But anions are located between such sheets.

Methyl groups of the TMTSF molecule also prevent intersheet electron transfer. Therefore, neglecting this process we consider the 2d electronic system of one sheet in a tight-binding approximation. The latter must be very good since the band width is moderate and the level of the relevant highest occupied molecular orbital of TMTSF is well separated from other levels [10].

The electronic sites, i.e., TMTSF molecules, are arranged in one sheet as in Fig. 1. Slight dimerization due to zigzag stacking in a column is neglected, since it makes little effect in the vicinity of the Fermi surface. So we take one molecule distance along the column as $\vec{a}$, although it is half of the real lattice constant. Another lattice constant vector $\vec{b}$ is as in Fig. 1. The transfer energy t_a between the neighbor molecules in the a direction has been known to be large, forming the band width of $4t_a \cong 1$ eV.

The magnitude of the transfer energy t_b in the b direction was controversial. The motivation at the start of the present study was to determine it from the features of the pressure-temperature (PT) phase diagram of SDW. With three transverse transfer energies, t_{bo}, t_{b1} and t_{b2}, shown in Fig. 1, the band energy $\varepsilon_{\vec{k}}$ is given by

$$\begin{aligned}\varepsilon_{\vec{k}} &= -2t_a\cos x - 2t_{bo}\cos y - 2t_{b1}\cos(y-x) - 2t_{b2}\cos(y-2x) \\ &= -2t_a\cos x - 2t_b(x)\cos(y-\phi(x)) ,\end{aligned} \tag{2.1}$$

where $x=\vec{a}\cdot\vec{k}$ and $y=\vec{b}\cdot\vec{k}$; the second equation is a trigonometric rewriting with

$$t_b(x) = [t_{bo}^2 + t_{b1}^2 + t_{b2}^2 + 2t_{b1}(t_{bo}+t_{b2})\cos x + 2t_{bo}t_{b2}\cos 2x]^{1/2} , \tag{2.2}$$

$$\phi(x) = \tan^{-1}[(t_{b1}\sin x + t_{b2}\sin 2x)/(t_{bo} + t_{b1}\cos x + t_{b2}\cos 2x)] . \tag{2.3}$$

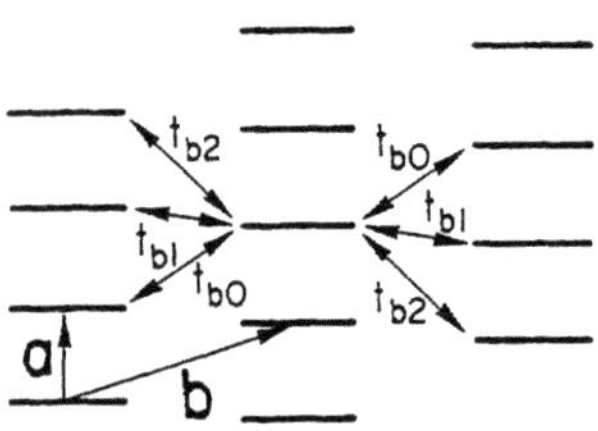

Fig. 1 Arrangement of TMTSF molecules in the ab sheet. t_{bi}, i=1,2,3, are important transverse transfer energies

Since the Fermi surface is nearly flat in the Bechgaard salts, x is nearly constant in the physically important region around the Fermi surface. So we can take $t_b(x)$ and $\phi(x)$ as constant. Further, the value of ϕ has been shown to be small by a band calculation [11] and also by the observation of the resistance anomaly accompanying the anion ordering [12]. Anyway it does not play an important role in the present problems. Then the band is reduced to the following simple one:

$$\varepsilon_{\vec{k}} = -2t_a\cos\vec{a}\cdot\vec{k} - 2t_b\cos\vec{b}\cdot\vec{k} - \mu . \tag{2.4}$$

Chemical potential μ is subtracted for the later convenience. In this model we can take $\vec{a}$ and $\vec{b}$ as perpendicular to each other.

The plasma frequency of this system in the a and b directions, $\omega_{pl,a}$ and $\omega_{pl,b}$, is given [13,7] by

$$\omega^2_{pl,a} = 16N_s e^2 a^2 t_a \sin ak_F ,$$

$$\omega^2_{pl,b} = 8N_s e^2 b^2 t_b^2/t_a \sin ak_F , \qquad (2.5)$$

where N_S is the density of electronic sites and k_F is the Fermi wave number in the limit of t_b=0. Applying these equations to the data obtained on the PF_6 system, we get [7]

$$t_a = 0.318 \text{ eV} = 3690 \ k_B , \quad t_b = 361 \ k_B , \quad t_b/t_a = 0.098 . \qquad (2.6)$$

The value of $t_b/t_a \cong 0.1$ is now widely accepted for all Bechgaard salts. The transfer energy t_c in the c direction is believed to be less than t_b by one order [3]. The band calculations [14,11] are in qualitative agreement with the above 2d band, but they still do not give the precise values of transfer energies due to difficulty in treating d orbitals properly.

Since two TMTSF molecules pass one electron to the electron acceptor X, there is one hole per two sites. For ease of treatment we make the electron-hole reversal and assume that there is one electron per two sites, i.e., we assume a quarter-filled band. We take t_a and t_b as positive without any loss of generality. Then in our model the Fermi wave number k_F is given by

$$k_F = \pi/4a . \qquad (2.7)$$

3. SDW Phase

3.1 Mean Field Theory

The mean field theory of SDW in the 2d system gives one remarkable feature which the 1d theory does not share. An upper bound is imposed on the transverse transfer energy t_b for the existence of the usual SDW state with the best nesting wave vector [6]. Such an upper bound is known for the quasi 1d Peierls system with 3d coupling [15].

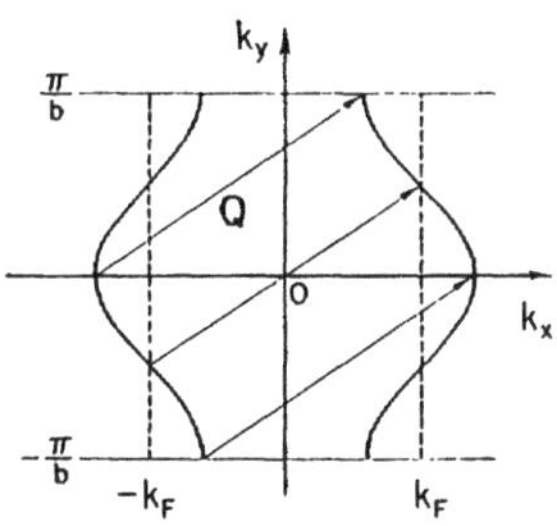

Fig. 2 Nesting Fermi surface of the 2d band (2.4)

It is well known that in spite of the warping of the Fermi surface due to t_b, as in Fig. 2, the pieces of the Fermi surface well nest with the nesting vector

$$\vec{Q} = (2k_F, \pi/b) \qquad (3.1)$$

in the present 2d model, when t_b is much smaller than t_a. With the aid of this nesting, the Coulomb interaction drives SDW. For simplicity, we employ the following Hubbard model as our system:

$$H = \sum_{\vec{k},\sigma} \varepsilon_{\vec{k}} c_{\vec{k}\sigma}^{\dagger} c_{\vec{k}\sigma} + 1/2 \cdot I \sum_{\vec{k},\vec{k}',\vec{q},\sigma,\sigma'} c_{\vec{k}\sigma}^{\dagger} c_{\vec{k}+\vec{q}\sigma} c_{\vec{k}'\sigma'}^{\dagger} c_{\vec{k}'-\vec{q}'\sigma'} \ . \tag{3.2}$$

For the polarization of the SDW magnetization there is no special direction. With the z direction chosen for it, the SDW gap parameter M is defined by

$$M = 1/2 \cdot I \sum_{\vec{k}} \left[\langle c_{\vec{k}-\vec{Q}\uparrow}^{\dagger} c_{\vec{k}\uparrow} \rangle - \langle c_{\vec{k}-\vec{Q}\downarrow}^{\dagger} c_{\vec{k}\downarrow} \rangle \right] . \tag{3.3}$$

Then the average Hamiltonian is given by

$$H_{aver} = \sum_{\sigma} \sum_{\substack{|k_x|<k_F \\ |k_y|<\pi/b}} \left(c_{\vec{k}+\vec{Q}\sigma}^{\dagger}, c_{\vec{k}\sigma}^{\dagger}, c_{\vec{k}-\vec{Q}\sigma}^{\dagger} \right) \begin{pmatrix} \varepsilon_{\vec{k}+\vec{Q}}, & -\sigma M & \\ -\sigma M, & \varepsilon_{\vec{k}}, & -\sigma M \\ & -\sigma M, & \varepsilon_{\vec{k}-\vec{Q}} \end{pmatrix} \begin{pmatrix} c_{\vec{k}+\vec{Q}\sigma} \\ c_{\vec{k}\sigma} \\ c_{\vec{k}-\vec{Q}\sigma} \end{pmatrix} . \tag{3.4}$$

Here M is assumed real without loss of generality. By diagonalizing the 3×3 matrix and transforming one-electron operators to the ones for the new bands, we get

$$H_{aver} = \sum_{\sigma} \sum_{i=1}^{3} \sum_{\vec{k}} E_i(\vec{k}) \alpha_{i\vec{k}\sigma}^{\dagger} \alpha_{i\vec{k}\sigma} \ . \tag{3.5}$$

Here $E_i(\vec{k})$, i=1,2,3, give the SDW-reorganized bands and $\alpha_{i\vec{k}\sigma}^{\dagger}$ ($\alpha_{i\vec{k}\sigma}$) is the creation (annihilation) operator of an electron in the new bands.

The reorganized bands are shown in Fig. 3 as functions of k_x for fixed values of k_y. $E_3(\vec{k})$ is not illustrated since it is approximately equal to $E_2(\vec{k}+\vec{Q})$. Both $E_1(\vec{k})$ and $E_2(\vec{k})$ have the maximum and minimum, respectively, at

$$k_x = k_F + t_b \cos bk_y / at_a \sin x_F \ , \tag{3.6}$$

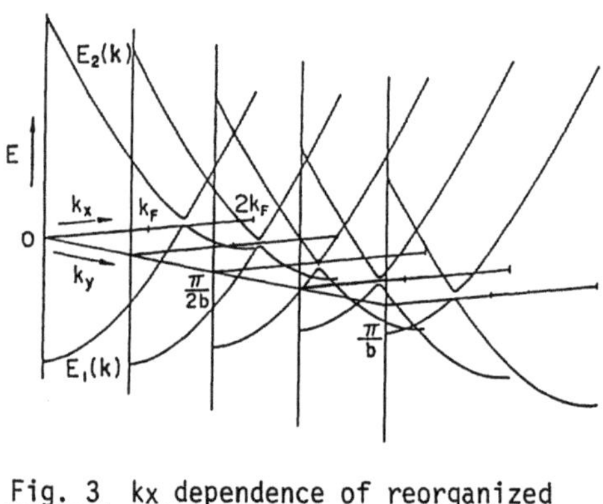

Fig. 3 k_x dependence of reorganized bands $E_1(\vec{k})$ and $E_2(\vec{k})$ for five values of k_y

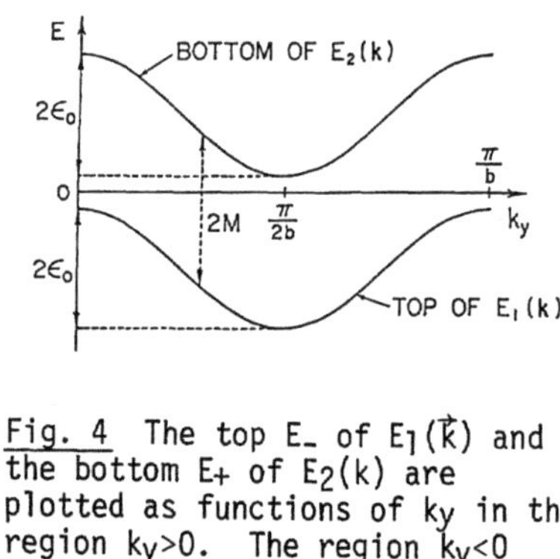

Fig. 4 The top E_- of $E_1(\vec{k})$ and the bottom E_+ of $E_2(k)$ are plotted as functions of k_y in the region $k_y>0$. The region $k_y<0$ is dropped since the figure is symmetric around the E axis

where $x_F = ak_F$. It is slightly shifted from k_F, depending on k_y. This is the position of the Fermi surface in the absence of SDW, which is shown in Fig. 2. The extremum values are given by

$$E_\pm \cong 2\varepsilon_0 \cos^2 bk_y \pm M - \mu - 2t_a \cos x_F \,, \tag{3.7}$$

where $\quad \varepsilon_0(t_b) = t_b^2 \cos x_F / 2t_a \sin^2 x_F \,.$ (3.8)

They are plotted in Fig. 4 as a function of k_y. The self-consistency equation for M is reduced to

$$\frac{1}{I} = \sum_{\substack{|k_x|<k_F \\ |k_y|<\pi/b}} \left(\frac{f(E_1(\vec{k})) - f(E_2(\vec{k}))}{\sqrt{(\varepsilon_{\vec{k}} - \varepsilon_{\vec{k}-\vec{Q}})^2 + 4M^2}} + \frac{f(E_1(\vec{k})) - f(E_3(\vec{k}))}{\sqrt{(\varepsilon_{\vec{k}} - \varepsilon_{\vec{k}+\vec{Q}})^2 + 4M^2}} \right) = \sum_{\substack{-k_F<k_x<3k_F \\ |k_y|<\pi/b}} \frac{f(E_1(\vec{k})) - f(E_2(\vec{k}))}{\sqrt{(\varepsilon_{\vec{k}} - \varepsilon_{\vec{k}-\vec{Q}})^2 + 4M^2}} \,. \tag{3.9}$$

When $\varepsilon_0 < M$, the upper and lower bands do not overlap; otherwise they overlap each other. At absolute zero of temperature, when $\varepsilon_0 < M$, the self-consistency equation is reduced to the well-known form

$$1/I = N(0) \cdot \ln(2D/M) \,, \tag{3.10}$$

where $\quad N(0) = N_s / 4\pi t_a \sin x_F$ (3.11)

is the state density for one k_x branch per spin, and

$$D = 4t_a \sin^2 x_F / \cos x_F \tag{3.12}$$

is the quantity of the order of the band width. Then we get

$$M = 2D \exp[-1/N(0)I] \,. \tag{3.13}$$

This is independent of t_b and we denote it by M_0. For $\varepsilon_0 > M$, i.e., when both the bands overlap each other, detailed calculation showed that M in (3.10) is replaced by ε_0 and that there is no solution which has the nesting vector (3.1) and can be continued to the above solution as a function of t_b. Therefore, as t_b increases, this type of solution suddenly disappears at an upper critical value, denoted by $t_{b,cr}$, which is defined by

$$\varepsilon_0(t_{b,cr}) = M_0 \,. \tag{3.14}$$

This result can be obtained also by calculating the energy difference between the SDW and normal states in the ground state in the absence of band overlap. It is given by expanding in powers of ε_0 and M_0 as

$$E_{SDW} - E_n = N(0)(\varepsilon_0^2 - M_0^2) \,. \tag{3.15}$$

The second term is due to the SDW condensation and the first term is due to the increase of the kinetic energy needed for the redistribution of electrons for the SDW formation. The difference vanishes as ε_0 approaches M_0. Equation (3.14) gives a slightly larger upper bound for transverse transfer energy than the result in [15].

The SDW transition temperature T_{SDW}, at which the SDW gap parameter begins to appear, was numerically obtained as in Fig. 5, where we used t_a=0.25 eV and IN_s=0.34 eV so that $T_{SDW}^{(0)}$, i.e., T_{SDW} for t_b=0, is equal to 11.5 K. We get in agreement with the value obtained in Fig. 5

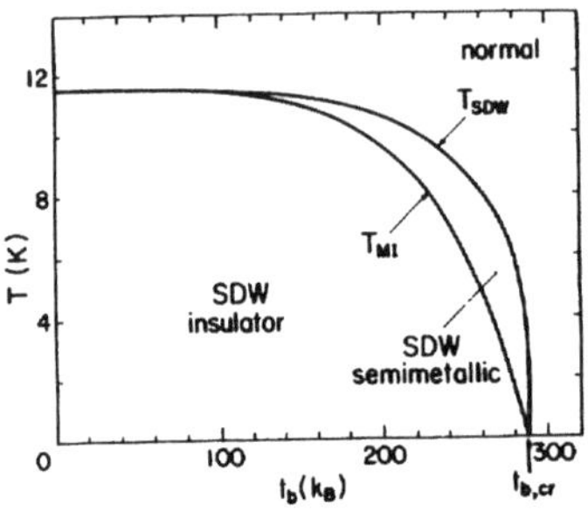

Fig. 5 Curve of SDW transition temperature T_{SDW} vs transverse transfer energy t_b. The curve of T_{MI} vs t_b divides the SDW phase into the semimetallic region where the upper and lower bands overlap and the insulator region where they do not

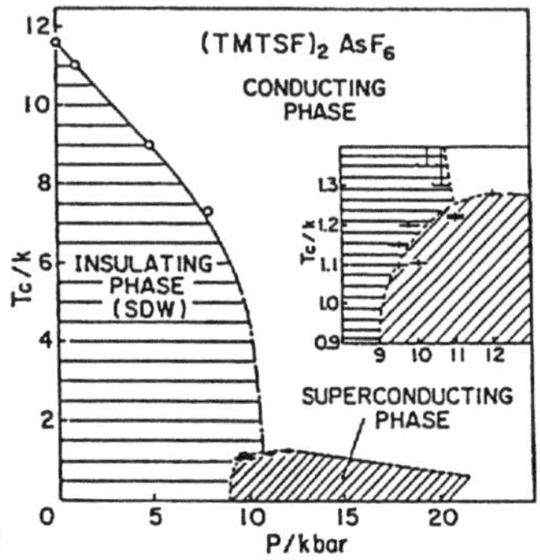

Fig. 6 PT phase diagram of $(TMTSF)_2AsF_6$ [16]

$$t_{b,cr} = (3.53\, T^{(o)}_{SDW}\, t_a \sin^2 x_F/\cos x_F)^{1/2} \cong 288\ k_B\ . \tag{3.16}$$

Although the c-axis direction is apparently neglected, we implicitly assume its existence. Supposing t_c is smaller than t_b, the condition for the decrease of T_{SDW} due to thermal fluctuations to be much smaller than the mean field 2d value above obtained proves to be reduced to

$$t_b \gg T_{SDW}\ , \tag{3.17}$$

which is amply fulfilled in the case of the Bechgaard salts. Therefore, we can rely on the 2d mean field calculation.

3.2 Comparison with the PT Phase Diagram

The above results have a good correspondence in the PT phase diagram of the Bechgaard salts. Figure 6 is for the AsF_6 system [16]. The phase diagrams for other systems are considered to be similar, if we move the origin of the pressure scale. Since the pressure decreases the lattice constants b and a, although we find only a preliminary report at room temperature on the pressure dependence of the lattice parameters [17], t_b and t_a are considered to increase under pressure. Therefore $\varepsilon_0 \propto t_b^2/t_a$ has the tendency to increase under pressure, while

$$M_0 \propto t_a \exp(-4\pi t_a \sin x_F/IN_s) \tag{3.18}$$

has the tendency to decrease. Then the condition

$$\varepsilon_0 < M_0 \tag{3.19}$$

for the existence of SDW has the tendency to be violated, even if it is satisfied under ambient pressure. On the other hand if we assume $T^{(o)}_{SDW}$=11.5 K and t_a=0.25 eV we get (3.16) for $t_{b,cr}$ and $t_{b,cr}/t_a$=0.099. If we employ t_a=3690 k_B given in (2.6), we get $t_{b,cr}$=325 k_B and $t_{b,cr}/t_a$=0.088. These values are amazingly close to the values of t_b and t_b/t_a in (2.6) obtained from the analysis of the plasma frequencies in view of uncertainties in $T^{(o)}_{SDW}$ and t_a, effect of fluctuations to T_{SDW}, mass enhancement effect,etc. Therefore, we are strongly led to the viewpoint that SDW disappears under pressure or when the electron acceptor is ClO_4 since condition (3.19) is violated. If we ascribe the strongest pressure dependence to t_b, then the t_b scale in the theoretical diagram can be taken as the pressure scale. Then the observed very steep descent of T_{SDW} is understood as the similar descent of T_{SDW} in the calculated curve. The critical pressure needed for suppressing SDW depends on the difference of t_b of the system from $t_{b,cr}$.

The pressure dependence of t_b has a subtle problem. Although the straightforward decrease of interstack Se-Se contacts apparently increases the calculated values of most transverse transfer energies corresponding to t_{bi} in Fig. 1 [18], t_b which is calculated by (2.2) is observed to decrease on the contrary. This is caused by the different signs of t_{b1} and $t_{bo}+t_{b2}$. But the pressure works more plausibly in a slightly different way. Since the interstack Se-Se contacts in the systems of PF_6, AsF_6 and ClO_4 at low temperatures are already very shortened, the TMTSF molecule on the nearest-neighbor column must have tendency to slip into the crack between TMTSF molecules of the column with the distance of the interstack Se-Se contacts not much shortened than to be simply transversely pushed to the neighbor molecule. This type of displacement decreases t_{b1} in Fig.1, since the nodal planes of the π orbitals of the relevant two molecules become more shifted, although it increases t_{bo} and t_{b2} . Therefore it increases t_b. t_{b1} should be taken equal to $(t_{I1}+t_{I2})/2$ in Grant's notation [11]. This conjecture could be checked by the increase of the crystallographic angle γ under pressure, which is already known to be easily changeable and also by the increased tilt of the molecular plane.

Some authors proposed that SDW vanishes under pressure due to the rapid decrease of the coupling constant g_3 for the umklapp process, applying the purely 1d g-ology theory [18]. They neglect the 2d nature of the band. When ϕ in (2.1) is small, this process surely contributes to SDW formation. Therefore the decrease of g_3 under pressure should be considered to contribute to the violation of condition (3.19). But it is plausibly not the main source of the pressure dependence since g_3 itself is very small.

Finally the value of IN_S=0.34 eV should be considered as the lower limit, since our calculation takes no account of the electron-phonon interaction which works against SDW formation.

3.3 Shift of SDW Wave Vector

In the present argument we have assumed that the SDW wave vector $\vec{Q}$ is fixed to the usually believed best nesting vector in (3.1). Consequently at T=0 K, the lower band is completely filled, while the upper band is empty, although there appear some small semimetallic carrier pockets at finite temperatures. At first the Shubnikov-de Haas (SdH) oscillation found on the PF_6 [19] and ClO_4 [20] systems was claimed to be related with these pockets [6]. Apparently the smallness of the pockets and the zero-temperature nature show difference from the observation. Kwak proposed that $\vec{Q}$ drastically changes from the best nesting vector to another commensurate vector under the magnetic field [19b]. But such a change would let the SDW condensation energy go away and would not be realized. Recently $\vec{Q}$ has been found to change slightly when t_b is close to $t_{b,cr}$ and gives a carrier pocket of appropriate magnitude [8].

As a function of the SDW wave vector written as

$$\vec{Q} = (\pi(1+s)/2a, \quad (\pi-2\phi)/b) \tag{3.20}$$

and M, we look for the minimum of the SDW state energy at T=0 K given by

$$E_{SDW} = \sum_{\sigma} \sum_{i=1}^{3} \sum_{\vec{k}} [E_i(\vec{k}) + \mu] f(E_i(\vec{k})) + 2M^2/I . \tag{3.21}$$

The result of numerical computation is shown in Fig. 7. The value t_b=3690 k_B in (2.6) is employed. I=0.4073 eV/N_s is chosen so that $T_{SDW}^{(0)}$ is equal to 11.5 K. E_{SDW} is shown in reference to the normal state energy E_n. The thick straight line shows the result with assuming s=ϕ=0, i.e., with the so far believed best nesting vector. In this case $t_{b,cr}$=325.6 k_B. Closed circles are the minimum when this restriction is lifted. They begin to deviate downward from the thick line at $t_b \cong t_{b,cr}$-1 k_B. There s and ϕ begin to increase in magnitude while M decreases. From $t_b \cong$326 k_B on, they take matured nearly constant values inscribed in Fig. 7. E_{SDW} increases as t_b increases and finally crosses the zero level. We denote this SDW state by SDW2 while we denote the so far known SDW state by SDW1. Open circles show another type of SDW state with nearly constant values of s, ϕ and M written in Fig. 7. We call this SDW3. Although M is very small in this case, it surely exists. It was found to have a negative value of E_{SDW} up to t_b =400 k_B. Perhaps the region where it exists extends up to t_b of order t_a. When t_b increases, SDW1 continuously changes to SDW2. Then SDW2 turns to SDW3 in a first-order transition. The extremum curves of the upper and lower bands are shown in Fig. 8 for SDW2. Since E_{SDW} is an even function of ϕ, we study only the case of $\phi > 0$. The finite value of ϕ lifts the left-hand side and lowers the right-hand side of the extremum curves in Fig. 8. The negative value of s reduces the phase space of the lower band and the electrons spill into the upper band. E_{SDW} gains the SDW condensation energy on the left-hand side but it neither gains nor loses much on the right-hand side in Fig. 8. In the case of SDW3, due to a larger value of ϕ the double peak in Fig. 8 becomes flattened with a much smaller gap. The phase space occupied by the electron pocket is given by

$$S_{el} = |s| \cdot S_{BZ}/2 \tag{3.22}$$

where S_{BZ} is the extent of the real Brillouin zone (BZ) in the a* and b* directions. Therefore S_{el}/S_{BZ} is equal to 0.26 % for SDW2 and 0.75 %

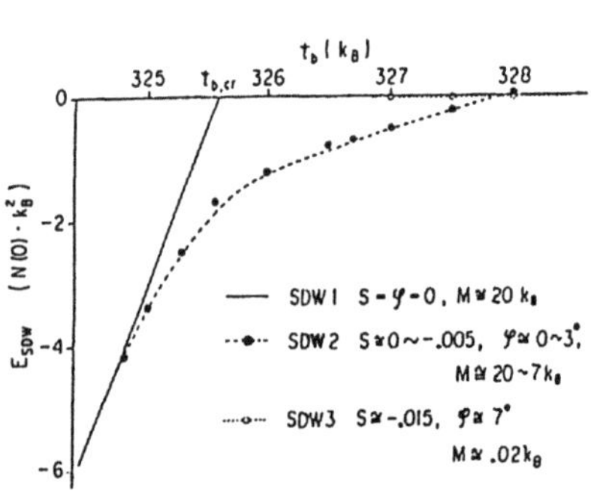

Fig. 7 Minimized energies for three types of SDW states in reference to the energy of the normal ground state

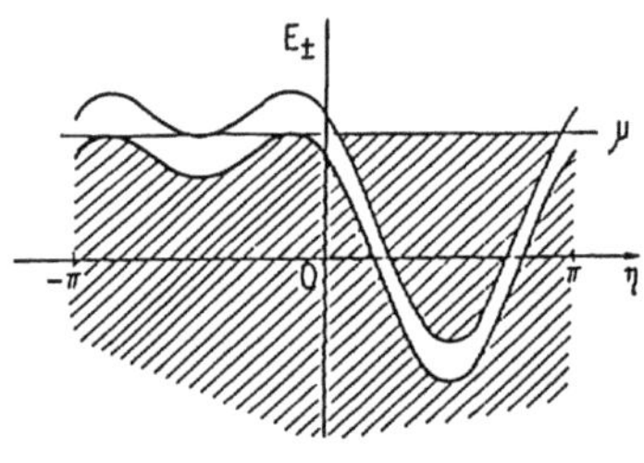

Fig. 8 The top and bottom of the reorganized bands are plotted as function of $\eta = bk_y + \phi$ for the SDW2 state with s=-0.0053, ϕ=3° and M=7 k_B. Non-zero value of ϕ makes the figure asymmetric around the E axis

for SDW3. The transition temperatures of SDW2 and SDW3 have not yet been calculated. They are presumably given as in Fig. 9.

The magnetic field dependence of T_{SDW} is crucial for the next argument. The perpendicular magnetic susceptibility $\chi_\perp$ in the SDW state is larger than the normal state one by a few tens of %. The introduction of the Zeeman energy was found not to cause the expected change. The increased susceptibility seems to originate from the increase of the enhancement factor due to the effective Coulomb interaction; the latter is caused by the weakened screening due to the decrease of the state density at the Fermi energy in the SDW state. Therefore, the transition temperature T_{SDW} must be elevated in the perpendicular magnetic field as shown by the dotted level line in Fig. 9, since $E_{SDW}-E_n$ is further lowered. This should be confirmed by a self-consistent calculation of the effective Coulomb interaction.

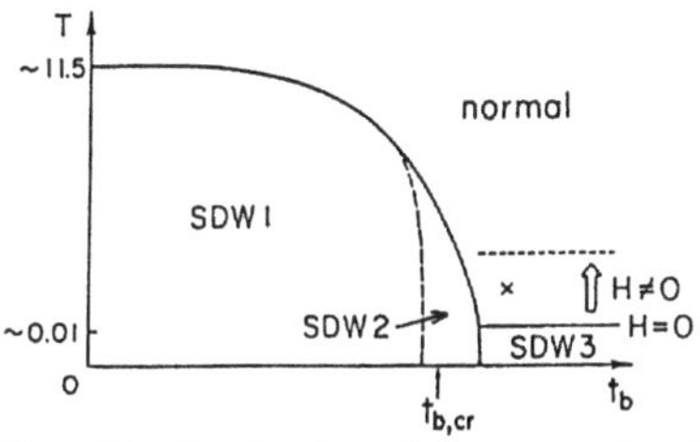

Fig. 9 Conjectured curves for T_{SDW2} vs t_b and T_{SDW3} vs t_b. The dotted line suggests the elevated T_{SDW3} in a perpendicular magnetic field. Both scales are extremely modified to expand the relevant parts

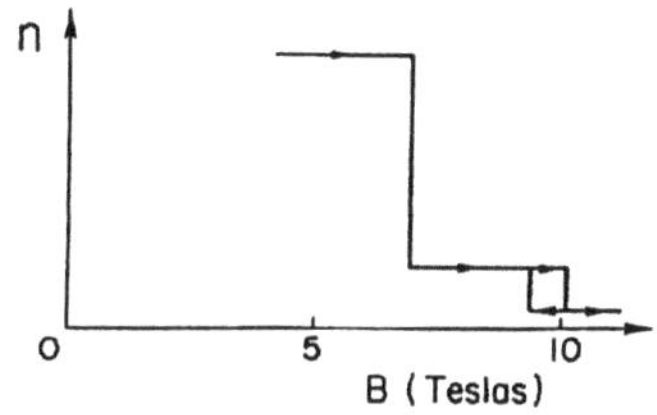

Fig. 10 Schematic figure showing the step-wise decrease of carrier density n of $(TMTSF)_2ClO_4$ in a perpendicular magnetic field. The scale gives only the order of magnitude

3.4 Comparison with SdH and Hall Effect Experiments

The SdH oscillation was observed above a significantly large threshold magnetic field in the temperature range of 0.03 ~ 2 K. The magnitude of the carrier pocket in the ClO_4 and PF_6 systems is 0.7 % and 1 % of the real BZ, respectively. These numbers are remarkably close to that of SDW3. Therefore, it is consistent to assume that the ClO_4 system in ambient pressure and the PF_6 system under pressure of 6.5 Kbar have their parameters located outside of the SDW region, e.g., at the X point in Fig. 9. When a magnetic field is applied perpendicular to the direction of magnetization, T_{SDW3} is considered to be elevated, since the free energy of SDW3 is decreased due to its larger $\chi_\perp$. The strength of the field at which T_{SDW} becomes equal to the measuring temperature should be the threshold field. With this reasoning the threshold field should increase as T increases, as has been already shown [21]. Recently three groups measured the Hall coefficient of the ClO_4 system in the perpendicular magnetic field [22]. Its sign is hole-sense [22a] in agreement with the present result since our model system is electron-hole reversed. The measurements showed step-wise decrease [22c] of the carrier density n as in Fig. 10. This can be well understood on the basis of the present results, if we assume that the system point moves from the X point in Fig. 9 to the right crossing the boundaries between SDW3 and SDW2 and then between SDW2 and SDW1. Incidentally, there appear electron pockets in SDW1 under a magnetic field. From this picture the first transition is expected to be a first-order-like one and the second one continuous, just as observed. Such a movement of the system point can be caused, e.g., by the increase of the Coulomb interaction due to screening weakened by the growing SDW gap and/or the magnetic field and/or by the decrease of t_b due to shrink of atomic orbitals.

4. Superconducting Phase

4.1 Problem of Interest

One major interest in the superconductivity in the Bechgaard salts has been the nature of the so-called reentrant region. In the PT phase diagram of the AsF_6 system in Fig. 6 the superconducting region intrudes into the region which must have been the SDW phase if the superconductivity did not appear. This region is called the reentrant region. Such intrusion is known also in the PF_6 system. MACHIDA and MATSUBARA theoretically pointed out the possibility of the microscopic coexistence of the superconducting and SDW orders in this region [23]. In this case superconducting order is asserted to contain both spin singlet and triplet pairing [24,4]. In this section [7], first we make a free energy consideration and find that this region can be well understood as intrusion of the purely superconducting region. Secondly, we discuss in more detail the possibility of the microscopic coexistence of both the orders and find it difficult. The possibility of triplet superconductivity is incidentally discussed. Finally the status of experiments is remarked.

4.2 Intrusion of Purely Superconducting Phase

Now we compare the free energies of the ordered states shown in the T vs t_b diagram in Fig. 5. The free energy F_{SDW} of the SDW state is equal to that F_n of the normal state along the curve of T_{SDW} vs t_b. It decreases continuously as the phase point moves into the SDW region off the phase boundary. When the normal state alongside and outside of the phase boundary changes into the superconducting state, the free energy F_{super} of the latter state is lower than F_n by the order of $N(0)\Delta^2$, with Δ being the superconducting gap parameter at 0 K. Therefore, F_{super} must be lower than F_{SDW} of the SDW state for the phase points inside and in the neighborhood of the phase boundary. This means that the SDW region in the vicinity of the boundary curve is replaced by the superconducting phase. The new first-order phase boundary between the intruding superconducting phase and the SDW phase is given by

$$F_{super} = F_{SDW} \ . \tag{4.1}$$

This boundary was calculated by assuming that SDW has the fixed wave vector, since the other possibility was not known. The dashed curve with closed circles labeled by $T_c^{(1)}$ in Fig. 11 is the result of this calculation. A BCS-type interaction introduced in the next subsection was added to our Hubbard model. The superconducting free energy was calculated in the mean field approximation. This approximation is justified by the observation that the coherence length in each direction is much larger than each lattice constant [25]:

$$\xi_a(0) = 600\ \text{Å} \gg a\ , \quad \xi_b(0) = 540\ \text{Å} \gg b\ , \quad \xi_c(0) = 60\ \text{Å} \gg c\ . \tag{4.2}$$

Features of the specific heat are also in accord with the mean field theory [26]. The parameters were chosen so that the SDW transition temperature reproduces the previous curve obtained without the BCS-type interaction.

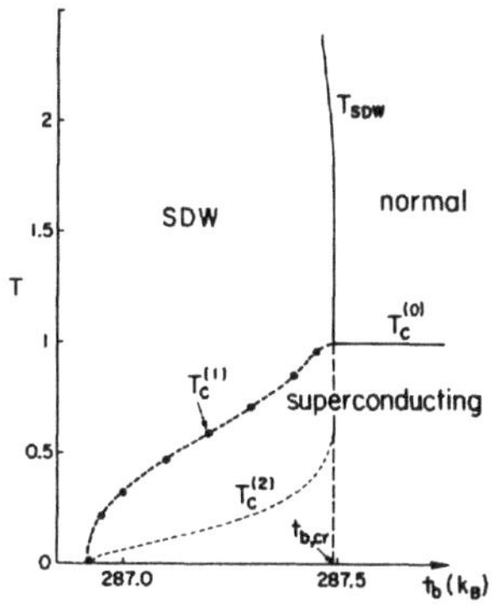

Fig. 11 The first-order transition temperature $T_c^{(1)}$ between the SDW1 and superconducting phases is plotted againt t_b. $T_c^{(2)}$ is the second-order transition temperature obtained in 4.5 when the superconductivity is of triplet spin pairing

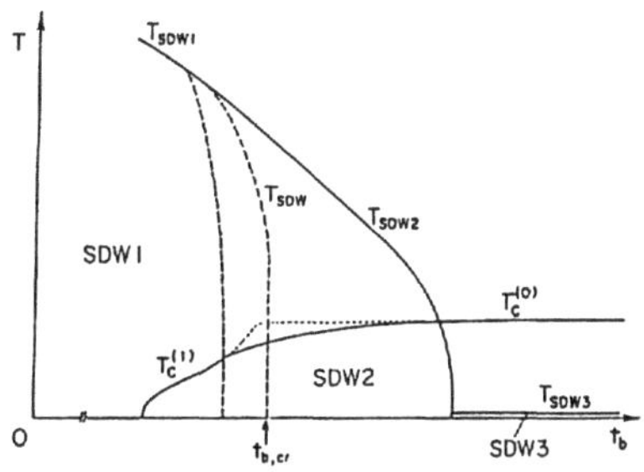

Fig. 12 Conjectured curve of the first-order transition temperature $T_c^{(1)}$ between the SDW and superconducting phases when all three types of SDW states are considered. The curve labeled by T_{SDW} gives the phase boundary when we assume only SDW1 exists. The dashed curve to the left of it is the boundary between SDW1 and SDW2. The dotted curve shows T_c when we take account of SDW1 only given in Fig. 11

The superconducting transition temperature $T_c^{(o)}$ not influenced by SDW is adjusted to 1 K. The result is almost independent of the type of interaction and the nature of superconductivity. The width of the intrusion in scale of t_b can be estimated by equating the following two quantities at T=0 K:

$$F_{super} - F_n = - N(0)\Delta^2 , \tag{4.3}$$

$$F_{SDW} - F_n = N(0)[\varepsilon_0^2(t_b) - \varepsilon_0^2(t_{b,cr})] . \tag{4.4}$$

Using the mean field relation Δ=1.76 $T_c^{(o)}$ k_B, we get

$$t_{b,cr} - t_b \cong 0.6[T_c^{(o)}]^2 k_B . \tag{4.5}$$

This is in good agreement with the result in Fig. 11.

Actually the SDW phase is more extended with the new SDW2 and SDW3 phases. The boundary taking account of the new SDW phases has not yet been calculated but it can be obviously given as in Fig. 12, if we overlap the plausible SDW2 and SDW3 regions over Fig. 11. The boundary is given by the previous curve $T=T_c^{(1)}$ in the deep region and is fixed to $T=T_c^{(o)}$ in the outmost part. The width of intrusion in scale of t_b is about 3 k_B.

4.3 BCS-Type Interaction

As the galvanomagnetic experiments and the present theory show, the SDW state can have current carriers. Transformation of them into a superconducting state may further lower the free energy of the SDW state. It may become lower than F_{super} for the purely superconducting state and the microscopic coexistence of the superconducting and SDW orders may be realized. Since the superconducting properties are qualitatively the same as for the ordinary superconductors as stated in 4.2, the phonon-mediated attraction between electrons is the most plausible interaction driving superconductivity in the Bechgaard salts, although we cannot specify which phonon mode works for this interaction for the moment. So we add to the previous Hubbard model the following BCS-type interaction:

$$H_{super} = -1/2\cdot\sum_{\vec{k},\vec{k}',\vec{q},\sigma,\sigma'}[g\theta(k_F-|q_x|)+g'\theta(|q_x|-k_F)]c_{\vec{k}\sigma}^{\dagger}c_{\vec{k}+\vec{q}\sigma}c_{\vec{k}'\sigma'}^{\dagger}c_{\vec{k}'-\vec{q}\sigma'}. \tag{4.6}$$

$\theta(x)$ is the step function. The related one-particle states are restricted, as usual, to those in the width of the Debye energy ω_D around the Fermi energy. g is the coefficient for the forward scattering in which the scattering electrons do not change momenta so much as to cross from one k_x branch to the other. g' is the coefficient for the backward scattering. Since the interaction is anisotropic, we get not only usual singlet superconductivity but also triplet superconductivity. From the examination of the parameter values, we find later that the singlet superconductivity is more plausible in the Bechgaard salts.

4.4 Case of Singlet Superconductivity

The mean field transition temperature for the usual singlet superconductivity is given by

$$T_c^{(s)} = 1.13\,\omega_D\,\exp[-1/N(0)(g + g' - 2I^*)]\ , \tag{4.7}$$

where $$I^* = I/[1 + 2N(0)I\cdot\ln(D/\omega_D)/\sin ak_F] \sim I/3\ . \tag{4.8}$$

I* is the effective Coulomb interaction. It is reduced by factor ~3. This is the reason why the superconductivity can arise alongside of the SDW phase which is realized by the sizable strength of Coulomb interaction.

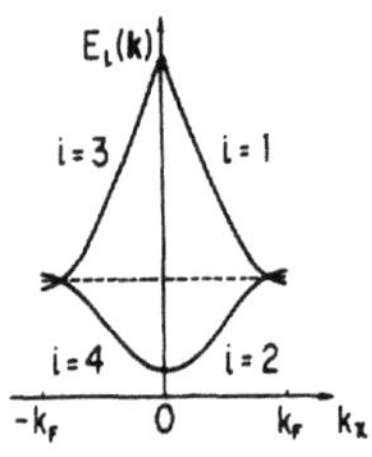

Fig. 13 Relabeled branches of the reorganized bands

We look for the appearance of superconductivity due to carriers, which exists in the SDW state of the present 2d system. Each branch is relabeled as in Fig. 13. The one-particle operators for the reorganized bands are given by the following orthogonal transformation:

$$\begin{pmatrix} c_{\vec{k}\sigma} \\ c_{\vec{k}-\vec{Q}\sigma} \end{pmatrix} = \begin{pmatrix} a(\vec{k}), & -\sigma b(\vec{k}) \\ \sigma b(\vec{k}), & a(\vec{k}) \end{pmatrix} \begin{pmatrix} \alpha_{2\vec{k}\sigma} \\ \alpha_{1\vec{k}\sigma} \end{pmatrix}\ , \tag{4.9}$$

where

$$a(\vec{k}) = [E_1(\vec{k}) - \varepsilon_{\vec{k}}]/\sqrt{[E_1(\vec{k}) - \varepsilon_{\vec{k}}]^2 + M^2}\ ,$$

$$b(\vec{k}) = M/\sqrt{[E_1(\vec{k}) - \varepsilon_{\vec{k}}]^2 + M^2} \ , \tag{4.10}$$

in the region $0<k_x<k_F$. In the region $-k_F<k_x<0$ the transformation is performed to get operators $\alpha_{3\vec{k}\sigma}$ and $\alpha_{4\vec{k}\sigma}$ by a similar matrix equation in which phase factors are chosen so that both operators become the same as $\alpha_{1\vec{k}+\vec{Q}\sigma}$ and $\alpha_{2\vec{k}+\vec{Q}\sigma}$, respectively, when $k_x=-k_F$. We assume that the superconducting order arises from the correlation of the type

$$<\alpha_{\bar{i}-\vec{k}\downarrow}\alpha_{i\vec{k}\uparrow}> \ . \tag{4.11}$$

Here the $\bar{i}$ band means the opposite band of i against $k_x=0$. We assume that our BCS-type interaction is invariant as an operator to the SDW formation, since the SDW gap, which is much smaller than the Debye energy, does not much modify the phonon-mediated electron-electron attraction. The operator can be transformed to the following effective one:

$$H_{super} = \sum_{i,j} \sum_{\vec{k},\vec{k}'} V_{ij}(\vec{k},\vec{k}')[<\alpha_{i\vec{k}\uparrow}^{\dagger}\alpha_{\bar{i}-\vec{k}\downarrow}^{\dagger}>\alpha_{\bar{j}-\vec{k}'\downarrow}\alpha_{j\vec{k}'\uparrow} + \alpha_{i\vec{k}\uparrow}^{\dagger}\alpha_{\bar{i}-\vec{k}\downarrow}^{\dagger}<\alpha_{\bar{j}-\vec{k}'\downarrow}\alpha_{j\vec{k}'\uparrow}>] \ . \tag{4.12}$$

The complicated explicit form of coefficient $V_{ij}(\vec{k},\vec{k}')$ is given in [7]. With the definition of the gap parameter as follows:

$$\Delta_i(\vec{k}) = - \sum_{j,\vec{k}} V_{ij} \ (\vec{k}, \vec{k}') <\alpha_{\bar{j}-\vec{k}'\downarrow}\alpha_{j\vec{k}'\uparrow}> \ , \tag{4.13}$$

the self-consistency equation for it is given by

$$\Delta_i(\vec{k}) = - \sum_{j=1}^{4} \sum_{\vec{k}'} V_{ij}(\vec{k},\vec{k}')F_i(\vec{k}')\Delta_j(\vec{k}') \ , \qquad i = 1,2,3,4, \tag{4.14}$$

where $$F_j(\vec{k}) = \tanh(\sqrt{E_j^2(\vec{k}) + |\Delta_j(\vec{k})|^2}/2k_BT)/2\sqrt{E_j^2(k) + |\Delta_j(\vec{k})|^2} \ . \tag{4.15}$$

On the other hand the self-consistency equation for the SDW gap parameter M is given by

$$I/4 \sum_i \sum_{\vec{k}} (-1)^i[(\varepsilon_{\vec{k}\pm Q}-\varepsilon_{\vec{k}})^2/4+M \ ]^{-1/2}[1-2E_i(\vec{k})F_i(\vec{k})] = 1 \ . \tag{4.16}$$

For the above system of equations we find two sets of solution as follows:

$$\Delta_{\bar{i}}(-\vec{k}) = \pm \ \Delta_i(\vec{k}) \ . \tag{4.17}$$

With the + sign chosen, we finally get

$$g_s \sum_{i=1}^{2} \sum_{o<k_x<k_F} \eta(\vec{k})[a^2(\vec{k}) - b^2(\vec{k})]^2F_i(\vec{k}) = 1 \ , \tag{4.18}$$

where $$\eta(\vec{k}) = \theta(\omega_D - |\varepsilon_{\vec{k}}|) \ ,$$

$$g_s = g + g' - 2I^* \ , \tag{4.19}$$

$$[a^2(\vec{k}) - b^2(\vec{k})]^2 = 1 - M^2/[(\varepsilon_{\vec{k}} - \varepsilon_{\vec{k}-\vec{Q}})^2/4 + M^2] \ . \tag{4.20}$$

This is reduced to the gap equation leading to (4.7) for the singlet superconductivity when M is reduced to zero. This shows that the effective superconducting coupling constant is diminished by (4.20). This is almost equal to zero in the vicinity of the Fermi surface. In addition, due to the small number of carriers in the SDW state, a finite transition temperature in the case of coexistence with SDW1 was not obtainable numerically as a finite value, although the case of SDW2 has not yet been examined. Evidently coexistence does not occur in the case of SDW3 [23], since its T_{SDW} is lower than $T_c^{(o)}$ in the absence of a magnetic field. With the similar situation in the case of SDW2 as in the case of SDW1, it looks unlikely that $T_c^{(2)}$ becomes higher than $T_c^{(1)}$ obtained in subsection 4.2. And so is the possibility that the coexistence state is the ground state. The present solution is the type whose existence was pointed out first by Baltensperger and Strässler [27]. And the diminishing factor mentioned (4.20) is in agreement with that given in the paper of J. Keller [28].

4.5 Case of Triplet Superconductivity

The other set of solution with the - sign in (4.17) gives the following reduced gap equation:

$$g_t \sum_{i=1}^{2} \sum_{o<k_x<k_F, k_y} \eta(\vec{k}) F_i(\vec{k}) = 1 , \tag{4.21}$$

where $g_t = g - g'$. (4.22)

In this case the diminishing factor in (4.20) does not appear. In the absence of the SDW gap, it gives the transition temperature

$$T_c^{(tr)} = 1.13 \exp[-1/N(0)g_t] . \tag{4.23}$$

Since in this case we have

$$<c_{-\vec{k}\downarrow} c_{\vec{k}\uparrow}> = - <c_{\vec{k}\downarrow} c_{-\vec{k}\uparrow}> , \tag{4.24}$$

this solution gives superconductivity with triplet spin pairing. Setting parameter values so that $T_c^{(tr)}$=1 K in the absence of SDW, the coupled equations for both the gap parameters were solved by assuming the fixed SDW wave vector. The second-order transition temperature $T_c^{(2)}$ at which we get the infinitesimal superconducting gap and a finite SDW gap is obtained as in Fig. 11. But the solution having a finite superconducting gap was found to exist only on the upper side of this temperature and that its free energy is larger than that of the pure SDW state. So this coexistence solution is unstable. The reaction on the SDW gap parameter from the finite superconducting gap seems to increase the free energy. Therefore in this case the coexistence is not realized, either. Incidentally this shows the necessity to obtain the free energy as well as T_c in looking for the coexistence solution.

Next we check the condition on the parameters for the triplet superconductivity to be realized. For the solution of triplet superconductivity to exist it is necessary

$$g > g' . \tag{4.25}$$

The condition for this state to dominate over the singlet one is

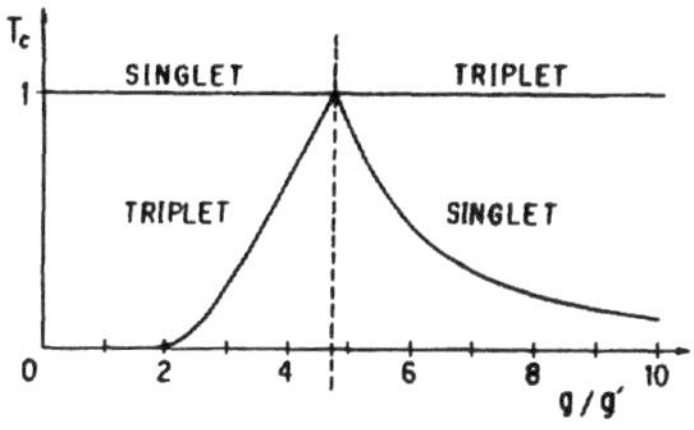

Fig. 14 The transition temperatures to the singlet and triplet superconducting states are plotted against the ratio g/g' of anisotropic BCS coupling constants. The higher one is fixed to 1 K

$$I^* > g' \ . \tag{4.26}$$

In Fig. 14 $T_c^{(s)}$ and $T_c^{(tr)}$ are shown as functions of ratio g/g'; they were calculated from two restrictions that the higher one between the two is 1 K and that T_{SDW} is fixed to 11.5 K with inclusion of the suppressive effect to SDW from the BCS Hamiltonian. $T_c^{(tr)}$ is higher than $T_c^{(s)}$ when

$$g/g' > 4.7 \ . \tag{4.27}$$

This anisotropy looks too large to be satisfied in the Bechgaard salts, although we cannot definitely deny the possibility until the main electron-phonon interaction driving superconductivity is clarified. But even in the case when the singlet one is realized, there is a possibility that the triplet one appears after the singlet one is artificially suppressed, e.g., by a magnetic field.

We have neglected the off-diagonal order of the type

$$\langle \alpha_{i-\vec{k}\downarrow} \alpha_{j\vec{k}\uparrow} \rangle \quad \text{with} \quad i \neq j \ . \tag{4.28}$$

This type of order is known to enhance superconductivity when one band is empty [29]. In the case of the inorganic magnetic superconductors the contribution of this type of correlation is known to be small [30].

Although the present analysis has still imperfection in the above point and in relation to SDW2, it is very plausible in the case of the Bechgaard salts that the coexistence solution has generally a lower transition temperature than the first-order transition temperature between the purely superconducting and SDW phases and that it is an unstable solution.

4.6 Comparison with Experiments

Although the resistive transition from the SDW to the superconducting phase has been observed to be gradual in most cases for the AsF_6 system under pressure, it is in some cases sharp enough to justify to take it as a first-order transition [31]. The apparent gradual transition may well occur due to inhomogeneity of pressure and also to the problems of the sample. This situation is reminiscent of the case of ClO_4, where the resistive transition is complicated with the problem of the anion ordering [21,32]. Here slowly cooled samples have the phase change between the normal and superconducting phases. Rapidly cooled ones have the SDW phase but no superconducting phase [33]. Intermediate samples have a gradual transition between the SDW and superconducting phases but it is now clear that both the phases are mixed, making small domains, and that the observed gradual transition is not intrinsic [34]. Since the randomness of anions in quenched samples is revealed [35] to arrest the contraction of the system at temperatures lower than the anion ordering temperature 24 K, the value of t_b of these samples must be smaller than that of annealed ones. The intruding region in Fig. 12 is considered to be situated between both values of t_b.

The resemblance of the curve of the first-order transition temperature vs t_b in Fig. 12 obtained in subsection 4.2 to the observed superconducting phase boundary for the AsF_6 system in Fig. 6 suggests to take t_b as representing the pressure. If we can attribute the pressure dependence mainly to t_b, as asserted here, the widths of intrusion in both scales give

$$dt_b/dP \sim 3\ k_B/\mathrm{Kbar}\ . \tag{4.29}$$

This is of the same order as the pressure dependence 14 k_B/Kbar expected from the extent of the related atomic orbitals [11] and by assuming the same compressibility as in TTF-TCNQ [36]. This is a fairly good agreement in view of the composite nature of t_b in (2.2). Thus the reentrant region is plausibly considered as a purely superconducting phase which just replaces the SDW phase.

Sensitivity to crystal defects of superconductivity in the Bechgaard salts is still a mystery [37,38]. This is a feature of triplet superconductivity. But there is an argument for singlet one based on the infrared presence of the Pauli limit to H_{c2} in the a direction [38]. Together with the triplet solution in the case where the singlet superconductivity is suppressed, the problem of triplet superconductivity seems to require more work.

5. Summary

On the basis of the 2d band model we obtained 3 types of SDW states, SDW1, SDW2 and SDW3. SDW1 and SDW2 impose an upper bound of transverse transfer energy t_b for their existence. Closeness of its estimated value to that of t_b of the Bechgaard salts strongly supported the viewpoint that SDW in the salts disappears under pressure or depending on the kind of anion since t_b exceeds the upper bound. The pressure dependence of T_{SDW} in the PT phase diagram of the SDW state was well understood from the theoretical results with a plausible assumption of sensitive pressure dependence of t_b. From a free energy consideration, the so-called reentrant region was found to be a purely superconducting phase intruding into the SDW phase. The possibility of the microscopic coexistence of the SDW and superconducting orders was examined with results showing its difficulty. The problem of triplet superconductivity was briefly discussed.

Acknowledgment The author heartily thanks Dr. J. Kondo and Dr. T. Ishiguro for helpful discussions and encouragement. He also acknowledges useful comments from and discussions with many other persons.

References

1. D. Jerome, A. Mazaud, M. Ribault and K. Bechgaard: J. Phys. (Paris) Lett. 41, L95 (1980)
2. K. Bechgaard: Mol. Cryst. Liq. Cryst. 79, 1 (1982)
3. D. Jerome: Mol. Cryst. Liq. Cryst. 79, 155 (1982)
4. E.W. Fenton: Lecture note on "Organic Superconductors" in these Proceedings
5. For recent developments see Proceedings of the Les Arcs Conference, 1982, published as J. Phys. (Paris) 44, C3 (1983)

6. K. Yamaji: J. Phys. Soc. Jpn. 51, 2787 (1982)
7. K. Yamaji: J. Phys. Soc. Jpn. 52, 1361 (1983)
8. K. Yamaji: talk at the autumn meeting of Phys. Soc. Jpn., Okayama, 1983; preprint
9. N. Thorup, G. Rindorf, H. Soling and K. Bechgaard: Acta Cryst. B 37, 1236 (1981)
10. N. Kinoshita, M. Tokumoto, H. Anzai, T. Ishiguro, G. Saito, T. Yamabe and H. Teramae: to be published in J. Phys. Soc. Jpn.; for the essentially similar level distribution of TTF see I.P. Batra, B.I. Benett and F. Herman: Phys. Rev. B11, 4927 (1975)
11. P.G. Grant: J. Phys. (Paris) 44, C3-847 (1983)
12. P.M. Grant: Phys. Rev. Lett. 50, 1005 (1983)
13. J.F. Kwak: Phys. Rev. B 26, 4789 (1982)
14. M.-H. Whangbo, W.M. Walsh, Jr., R.C. Haddon and F. Wudl: Solid State Commun. 43, 637 (1982)
15. B. Horovitz, H. Gutfreund and M. Weger: Phys. Rev. B 12, 3174 (1975)
16. B. Brusetti, M. Ribault, D. Jerome and K. Bechgaard: J. Phys. (Paris) 43, 801 (1982)
17. B. Morosin, J. E. Schirber, R. L. Greene and E. M. Engler: Phys. Rev. B26, 2660 (1982)
18. (a) B. Horovitz, H. Gutfreund and M. Weger: Solid State Commun. 39, 541 (1981); (b) V.J. Emery, R. Bruinsma and S. Barisic: Phys. Rev. Lett. 48, 1039 (1982)
19. (a) J.F. Kwak, J.E. Schirber, R.L. Greene and E.M. Engler: Phys. Rev. Lett. 46, 1296 (1981); (b) J.F. Kwak: J. Phys. (Paris) 44, C3-839 (1983)
20. (a) H. Bando, K. Oshima, M. Suzuki, H. Kobayashi and G. Saito: J. Phys. Soc. Jpn. 51, 2711 (1982); (b) K. Kajimura, H. Tokumoto, M. Tokumoto, K. Murata, T. Ukachi, H. Anzai, T. Ishiguro and G. Saito: Solid State Commun. 44, 1573 (1982); (c) R. Brusetti, K. Bechgaard, G.G. Lonzarich and R.H. Friend: J. Phys. (Paris) 44, C3-1055 (1983)
21. T. Takahashi, D. Jerome and K. Bechgaard: J. Phys. (Paris) 44, C3-805 (1983)
22. (a) P.M. Chaikin, M.-Y. Chai, J.F. Kwak, J.S. Brooks, K.P. Martin, M.J. Naughton, E.M. Engler and R.L. Greene: preprint; (b) M. Ribault, D. Jerome, J. Tuchendler, C. Weyl and K. Bechgaard: J. Phys. (Paris) Lett. 44, 953 (1983); (c) K. Oshima, M. Suzuki and T. Takayama: talk at the autumn meeting of Phys. Soc. Jpn., Okayama, 1983
23. K. Machida: J. Phys. Soc. Jpn. 50, 2195 (1981); K. Machida and T. Matsubara: ibid. 3231
24. E.W. Fenton and G.C. Psaltakis: Solid State Commun. 45, 5 (1982)
25. K. Murata, H. Anzai, K. Kajimura, T. Ishiguro and G. Saito: Mol. Cryst. Liq. Cryst. 79, 283 (1982)
26. P. Garoche, R. Brusetti and K. Bechgaard: Phys. Rev. Lett. 49, 1346 (1982)
27. W. Baltensperger and S. Strässler: Phys. Kondens. Mat. 1, 20 (1963)
28. J. Keller: J. Magn. Magn. Mat. 28, 193 (1982)
29. J. Kondo: Progr. Theor. Phys. 29, 1 (1963)
30. K. Levin: these Proceedings
31. D. Jerome: private communication
32. J.P. Pouget, G. Shirane, K. Bechgaard and J.M. Fabre: Phys. Rev. B 27, 5203 (1983)
33. M. Ribault: J. Phys. (Paris) 44, C3-827 (1983)
34. T. Ishiguro, K. Murata, K. Kajimura, N. Kinoshita, H. Tokumoto, M. Tokumoto, T. Ukachi, H. Anzai and G. Saito: J. Phys. (Paris) 44, C3-831 (1983); H. Schwenk, K. Andres and F. Wudl: Phys. Rev. B27, 5846 (1983)

35. S. Kagoshima, T. Yasunaga, T. Ishiguro, H. Anzai and G. Saito: Solid State Commun. 46, 867 (1983)
36. D. Debray, R. Millet, D. Jerome, S. Barisic, L. Giral and J.M. Fabre: J. Phys. (Paris) Lett. 38, 227 (1977)
37. S. Bouffard, M. Ribault, R. Brusetti, D. Jerome and K. Bechgaard: J. Phys. C15, 2951 (1982); S. Tomic, D. Jerome, D. Mailly, M. Ribault and K. Bechgaard: J. Phys. (Paris) 44, C3-1075 (1983)
38. M.-Y. Choi, P.M. Chaikin, P. Haen and R.L. Greene: Solid State Commun. 41, 225 (1982)

Theory of Infrared Conductivity from Density Waves: $(TMTSF)_2ClO_4$

E.W. Fenton

National Research Council of Canada, Ottawa, Canada K1A 0R6

In Fig. 1 I show schematically the real part of the conductivity below about 50 wavenumbers obtained by NG, TIMUSK, and BECHGAARD [1] for $(TMTSF)_2ClO_4$ at 2K and zero pressure. Similar results were obtained by CHALLENER, RICHARDS and GREENE [2]. When I first saw these results late in 1982, we believed that at this temperature the specimen was in a spin density wave (SDW) state. The sharp peak at 7 cm^{-1} and the asymmetric peak commencing slightly above 20 cm^{-1} suggest the sharp phonon peaks and asymmetric single-particle peak in Re$\sigma(\omega)$ for $TEA(TCNQ)_2$ explained by the theory of Michael RICE and colleagues [3,4]. In the Rice theory, intramolecular vibrational modes which are infrared inactive for a normal metallic or insulating state become active in the presence of a charge density wave (CDW), and the phonon states with wavevector equal to the CDW wavevector $\underline{Q}$ and with frequency in the gap for electron states are very sharp. The problem appeared then to be that we required a theory for a similar effect arising from a SDW in $(TMTSF)_2ClO_4$.

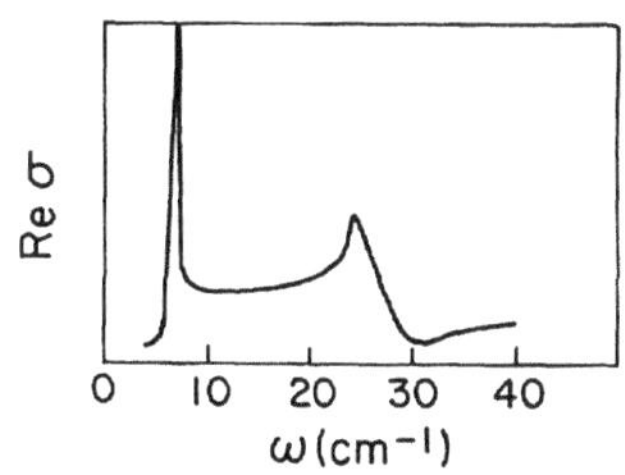

Fig. 1 Conductivity versus wavenumber (schematic) observed for $(TMTSF)_2ClO_4$ at 2K with $\underline{E} \parallel \hat{a}$

For a SDW or for a hypothetical CDW in a rigid crystal lattice (Overhauser CDW), the single-particle optical absorption commencing at the gap for electron states due to the density wave, 2Δ, is exactly cancelled by the absorption due to the collective mode effects of the electromagnetic field on the density wave, according to general arguments by HOPFIELD [5] and by FEDDERS and MARTIN [6]. The response diagrams for the single-particle conductivity $\sigma_{s.p.}$ and collective mode conductivity $\sigma_{c.m.}$ are shown in Figs. 2(a) and 2(b). For a CDW in a non-rigid crystal lattice, LEE, Maurice RICE and ANDERSON [7] showed that this collective-mode contribution to the absorption does not completely cancel the single-particle absorption and an asymmetric absorption peak commences at 2Δ. Later Michael Rice showed that many normal-mode distortions (phonon states) of the lattice at wavevector $\underline{Q}$ of the CDW are involved in the CDW state [3]. For organic conductors, where the number of atoms per unit cell is very large, the number of such normal modes is large. If there is an "external" potential $V(\underline{Q})$ due to ordering of counter-ions or a commensurability potential $V(\underline{Q})$ for higher order commensurability $\underline{G} = n\underline{Q}$, where $\underline{G}$ is a reciprocal lattice vector of the crystal in the absence of the density wave, and n is an integer greater than two, then sharp phonon peaks may occur in the gap, $0 < \omega < 2\Delta$. Roughly speaking, if the frequency $\omega_Q^{(u)}$ for the uth phonon mode of the normal metal with wave vector $\underline{Q}$ is in the gap, then it becomes optically active and appears in $\mathrm{Re}\sigma(\omega)$ for the CDW state at a still lower position within the gap, depending on the magnitude of $\frac{V}{\Delta} < 1$. It has narrow linewidth because electron-hole emission broadening does not occur within the gap 2Δ.

A SDW does not exist without $n\underline{Q}$-wavevector harmonics, simply because when an electron state at wavevector $\underline{k}$ is correlated with another state at wavevector $\underline{k} + \underline{Q}$ by the SDW ordering, then in general an electron state at wavevector $\underline{k} - \underline{Q}$ is also correlated with state k. Since any two states correlated with a third state are also correlated with each other, the states at wavevectors $\underline{k} - \underline{Q}$ and $\underline{k} + \underline{Q}$ must also be correlated with each other and this correlation is at wavevector $2\underline{Q}$, and so on for $n\underline{Q}$. A $\underline{Q}$-wave SDW correlation with amplitude and correlation energy means a $2\underline{Q}$ (and $n\underline{Q}$) correlation must occur with amplitude but not necessarily with any significant correlation energy, depending on whether or not a non-negligible correlation interaction exists for $2\underline{Q}$ (and $n\underline{Q}$).

The $n\underline{Q}$ harmonics of a SDW where n is even are in fact CDW amplitudes, which can be seen by essentially just squaring the SDW spin-space function to yield a diagonal matrix in spin space. For each such CDW harmonic, electron-phonon coupling exists for a non-rigid lattice, and the same general lattice relaxation mechanism causing infrared conductivity from a CDW fundamental must also cause infrared conductivity from a SDW fundamental plus harmonics, although it may be of quite different form depending on the ratio of harmonic CDW amplitudes to the SDW fundamental amplitude.

We now require the electron Green function including effects of harmonics of the SDW. (In the theories of LEE, RICE, and ANDERSON [7] and RICE [3], only the effects of the fundamental of the density wave were considered.) When the SDW is incommensurate, then with no point $\underline{r}$ in the crystal exactly the same as point $\underline{r}'$, the unit cell is the entire volume of the crystal. In this case, repeated Brillouin zones in k space are all inequivalent and the electron Green function compared to that of the normal metal state is an infinite-order matrix on a two-fold wavevector basis $k + n\frac{Q}{2}$, $k + m\frac{Q}{2}$, where n and m are odd positive or negative integers. We represent the components of this matrix by $G_{mn}(k,\omega)$, where $|k| < k_f$ in a quasi-one-dimensional band structure and m,n = ±, 3±, 5±, ... represents $\frac{\pm Q}{2}$, $\frac{\pm 3Q}{2}$, $\frac{\pm 5Q}{2}$, $\frac{\pm 7Q}{2}$, etc., in $G(k + \frac{Q}{2}; k + \frac{Q}{2}; \omega)$, $G(k + \frac{Q}{2}; k - \frac{Q}{2}; \omega)$, etc. The matrix Green function is of the form

$$\underline{\underline{G}} = \begin{bmatrix} G_{-\infty,-\infty} & \cdots & & \cdots \\ \vdots & \ddots & G_{--}\,G_{-+} & \vdots \\ \vdots & & G_{+-}\,G_{++} & \vdots \\ \vdots & & \ddots & \vdots \\ \cdots & & \cdots & G_{\infty,\infty} \end{bmatrix}$$

where the central 2×2 block is the only part non-zero in earlier theories for $\sigma(\omega)$ [3,7].

For a number of problems discussed in the literature, effects of harmonics of a density wave condensate have been considered by allowing more than the central 2×2 block of the matrix $\underline{\underline{G}}(k,\omega)$ to be non-zero, e.g. 3×3, 4×4, or 5×5 matrices for $\underline{\underline{G}}$ have been used. As we have discussed, at least for crystals where the normal state has central symmetry, any odd-order square matrix is unphysical [8]. In addition, to our knowledge no justification or consistent scheme has been developed for ignoring all other m, n elements outside of the central square. I have tried to deal with a $\underline{\underline{G}}$ matrix restricted to 4×4 or 6×6 blocks with all other elements arbitrarily set to zero. The resulting theory is far too cumbersome to be used for calculating the infrared conductivity, and moreover the block chosen is purely an arbitrary choice.

A simple scheme is possible for including effects of SDW harmonics which is exact in the limit where the ratio of CDW $2\underline{Q}$ harmonic amplitude (modulus) to the $1\underline{Q}$ SDW fundamental amplitude (modulus) goes to zero, $\frac{<2Q>}{<1Q>} \to 0$, and which is representative of the physics for an actual system when $\frac{2\Delta}{\varepsilon_f} \ll 1$. We take an electron band structure with the nesting condition $\varepsilon_{k+\frac{Q}{2}} = -\varepsilon_{k-\frac{Q}{2}} = v_f k$ with $Q = 2k_f$ for a quasi-one-dimensional band structure or for region 1 of a two-region Fermi surface in which the SDW annihilates or gaps all of a flat region 1 at temperatures far below the SDW transition temperature T_{SDW}. We take $\varepsilon_{k\pm\frac{3Q}{2}} = \varepsilon_{k\pm\frac{Q}{2}} + \varepsilon_Q \equiv \varepsilon_{k^\pm} + \varepsilon_Q$, where $\varepsilon_Q > 2\varepsilon_f$, and we take $\varepsilon_{k+n\frac{Q}{2}}$ infinite for $n = \pm 5, \pm 7, \pm 9 \ldots$ The usual theories with no harmonics present are equivalent to also taking $\varepsilon_Q \to \infty$ and $\varepsilon_{k\pm\frac{3Q}{2}} \to \infty$. As will be seen, the $G_{mn}(k,\omega)$ are all zero if either one of $|m|$ or $|n|$ is greater than 3. As will be seen, all of the $G_{mn}(k,\omega)$ with $|m|, |n| < 3$ can be expressed in terms of the $G^{(0)}_{mn}(k,\omega)$ of Lee, Rice, and Anderson, where $|m|$ and $|n|$ are both unity (m or n taken as simply + or −).

The Hamiltonian is

$$H = \sum_{k,\alpha} \varepsilon_k c^+_{k\alpha} c_{k\alpha} - \frac{V_{SDW}}{2N} \sum_{\substack{k,k',q \\ \alpha,\beta}} c^+_{k-\frac{q}{2}\beta}\, c_{k+\frac{q}{2}\alpha}\, c^+_{k'+\frac{q}{2}\alpha}\, c_{k'-\frac{q}{2}\beta}$$

$$+ \sum_{q,u} \omega^{(u)}_q b^+_{uq} b_{uq} + \sum_{q,k,\alpha,u} \frac{ig_u}{N^{\frac{1}{2}}} c^+_{k+q\alpha} c_{k\alpha} (b_{uq} + b^+_{u-q}) \tag{1}$$

where N is the number of atoms, k, k' and q are wavevector components parallel to the highly conducting axis, α and β denote spin states, and u is the index of the phonon band. The direct coulomb interaction appears in screening of the exchange potential V_{SDW}, averaged over k and k', and in renormalization of coupling g_u and frequency $\omega^{(u)}_q$ for the phonons, which are discussed by Michael RICE [3]. For a SDW state, there is a mean-field part of H given by

$$H_{m.f.} = -\sum_{k,\alpha,\beta} M(\underline{\sigma}\cdot\underline{n})_{\alpha\beta}\, c^+_{k-\frac{Q}{2}\alpha}\, c_{k+\frac{Q}{2}\beta}$$

$$- \sum_{k,\alpha} \Gamma\left(c^+_{k-\frac{3Q}{2}\alpha}\, c_{k+\frac{Q}{2}\alpha} + c^+_{k-\frac{Q}{2}\alpha}\, c_{k+\frac{3Q}{2}\alpha}\right) + \text{h.c.} \tag{2}$$

where

$$M = \frac{V_{SDW}}{2N} \sum_{k',\alpha,\beta} (\underline{\sigma}\cdot\underline{n})_{\alpha\beta} \langle c^{+}_{k'+\frac{Q}{2}\,\alpha}\, c_{k'-\frac{Q}{2}\,\beta} \rangle \quad \text{and} \tag{3}$$

$$\Gamma = \frac{1}{N} \sum_{k',\alpha,u} \frac{g_u^2}{\omega_{2Q}^{(u)}} \langle (c^{+}_{k'+\frac{3Q}{2}\,\alpha}\, c_{k'-\frac{Q}{2}\,\alpha} + c^{+}_{k'+\frac{Q}{2}\,\alpha}\, c_{k'-\frac{3Q}{2}\,\alpha}) \rangle . \tag{4}$$

Including only the kinetic energy and $H_{m.f.}$ parts of H, equations of motion for the electron field operators with electron energy $|\omega| < E_f$ lead to electron Green functions $G_{mn}(k,\omega)_{\alpha\beta} \equiv \langle\langle c_{k+\frac{mQ}{2}\,\alpha}; c^{+}_{k+\frac{nQ}{2}\,\beta}\rangle\rangle_{\omega}$ which are zero unless $m,n = \pm1, \pm3$. We define $G_{--}(k,\omega)_{\alpha\beta} = (\underline{1})_{\alpha\beta} G^{DIA}_{--}(k,\omega)$; $G_{+-}(k,\omega)_{\alpha\beta} = (\underline{\sigma}\cdot\underline{n})_{\alpha\beta} G^{SDW}_{+-}(k,\omega)$; $G_{3+,-}(k,\omega)_{\alpha\beta} = (\underline{1})_{\alpha\beta} G^{CDW}_{3+,-}(k,\omega)$; $G_{3+,3-}(k,\omega)_{\alpha\beta} = (\underline{\sigma}\cdot\underline{n})_{\alpha\beta} G^{3QSDW}_{3+,3-}(k,\omega)$; etc., and in the equations of motion expand

$$\frac{M}{\omega-\varepsilon_{k^{\pm}}-\varepsilon_Q} = -\frac{M}{\varepsilon_Q} - \frac{M(\omega-\varepsilon_{k^{\pm}})}{\varepsilon_Q^2} + \ldots \qquad \frac{\Gamma}{\omega-\varepsilon_{k^{\pm}}-\varepsilon_Q} = \frac{-\Gamma}{\varepsilon_Q} + \ldots$$

including terms only to order ε_Q^{-2} near the limit $\varepsilon_Q \to \infty$. ($\Gamma \propto \varepsilon_Q^{-1}$). We obtain:

$$G^{DIA}_{--} = \left(1 + \frac{|M|^2}{\varepsilon_Q^2}\right)^{-1} \frac{(\omega+\varepsilon_{k^-})}{(\omega-\varepsilon_{k^-})(\omega+\varepsilon_{k^-})-|\Delta|^2} \equiv \left(1 + \frac{|M|^2}{\varepsilon_Q^2}\right)^{-1} G^{(0)}_{--}\Big|_{M\to\Delta}$$

$$G^{SDW}_{+-} = \left(1 + \frac{|M|^2}{\varepsilon_Q^2}\right)^{-1} \frac{(-\Delta^*)}{(\omega-\varepsilon_{k^-})(\omega+\varepsilon_{k^-})-|\Delta|^2} \equiv \left(1 + \frac{|M|^2}{\varepsilon_Q^2}\right)^{-1} G^{(0)}_{+-}\Big|_{M\to\Delta}$$

$$G^{SDW}_{3-,-} = \frac{M}{\varepsilon_Q} G^{DIA}_{--} \quad ; \quad G^{CDW}_{3+,-} = \frac{M^*}{\varepsilon_Q} G^{SDW}_{+-}$$

$$G^{DIA}_{3-,3-} = \frac{|M|^2}{\varepsilon_Q^2} G^{DIA}_{--} \quad ; \quad G^{3QSDW}_{3+,3-} = \frac{|M|^2}{\varepsilon_Q^2} G^{SDW}_{+-} \tag{5}$$

and similar relations for other G_{mn}. $G^{(0)}_{mn}$ is the (standard) solution with $\varepsilon_Q \to \infty$ and no harmonics present [7]. In (5), an energy term $|M|^2/\varepsilon_Q$ has been incorporated into a shifted chemical potential. The SDW gap function with harmonics present is

$$\Delta \equiv M\left(1 + \frac{2|\Gamma|}{\varepsilon_Q}\right)\left(1 + \frac{|M|^2}{\varepsilon_Q^2}\right)^{-1} = \left(V_{SDW} + \sum_u \frac{2g_u^2 a^2}{\omega_{2Q}^{(u)}}\right) \frac{k_B T}{\left(1 + \frac{a^2}{4}\right)} \sum_{\omega,k} G_{-+}(k,\omega) \tag{6}$$

where M is the magnitude of Δ in the limit $\varepsilon_Q \to \infty$ where the 2Q CDW amplitude is zero. To first order in ε_Q^{-1}, the ratio of 2Q CDW amplitude to 1Q SDW amplitude is the harmonic amplitude parameter $a = \frac{2|M|}{\varepsilon_Q}$. At T = 0, the gap equation (6) becomes

$$1 = \lambda \, \ell n \frac{2E_f}{|\Delta|} \qquad \text{where}$$

$$\lambda = \left(1 + \frac{a^2}{4}\right)^{-2} N(0)V_{SDW} + \left(1 + \frac{a^2}{4}\right)^{-2} N(0) \sum_u \frac{2g_u^2 a^2}{\omega_{2Q}^{(u)}}$$

$$\equiv \lambda_{SDW} + \sum_u \lambda_u \tag{7}$$

with N(0) the one-spin density of electron states at the Fermi surface (f.s.).

The electron number n obtained using the $G_{mm}(k,\omega)$ of (5) with m = ±1,±3 is exactly $n^{(o)}$ of the usual theory with harmonics absent, with $M \rightarrow \Delta$. The single-particle contribution to the conductivity near T = 0 using all the diagrams of Fig. 2(a) with m,n = ±1,±3 and the G_{mn} of (5) is exactly

$$\sigma_{s.p.}(q{=}0,\omega) = \frac{ne^2}{i\omega m}\left[f\left(\frac{\omega}{2\Delta}\right) - f(0)\right]$$

$$= \sigma_{s.p.}^{(0)}(0,\omega)\Big|_{M \rightarrow \Delta} \qquad \text{where} \tag{8}$$

$$f(\tilde{\omega}) = \left\{\pi i + \ell n[(1-S)/(1+S)]\right\}/(2S\tilde{\omega}^2)$$

in which $S = (1-\tilde{\omega}^{-2})^{\frac{1}{2}}$ and f(0) = 1. m is the electron band-structure mass. The same expression for $\sigma_{s.p.}$ was obtained for the CDW by LEE, RICE, and ANDERSON [7] and by Michael RICE [3].

The fact that $G_{mn}(k,\omega)$ are functions only of the harmonic amplitude parameter a discussed above and the $G_{mn}^{(0)}(k,\omega)$ functions for the SDW without harmonics will ultimately mean that the collective mode conductivity represented by Fig. 2(b) with m,n',p',r = ±1,±3 and n,p = ±1,±2 can be determined in analytic form similar to existing theory for the CDW. The Dyson equation for the collective-mode propagator $D_{np}^{uv}(q,i\omega_0)$ at centre in the diagram of Fig. 2(b) (with wavevector q+nQ at one end and q+pQ at the other end) can be expressed as

$$\sum_{u,v} g'_u g'_v D_{mn}^{uv}(q,i\omega_0) = \delta_{m,n}\left\{\sum_{u=0}^{max} g'^2_u D_0^{uu}(i\omega_0)\right\} + \left\{\sum_{u=0}^{max} g'_u D_0^{uu}(i\omega_0)\right\} \times \left\{\left[\sum_r \pi_{m,-m;r,-r}(q,i\omega_0)\right]\left[\sum_{w,v=0}^{max} g'_w g'_v D_{rn}^{wv}(q,i\omega_0)\right]\right\}\frac{1}{g^2} \tag{9}$$

(a)

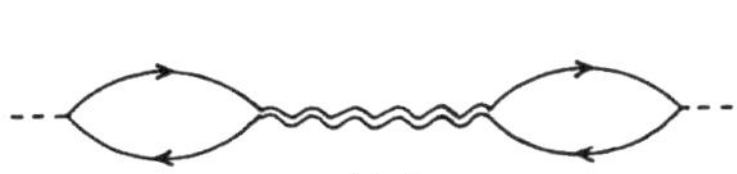

(b)

Fig. 2 Conductivity due to (a) single-particle excitations and (b) collective mode excitations

where

$$\pi_{m',m'';n',n''}(q,i\omega_0) = \frac{2g^2k_BT}{N} \sum_{k,\omega_n} G_{m'n'}(k+q,i\omega_n+i\omega_0)G_{n''m''}(k,i\omega_n)$$

and

$$g_0' = 1;\; g'_{u\neq 0} = ag_u;\; g^2 = \sum_{u=0}^{max} g_u'^2 \;.$$

The bare collective-mode propagators are

$$D_0^{00}(i\omega_0) = \frac{V_{SDW}}{2}\;;$$

$$D_0^{uu}(i\omega_0)\Big|_{u\neq 0} = \frac{\omega_{2Q}^{(u)}}{(i\omega_0)^2-\omega_{2Q}^{(u)2}+(i\omega_0)i\gamma_u}$$

for the SDW mode and for phonons with natural-linewidth parameters γ_u. The spin-space structure in the problem results in $V_{SDW} \to \frac{1}{2}V_{SDW}$ for D_0^{00} and otherwise can be incorporated in the g_u', and in Fig. 2(b) results finally in diagonal matrices in Pauli spinor space.

We define $\underline{D}$, $\underline{D}_0$, and $\underline{\pi}$ matrices in a 2×2 space for m,n = ±1, representing the entire ±1,±3 space with $g'_{u\neq 0}$ for ±3 different from g_0' for ±1. The matrix elements are

$$D_{mn} = \sum_{u,v=0}^{max} g_u'g_v'D_{mn}^{uv};\; (D_0)_{mn} = \delta_{m,n}\sum_{u=0}^{max} g_u'^2 D_0^{uu};\; \pi_{mn} = \pi_{m,-m;n,-n}\;.$$

Defining

$$\pi^{\pm} = \pi_{+-,+-} \pm \pi_{+-,-+} \quad \text{and}$$

$$g^2D^{\pm} = \sum_{u,v=0}^{max} g_u'g_v'(D_{++}^{uv} \pm D_{-+}^{uv}) \tag{10}$$

we solve the Dyson equation $\underline{D} = \underline{D}_0 + \underline{D}_0(\frac{1}{g^2}\underline{\pi})\underline{D}$ to obtain

$$D^-(q=0,i\omega_0) = \frac{\frac{\lambda}{g^2N(0)}\left(1+\frac{a^2}{4}\right)^2}{\left[\frac{-\lambda_{SDW}}{\lambda} + \sum_{u=1}^{max}\left(\frac{\lambda_u}{\lambda}\right)\frac{\omega_{2Q}^{(u)2}}{(i\omega_0)^2-\omega_{2Q}^{(u)2}+(i\omega_0)(i\gamma)}\right]^{-1} + 1 + \lambda\left(\frac{i\omega_0}{2\Delta}\right)^2 f\left(\frac{i\omega_0}{2\Delta}\right)} \tag{11}$$

where we have dropped the modulus on $|\Delta|$. The mode D^- is a phase fluctuation mode and is optically active. The mode D^+ is an amplitude fluctuation mode which we have shown to be optically inactive.

For each (u,v) component of D^- in (10), four electron-line pairs in Fig. 2(b) couple either the u or v end to one of the external photon dotted lines, because for both lines of an electron-line pair mQ/2 occurs at the photon with m = ±1,±3. Summation over the (u,v) components with the total coupling-to-photon for each u and v (using the G_{mn} of (5)) leads to an expression for the collective mode conductivity at T = 0:

$$\sigma^{SDW}_{c.m.}(q=0,\omega) = \frac{ine^2}{\omega m}\, g^2 N(0)\left(1+\frac{a^2}{4}\right)^{-2}\left(\frac{\omega}{2\Delta}\right)^2 f^2\left(\frac{\omega}{2\Delta}\right) D^-(0,\omega)$$

$$= \frac{-ne^2}{i\omega m}\left[\frac{\lambda\left(\frac{\omega}{2\Delta}\right)^2 f^2\left(\frac{\omega}{2\Delta}\right)}{\left[\frac{-\lambda_{SDW}}{\lambda} + \sum_{u=1}^{max}\frac{\lambda_u}{\lambda}\frac{\omega_{2Q}^{(u)2}}{\omega^2-\omega_{2Q}^{(u)2}+i\omega\gamma_u}\right]^{-1} + 1 + \lambda\left(\frac{\omega}{2\Delta}\right)^2 f\left(\frac{\omega}{2\Delta}\right)}\right] \tag{12}$$

where $\lambda = \lambda_{SDW} + \sum_{u=1}^{max}\lambda_u$. The SDW coupling constant λ_{SDW} and λ_u coupling constants for the phonons are defined in (7). The effective coupling constant for phonons, $\lambda_u = \lambda^{ph}_{u,2Q}a^2(1+\frac{a^2}{4})^{-2}$, is generally reduced from $\lambda^{ph}_{u,Q} = 2N(0)g_u^2/\omega_Q^{(u)}$ for the CDW case considered by Rice.

The expression for $\sigma^{SDW}_{c.m.}$ cancels exactly with $\sigma_{s.p.}$ of (8) for $\omega \neq 0$ if the harmonic amplitude parameter a is zero and/or all the $\lambda^{ph}_{u,2Q}$ are zero. The result we have obtained for the SDW with small-amplitude harmonics in a clean crystal is also obtained including Umklapp vectors of the host crystal, and also for large harmonic amplitude in an approximation where 2Q electron correlations are distributed among electron states in the same way as the 1Q correlations, corresponding to $G^{CDW} \propto a\, G^{SDW}$ in (5). However the theory is in all cases a weak-coupling theory, strictly speaking valid only when the coulomb interaction u and electron bandwidth 4t satisfy $\frac{u}{4t} \ll 1$. (For $(TMTSF)_2AsF_6$ at least [9], $\frac{u}{4t} < 1$.)

The result for the CDW at T = 0 obtained by RICE [3], including effects of an external potential with amplitude V and wavevector Q due to counter-ion ordering, is

$$\sigma^{CDW}_{c.m.}(q=0,\omega) = \frac{-ne^2}{i\omega m}\left[\frac{\lambda\left(\frac{\omega}{2\Delta}\right)^2 f^2\left(\frac{\omega}{2\Delta}\right)}{\left[\sum_{u=1}^{max}\frac{\lambda^{ph}_u}{\lambda}\frac{\omega_Q^{(u)2}}{\omega^2-\omega_Q^{(u)2}+i\omega\gamma_u}\right]^{-1} + 1 - \frac{V}{\Delta} + \lambda\left(\frac{\omega}{2\Delta}\right)^2 f\left(\frac{\omega}{2\Delta}\right)}\right] \tag{13}$$

with in this case $\lambda = \sum_{u=1}^{max}\lambda^{ph}_{u,Q}$. The gap equation for Δ in the CDW and $\frac{V}{\Delta} < 1$ are discussed by Rice. The total density-wave conductivity is $\sigma = \sigma_{s.p.} + \sigma_{c.m.}$ using (8) with (12) or (13).

Collective mode frequencies with only one phonon coupled to the density wave, obtained from zeros of the denominator in (11) (and of an analogous function for the CDW case), are

SDW: $\omega_1^2 = 0$

$$\omega_2^2 = \omega_{2Q}^2\left(\frac{\lambda}{\lambda_{SDW}}\right)+4\Delta^2\left(\frac{a^2\lambda_{ph}(1 + \frac{a^2}{4})^{-2}}{\lambda\lambda_{SDW} f\left(\frac{\omega_2}{2\Delta}\right)}\right) > \omega_{2Q}^2$$

$$\text{CDW: } \omega_1^2 = \omega_Q^2\left[\frac{V/\Delta}{1 + \frac{\lambda^{ph}\omega_Q^2}{4\Delta^2} f\left(\frac{\omega_1}{2\Delta}\right)}\right] < \omega_Q^2 .$$

The inequalities apply for $\omega_i < 2\Delta$ where f is real and positive. In Fig. 3(a) we show the infrared conductivity for the incommensurate CDW with counter-ion ordering and one low-frequency Q-wave phonon coupled to the density wave, for parameters $\frac{\omega_Q}{2\Delta} = 0.5$, $\frac{\gamma}{2\Delta} = 10^{-2}$, $\lambda = \lambda^{ph} = 0.2$, $\frac{V}{\Delta} = 0.5$ and with $f\left(\frac{\omega}{2\Delta}\right) \rightarrow f\left(\frac{\omega}{2\Delta} + 10^{-2}i\right)$(to represent some smearing of f due to inhomogeneity of the imperfect crystal and other effects on the electron lifetime). The absorption peak at $\frac{\omega}{2\Delta} \approx 0.3$ is due to the phonon, and the single-particle peak commences at $\frac{\omega}{2\Delta} = 1.0$. In Fig. 3(b) we show conductivity for the incommensurate SDW with one 2Q phonon coupled and for $\frac{\omega_{2Q}}{2\Delta} = 0.5$, $\frac{\gamma}{2\Delta} = 10^{-2}$, $\lambda = 0.2$, $\lambda_{ph}a^2(1 + \frac{a^2}{4})^{-2} = 0.05$, with $f\left(\frac{\omega}{2\Delta} + 10^{-2}i\right)$.

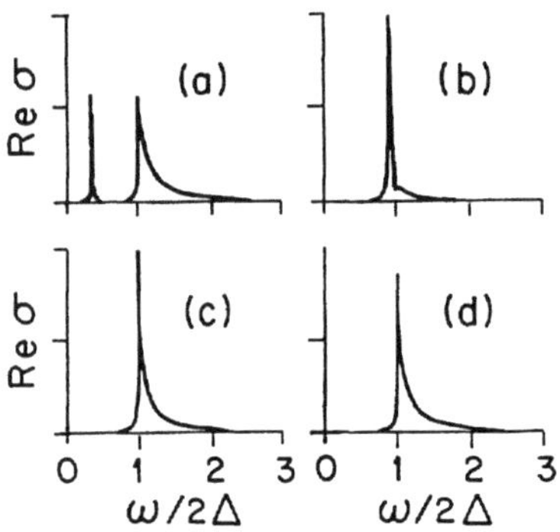

Fig. 3 Real part of the conductivity versus frequency. Examples for: (a) incommensurate charge density wave; (b) incommensurate spin density wave; (c) incommensurate spin density wave with larger harmonic amplitude; and (d) commensurate spin or charge density wave with rigid density wave locked to the host lattice. Conductivity is in units of $\frac{ne^2}{2\Delta m}$, and for parameters of the solid discussed in the text

The phonon peak and the single-particle peak are almost merged together at $\frac{\omega}{2\Delta} = 1$. The phonon linewidth parameter γ also broadens the peak at $\omega = 0$ (including the case with $f(\frac{\omega}{2\Delta})$). In Fig. 3(c) we show conductivity for the incommensurate SDW with the same parameters as in Fig. 3(b) except that $\lambda_{ph}a^2(1 + \frac{a^2}{4})^{-2} = 0.1$. Figure 3(c) is considerably different from the single-particle conductivity alone in Fig. 3(d) (for $f(\frac{\omega}{2\Delta} + 10^{-2}i)$), which represents the total conductivity for a commensurate SDW or CDW locked to the host lattice. We note that a peak at $\frac{\omega}{2\Delta} = 1$ of similar form to Fig. 3(c) for the SDW occurs also for the CDW if ω_Q for the phonon is in the low frequency range above 2Δ. However, absorption of the form in Fig. 3(a) distinguishes the incommensurate CDW with ordering of counter-ions, e.g. the relaxed state of $(TMTSF)_2ClO_4$, because for the incommensurate SDW a phonon peak this low in the gap below 2Δ (with $\lambda_{ph}a^2(1 + \frac{a^2}{4})^{-2}$ very small) does not occur along with an asymmetric single-particle peak of comparable oscillator strength commencing at $\frac{\omega}{2\Delta} = 1$.

$\sigma(q,\omega)$ with $q = 0$ discussed above and shown in Fig. 3 is appropriate for highly anisotropic conductors with the optical beam directed perpendicular to the highly conducting axis of the crystal and polarized parallel to that axis. For this $q = 0$ case, no infrared conductivity arises from a superconductor: there is no absorption edge at 2Δ. When $2\Delta \ll \omega_{Debye}$ and at frequencies much higher than 2Δ, optical absorption processes characteristic of the normal metal and not included in the theory above contribute strongly, e.g. the Holstein mechanism [10]. The conductivity derived above is applicable only for $\omega \lesssim 2\Delta$ (and of course with reduced oscillator strengths). When the f.s. is not entirely gapped by the SDW or CDW, (8), (12), and (13) describe the conductivity only for the gapped part of the f.s. with density of states $N_1(0) < N(0)$, and in this case these three conductivity expressions must each be reduced by a factor $N_1(0)/N(0)$. The remaining non-gapped fraction $N_2(0)/N(0)$ of the f.s. will contribute a conductivity term at low frequencies of the form

$$\sigma_2(\omega) = \frac{ne^2}{(\frac{1}{\tau} - i\omega)m} \frac{N_2(0)}{N(0)}$$

with $\tau \gg \tau_{normal\ metal}$ when $N_2(0) \ll N(0)$ because of the reduced number of final states at E_f for electron scattering.

When Dr. G.C. Psaltakis and I had finally obtained the results of (8) and (12) for the infrared conductivity due to a SDW with a CDW harmonic, we immediately asked Dr. G.C. Aers if he would put these results on our computer, which produces graphs of $Re\sigma(\omega)$ versus $\omega/2\Delta$ directly as in Fig. 3. No combination of parameters for the solid would produce a phonon peak below halfway in the gap with at the same time any appreciable amplitude remaining for the asymmetric peak commencing at $\omega/2\Delta = 1$. About this time we learned that Fig. 1 represents the infrared conductivity for a $(TMTSF)_2ClO_4$ specimen which was cooled slowly from room temperature to 2K, and in this case the SDW does not appear but instead the ClO^-_4 anions order as T decreases through a transition temperature of roughly 22K. In this case we expect the theory by Michael Rice for the CDW (no SDW) in the presence of counter-ion ordering to be descriptive, and of course results from his theory corresponding to Fig. 3(a) look much like the experimental $Re\sigma(\omega)$ spectrum in Fig. 1. (However Fig. 3(a) is relevant only to the region of the Fermi surface gapped by anion ordering, if such gapping occurs.) The wavevector of the anion ordering is $(0,\pi/b,0)$ [11]. Subsequently experiments were done by Tom Timusk and Hon Kie Ng at McMaster University for $(TMTSF)_2ClO_4$ specimens which were quench-cooled at the rate of 40-50K/minute. The experimental $Re\sigma(\omega)$ spectrum for the quench-cooled specimen looks very much like our theoretical result for the conductivity due to a SDW with a CDW harmonic, shown in Fig. 3(b). Of course it is now known that the SDW occurs at zero pressure in $(TMTSF)_2ClO_4$ only for specimens which have been rapidly cooled. If the experimental spectrum in Fig. 1 for the slowly cooled specimen arises from a CDW associated with anion ordering, this may explain a mysterious feature, which is that the two peaks are observed at temperatures as high as 30K, compared to 22K for the anion ordering transition. The explanation for this is as follows. The fluctuation temperature regime above a phase transition at T_c has magnitude $\frac{\delta T}{T_c} \sim z^{-1}$, where z is the coordination number (or number of nearest neighbours) according to arguments given long ago by V.L. Ginzburg. The coordination number for a superconductor in three dimensions with Cooper pair size so large that 10^7-10^9 pairs overlap is $z = 10^7$-10^9. The electron-hole pair of the itinerant-electron SDW condensate is also large and so for the SDW z is also large. However for anion ordering, the anion coordination number is $z < 10$ and we expect a large δT, of order unity times T_c. The $Re\sigma(\omega)$ spectra at 30K with peaks similar to those in Fig. 1 but with smaller amplitude may be due to the effects of short-range anion ordering at that temperature, precursor to the long-range ordering below T_c at 22K.

References

1. H.K. Ng, T. Timusk, and K. Bechgaard: J. Physique, to be published
2. W.A. Challener, P.L. Richards, and R.L. Greene: Bull. Am. Phys. Soc. 28, 445 (1983)
3. M.J. Rice: Phys. Rev. Lett. 37, 36 (1976)
4. M.J. Rice, L. Pietronero, and P. Bruesch: Solid State Comm. 21, 757 (1977)
5. J.J. Hopfield: Phys. Rev. 139, A419 (1965)
6. P.A. Fedders and P.C. Martin: Phys. Rev. 143, 245 (1966)
7. P.A. Lee, T.M. Rice, and P.W. Anderson: Solid State Comm. 14, 703 (1974)
8. E.W. Fenton and C.R. Leavens: J. Phys. F 10, 1853 (1980)
9. J.B. Torrance, H.J. Pedersen, and K. Bechgaard: Phys. Rev. Lett. 49, 881 (1982)
10. P.B. Allen: Phys. Rev. B3, 305 (1971)
11. J.P. Pouget, R. Moret, R. Comes, G. Shirane, K. Bechgaard and J.M. Fabre: J. Physique, to be published

Part IV

$BaPb_{1-x}Bi_xO_3$ and Some Metal Hydrides

$BaPb_{1-x}Bi_xO_3$ – An Example of the Strong Coupling Limit of Superconductivity?

T.M. Rice
Theoretische Physik ETH-Hönggerberg, CH-8093 Zürich, Switzerland

In the alloy series $Ba\,Pb_{1-x}\,Bi_x\,O_3$ the superconducting transition temperature rises with increasing x but for $x \gtrsim 0.3$ the alloys are semiconducting. An explanation of this transition at $x \simeq 0.3$ as a charge disproportionation into Bi^{3+} and Bi^{5+} ions, because of strong electron-phonon coupling, has been proposed but recent a priori electronic structure calculations question this explanation.

1. Introduction

The study of unusual superconductors has been an exciting topic in recent years with the discovery of ferromagnetic superconductors, heavy fermion superconductors and materials which are poor metals in the normal state but still high transition temperature superconductors. $Ba\,Pb_{1-x}\,Bi_x\,O_3$ with $x \simeq 0.25$ is such a material. The normal state resistivity is very high and it also shows unusual values of the parameters which characterize the superconducting state. In general high transition temperatures occur in metals with high values of the electronic density of states at the Fermi level. $Ba\,Pb_{1-x}\,Bi_x\,O_3$ is a clear exception with a low density of states although the compound with $x = 0.25$ has a value of T_c ($\simeq$ 12 K), which is the highest value for any material with an s-p derived valence band. Further, if x is increased beyond this value, instead of a continuing rise in T_c, there is an abrupt drop in $T_c(x)$ accompanied by a rise of the resistivity and a transition to insulating or semiconducting behavior which persists in the range $0.3 \leq x \leq 1$. There are other known examples of materials where crystallographic phase transitions cause abrupt drops in T_c. In this case there is a crystallographic phase transition but a priori electronic structure calculations show that the Fermi surface behavior is only slightly altered so that there is no obvious explanation for the disappearance of superconductivity. These calculations give a broad set of conduction bands and the coupling in these bands to the optic phonons is not particularly strong and the energy gaps, which they find in the distorted phase of $Ba\,Bi\,O_3$, are much smaller than the bandwidth. The charge disproportionation between inequivalent Bi atoms in $Ba\,Bi\,O_3$ is very small in these calculations. By contrast, several authors have proposed strong coupling models which assume a substantial charge disproportionation into Bi^{3+}-Bi^{5+} ionic states and propose to explain the origin of the insulating behavior over such wide range of x values as arising from such a local disproportionation.

In this lecture first the salient experimental features will be reviewed. Then the strong coupling models that have been proposed will be discussed, followed by a review of the results of the a priori electronic structure calculations. The last section summarizes the dilemmas one faces in trying to propose a theoretical model which could plausibly explain the unusual and intriguing behavior of these compounds.

2. Summary of Experimental Properties of $Ba\,Pb_{1-x}\,Bi_x\,O_3$

$Ba\,Pb_{1-x}\,Bi_x\,O_3$ apparently forms as a homogeneous alloy series over the entire range $0 < x < 1$ in perovskite-derived structures. Superconductivity was discovered in the range $0.05 \leq x \lesssim 0.3$ by SLEIGHT et al.[1] who found that the transition temperature, $\overline{T}_c$ reached a maximum value of $T_c \simeq 13$ K for values of x in the upper end of this range. Increasing values of x beyond this range led to a rise in the resistivity and a transition to semiconducting behavior. Subsequent more detailed examinations by KHAN et al.[2] and THANH et al.[3] extended and confirmed the earlier results. The maximum values of T_c occurred in samples with unusually large values of the normal state resistivity which also had an unusual temperature dependence, increasing as the temperature decreased. However, these results were obtained on pressed samples - a fact which could have a substantial influence on the transport properties.

Recently the first single-crystal results have been obtained by BATLOGG, REMEIKA et al.[4]. Their results show similar behavior to that found by previous workers on the pressed samples. In Fig. 1 the values of $T_c(x)$ are shown and in Fig. 2 the resistance at the concentration with the highest value of T_c, i.e., $x = 0.25$. This resistivity is high and rises by a factor ~ 2 between room temperature and T_c - a highly unusual behavior for a metallic alloy with a broad complex of conduction bands.

Crystals with Bi concentrations $x \gtrsim 0.3$ show a marked increase in the room temperature resistivity with values in the range of $10^0 - 10^2$ Ω cm (see Fig. 1). They are not superconducting. Rather they display an activated temperature dependence of the resistivity with activation energies in the range 0.2 eV to 0.3 eV. The concentration $x = 0.3$ is at the boundary between metallic, or a least better conducting, behavior for $0 \leq x \leq 0.3$ and semiconducting behavior for $1 \geq x \geq 0.3$.

COX and SLEIGHT [5] used x-ray and neutron techniques to determine the phase diagram of the complete alloy series at room temperature. They found a sequence of phases; orthorhombic $\rightarrow$ tetragonal $\rightarrow$ orthorhombic $\rightarrow$ monoclinic with inceasing values of x. The second boundary from tetragonal to orthorhombic occurs at $x \simeq 0.3$ and coincides with the loss of superconductivity and the rise in the room temperature resistivity. Both phases are distortions from the cubic perovskite structure which arise due a rotation of the oxygen octahedra in these structures. In each case the primitive unit cell contains two $Ba\,Pb_{1-x}\,Bi_x\,O_3$ formula units but in both the tetragonal and orthorhombic phases the nearest-neighbor $Pb_{1-x}\,Bi_x$ sites remain crystallographically equivalent. In both structures the rotations of the oxygen octahedra are through angles $\sim 9^0$ and are in the opposite sense in neighboring octahedral units. The difference between the tetragonal and orthorhombic phases is only in the axis about which the rotation takes place. A quite complete description of the crystallographic structures can be found also in the paper by MATTHEISS and HAMANN [6] on the electronic structure, to which we will return later.

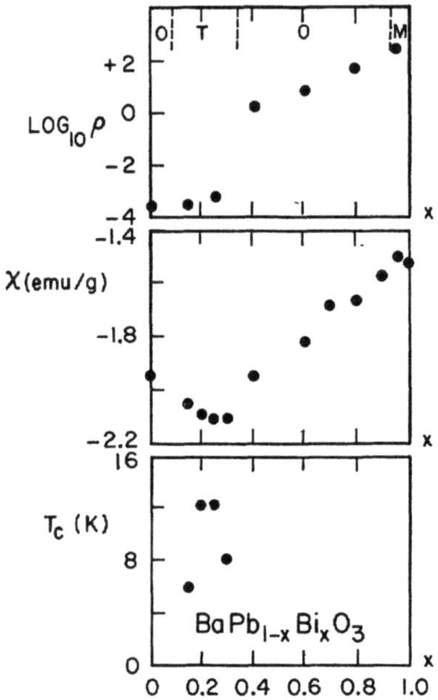

Fig. 1: Properties of $Ba\ Pb_{1-x}\ Bi_x\ O_3$ from BATLOGG et al. [4]

Top: Logarithm of resistivity (Ωcm) at T = 300 K
Crystal structure (T = 300 K)

O: Orthorhombic

T: Tetragonal

M: Monoclinic

from COX and SLEIGHT [5]

Middle: Magnetic Susceptibility (T = 15 K)

Bottom: Superconducting Transition Temperature

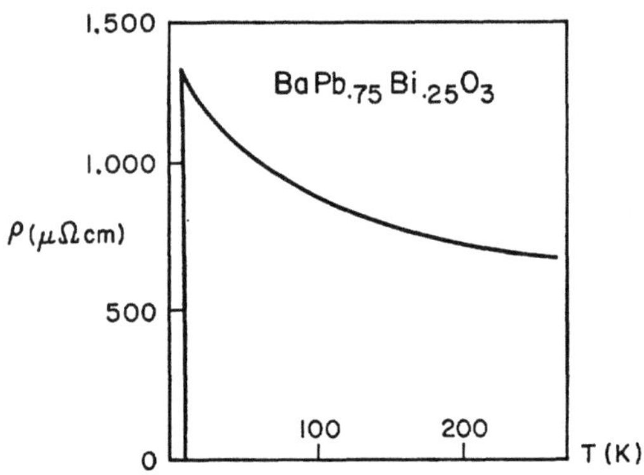

Fig. 2: Resistivity rs. Temperature on single crystal BATLOGG et al. [4]

In the region near to Ba Bi O_3, i.e., $0.95 \leq x \leq 1$, a monoclinic phase was found by COX and SLEIGHT [7,8]. In this phase there is also a breathing mode displacement of the oxygen lattice, i.e., the oxygen octahedra collapse around one set of Bi ions while around the nearest-neighbor Bi ions the oxygen octahedra are expanded. The result is two inequivalent sets of Bi ions. On the basis of the change in average Bi - O separations from an average of 4.0 a.u. in the contracted octahedra to an average value of 4.3 a.u. in expanded octahedra, COX and SLEIGHT [7,8] suggested a charge disproportionation into Bi^{5+} and Bi^{3+} ions.

The magnetic susceptibility has been measured by BATLOGG et al. [4] (see Fig. 1). They find diamagnetic values throughout the alloy series with, surprisingly, the more diamagnetic values observed in the metallic alloys

($\chi \sim -2.1$ emu/g). The onset of the semiconductor behavior coincides with a decrease in the magnitude of χ and a roughly linear behavior with x is observed in the region $0.4 < x < 1$ with χ varying from -2.0 to -1.5 emu/g. These data indicate a low value of the electronic density of states at the Fermi energy in the metallic alloys. Recent measurements [BATLOGG and OTT [9]] show evidence of a small specific heat discontinuity at the superconducting transition ($\sim$ 2.5%) indicating a value of the specific heat ($\sim 4 \times 10^{-1}$ states / eV - spin - $Pb_{0.75}Bi_{0.25}$) for the alloy with maximum T_c. No linear specific heat was observed in the superconducting state. Hall measurements at low temperatures give a density of carriers of $\sim 2.4 \times 10^{21}$ cm^{-3} or approximately one per Bi atom. Both these values are abnormally small for a material in which $T_c > 10$ K.

The nature of the superconducting state has been investigated by magnetic measurements and by tunneling spectroscopy. The tunneling spectra reported by BATLOGG et al.[4,9] show a wealth of low frequency phonon modes extending over the range up to 10 meV. A comparison with Pb shows the phonon energies are in the same energy range in the oxides and in pure Pb. Therefore the value of electron-phonon coupling constant must be larger (i.e., $\lambda > 1.5$ $(= \lambda_{Pb})$) since the superconducting transition temperature is substantially higher (T_c = 12 K vs. 7 K). The tunneling spectra show that it is a strong coupling superconductor and that the superconductivity is similar to that of standard BCS superconductors rather than a Bose condensation of charge bipolarons - a model considered by several authors and to which we will return later. One experimental result, that is rather puzzling, is the value of the superconducting energy gap. BATLOGG et al.[4] report a value of $2\Delta \simeq 3.3 \pm .1$ meV, which gives a value of the ratio ($2\Delta/k_B T_c \simeq 3.5$), i.e., exactly the BCS weak coupling value, rather than the enhanced value that should occur in a strong coupling superconductor (e.g., in Pb the ratio is 4.3).

The alloys because of their very large resistivity in the normal state, and therefore very short mean free path, are strongly type-II superconductors. The temperature dependence of H_{c2} has been measured on pressed powder samples by THANH et al.[3] and on single crystals by BATLOGG et al. [4, 9]. Both groups find very similar results with H_{c2} rising linearly with decreasing T. From their measurements they deduce values of the BCS coherence length $\xi_o \simeq 700$ Å while the penetration length $(\xi_o \ell)^{1/2}$ is ~ 65 Å. [ℓ: mean free path $\sim$ 6Å]. Again these values confirm the proposal that the superconductivity in $Ba\,Pb_{1-x}Bi_xO_3$ is of the conventional BCS kind.

BATLOGG et al.[4] conclude from their tunneling and magnetic measurements that the $Ba\,Pb_{1-x}Bi_xO_3$ alloys are bulk superconductors, i.e., that the superconductivity is not due to a network of filaments of a minority phase.

3. Strong Coupling Models

The Bardeen-Cooper-Schrieffer theory has been very successful in describing superconductors in which the electron-phonon coupling is weak, i.e., the dimensionless electron-phonon coupling constant, $\lambda < 1$. It has also been successfully generalized to the regime $\lambda \gtrsim 1$ by Eliashberg and others [10]. Superconductors with $\lambda \gtrsim 1$ are known as strong coupling superconductors, e.g., Pb with $\lambda \gtrsim 1.5$. However the Eliashberg theory is based on the existence of a small parameter ω_D/E_F where ω_D is the characteristic phonon energy and E_F is the Fermi energy characteristic of the electrons. As first pointed out by Migdal, the electron-phonon problem simplifies when $\omega_D/E_F \ll 1$. One way of seeing this is to recognize that only electron states within an energy range ω_D of E_F are strongly coupled to the phonons. For the majority of electron states, which are much further in energy from the Fermi level, the Pauli exclusion principle prevents any strong coupling to the phonon. However if $\lambda \gg 1$ then the chracteristic electron energy is $E_F/(1+\lambda)$, due to the mass renormalization, and in this case ω_D may be $\simeq E_F(1+\lambda)^{-1}$. In this case the Migdal approximation breaks down and all the electron states are strongly coupled to the phonons. By analogy with the corresponding limit in semiconductors this limit is called a small polaron metal or small polaron Fermi liquid.

Recently there have been several studies of such a system, which I will refer to as very strong coupling systems. ALEXANDROV and RANNINGER [11] have studied this problem in the case where there is an attraction between the small polarons. They compare two possible ground states for this case, one a Bose condensation of already paired polarons, or bipolarons, and the other a crystallization of the bipolarons. They find a critical concentration which separates the two phases with the Bose fluid at low concentrations and the crystal or charge density wave at higher concentrations. Similar results have also been obtained on a similar model by KULIK and PEDAN [12]. A more detailed study by KUBO and TAKADA [13] led to a more complicated phase diagram with the possibility of a condensed charge density wave state which has both diagonal and off-diagonal long range order characteristic of the usual charge density wave states and of Bose condensation, respectively. It should be stressed that all of the above theories use idealized models and neglect such features as the long range Coulomb interaction between the bipolarons which can be expected to favor crystallization.

The original impetus for the study of bipolarons arose from the experimental study of certain oxides, Ti_4O_7 in particular, where a fluid of bipolarons could be identified. In this case the bipolarons are a nearest-neighbor pair of Ti^{3+} ions which strongly polarize their oxygen neighbors. Ti_4O_7 does not remain in this fluid state down to low temperatures but has a crystallization transition, CHAKRAVERTY and SCHLENKER [14]. There is then a question as to whether such a model applies to the $BaPb_{1-x}Bi_xO_3$ alloy system in this case with an onsite attraction strong enough to form a mobile Bi^{3+} bipolaron. It seems unlikely. The tunneling spectrum of BATLOGG et al. [4], discussed in Sect. 2, is contrary to what one expects in a bipolaron model, where the single-particle energy gap is related to the bipolaron binding energy and not to the Bose condensation. For this and other reasons I believe one can confidently rule out the possibility that such very strong coupling models apply to the $BaPb_{1-x}Bi_xO_3$ system.

There is a very large range of values of λ between those in the conventional strong coupling regime $\lambda \simeq 1$ and the very strong coupling limit $\lambda \simeq E_F/\omega_D \gg 1$. The key question is whether for values of λ in between these two limits we should expect further instabilities. The standard answer to this question is that values of λ much larger than 1 are unobtainable because the crystal structure will become unstable and transform to a new structure with a smaller value of λ. This hypothesis is supported by an analysis of the stability against charge density wave formation. For example the CHAN-HEINE [15] criterion for a CDW instability in an ordered crystal will generally be satisfied for values of λ not too much greater than 1. However, if the system is disordered, i.e., an alloy, then the CDW instability may show itself as a local instability rather than an instability of a phonon with a well-defined wavevector. This is the essence of the model proposed by RICE and SNEDDON [16].

They examined a model of positive ions (Pb, Bi) placed on a simple cubic lattice with a negative O ion placed between each near-neighbor (Pb, Bi) pair. This is a model of the ideal perovskite structure which however assumes the Ba ions can be safely ignored. The (Pb, Bi) ions are assumed to have only one relevant state which is represented by a non-degenerate tight-binding 6s band. The position of each O ion is coupled linearly to the charge difference on the two neighboring ions. Eliminating the O coordinates leads to a Hamiltonian

$$H = \sum_{<ij>\sigma} t\, c^{+}_{i\sigma} c_{j\sigma} - \frac{g^2}{2C} \sum_{<ij>} (q_i - q_j)^2 + \sum_i U_c\, n_{i\uparrow} n_{i\downarrow} + \sum_{i>j} \frac{e^2}{\varepsilon r_{ij}} q_i\, q_j + \sum_{i\varepsilon P} t_o\, n_{i\sigma} .$$

Then $n_{i\sigma}$ are the electron numbers on each site and the net charge on a site is $q_i = 1 - n_{i\uparrow} - n_{i\downarrow}$. The first term describes the hopping of electrons between nearest-neighbor sites $<ij>$. The second term is the effective interaction between nearest-neighbor sites when the O coordinate is eliminated [g is the coupling constant between the O coordinate and the charge difference and C the spring constant for an O displacement]. The third and fourth terms are the intrasite and intersite Coulomb interactions. Rice and Sneddon assumed the valence band was the 6s band of the (Pb, Bi) ions and t_o (> 0) represents the energy difference between the Pb 6s level and Bi 6s level. The set of site {P} is occupied by Pb ions and the other sites by Bi ions. The number of electrons is equal to the number of Bi ions.

Consider first the case of Ba Bi O_3. In this case the 6s band is exactly half-filled and, as is well known, a half-filled tight binding s band in a s.c. lattice is perfectly nesting so that the metallic state is unstable. If the electron-phonon coupling constant g^2/C is large enough ($24g^2/C - U_c + 2\, e^2\alpha/\varepsilon a > 0$; α, Madelung const. (1.75 for s.c.); a, lattice constant) then the instability is towards a CDW state with excess charge on half of the sites and reduced charge on the others.

In the limit of strong coupling (i.e., small t) RICE and SNEDDON [16] showed that the model could be rewritten as a Blume-Emery-Griffiths model by introducing a "spin variable $m_i = n_{i\uparrow} + n_{i\downarrow} - 1$ on each site, leading to

$$H = \Delta \sum_i m_i^2 + J \sum_{<ij>} m_i\, m_j$$

with $\Delta = \frac{1}{2}(U_c - 6g^2/C)$ and $J = g^2/C + e^2\alpha/6\varepsilon a$. [The long range Coulomb interaction has been replaced by a short range interaction]. In this model the parameter Δ acts as an effective intrasite Coulomb interaction and an attractive value ($\Delta < 0$) occurs for strong electron-phonon interaction. However the CDW ground state (in the BEG representation, an antiferromagnetic ground state) occurs even if $\Delta > 0$ due to the "exchange" term J.

Rice and Sneddon studied the effect of diluting pure Ba Bi O_3 with Pb. They assumed that Pb sites would be always in the $m_i = 0$ state (corresponding to a valence state Pb^{4+}) and so their presence weakens the "exchange" term. They found a critical value of dilution, depending on the Δ/J ratio, for a transition to the uniform charged state (or paramagnetic state). They identified this as a transition to a metallic state.

The important question is how this phase diagram is modified at finite values of t. Clearly for small values of t where a perturbation theory may be used, the modification is not essential. In this strong coupling regime, Rice and Sneddon proposed that the metallic state would be a strong coupling superconductor which then became locally unstable against Bi^{3+} - Bi^{5+} formation leading to an insulating ground state composed of Pb^{4+}, Bi^{3+} and Bi^{5+} site in a wide range of Pb concentrations. This state they called a state with electron pairing in real space as opposed to the superconducting state in which electrons are paired in $\vec{k}$ space. This transition between $\vec{k}$-space order and real space order would be Fermion analog of the transition for Bose systems between superfluid and crystalline order. However if t is large so that the band width (12 t) is much larger than the Pb-Bi energy splitting (t_o) or the energy gaps associated with the charge density wave, then one would expect the CDW state to occur only in a limited range of concentrations near Ba Bi O_3 and for the rest of the range to have an essentially free electron metal. Further work is required to establish the limits on the bandwidth and electron-phonon coupling constant within which the Rice-Sneddon theory is applicable. They suggested that the value of λ required would be much closer to 1 than that required for the bipolaron theory where the value of λ required ($\lambda = E_F/\omega_D$) is $\sim$ 100.

4. A Priori Electronic Structure Calculations

MATTHEISS and HAMANN [6] have carried out an extensive series of calculations of the electronic structure of the Ba Pb_{1-x} Bi_x O_3 alloy series. They used a self-consistent, scalar- relativistic, linear-augmented planewave (LAPW) method. Thus their calculations are state-of-the-art band structure calculations. The energy bands near the Fermi level are a hybridized set of bands arising from the 6s levels of the Pb or Bi ions and the 2p levels of the O ions. Since each O atom has only two Pb/Bi nearest neighbors the 2p states are split between those $2p_z$ states (z along the Pb/Bi - O - Pb/Bi axis) which are strongly hybridized with the 6s states and the $2p_{x,y}$ and other $2p_z$ states which are essentially unhybridized. In their calculations they find the 6s level lies below the $2p_z$ level by some 2 eV, thus the lower half of the 6s-$2p_z$ complex has more 6s character while the upper half has more $2p_z$ character. The $2p_x$-$2p_y$ states form a narrow complex of bands in the middle. The general form of the bands is shown in Fig. 3. The $2p_x$-$2p_y$ complex is narrow with a total width $\gtrsim$ 2 eV. However the 6s-$2p_z$ complex is very broad with a total width of no less than 16 eV.

In the end members of the alloy series the Fermi level is in the middle of the upper $2p_z$-6s band (Ba Bi O_3) while in Ba Pb O_3 it lies at the lower boundary of upper $2p_z$-6s band where it overlaps the $2p_x$-$2p_y$ complex. Thus Ba Pb O_3 is found to be a semi-metal due this small overlap in agreement with experiment. As the Bi concentration rises the Fermi energy rapidly moves above the $2p_x$-$2p_y$ complex and then rises in the upper $2p_z$-6s band. This subband is well approximated by a canonical s band for a sc lattice.

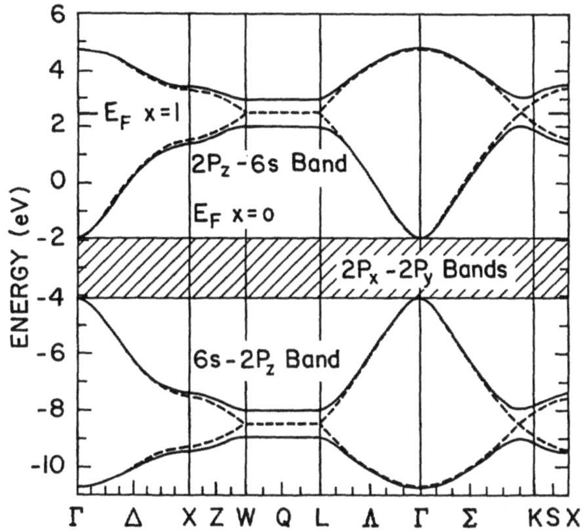

Fig. 3: Band Structure of Ba Pb_{1-x} Bi_x O_3 represented as a tight-binding fit to a priori calculations by MATTHEISS and HAMANN [6]. The bands are drawn in the doubled unit cell resulting from the breathing mode displacement of the O atoms in the monoclinic phase of Ba Bi O_3 (solid lines). The dashed lines are the bands in the undistorted perovskite structure

There is some distortion but it is not too large. The result is a simple Fermi surface which expands and reaches the zone boundary only as $x \to 1$. The width of the upper $2p_z$-6s band is quite large (~ 6 eV) so the density of states is quite low for a high T_c material (they calculate 0.16 states/eV-spin-$Pb_{0.7}$ $Bi_{0.3}$ at x = 0.3). This value is in reasonable agreement with that deduced from the specific heat discontinuity at T_c by BATLOGG and OTT [9] when one allows for electron-phonon renormalization and strong coupling corrections. The number of carriers obtained is one per Bi atom in agreement with the Hall effect data.

MATTHEISS and HAMANN [6] do not find a large splitting between the Pb and Bi 6s levels. They find a value ~ 3 eV for this splitting which is much smaller than the total bandwidth of the 6s-$2p_z$ hybridized bands (~ 16 eV). From this they conclude that the virtual crystal approximation gives a good description of the alloys series and that the results of coherent potential calculation would not be essentially different. Therefore the states at the Fermi energy in the alloys should not be near to localization and presumably the resistivity of the alloys due to the Pb/Bi disorder should not be close to the Mott value. The experimental fact that the mean free path is small and essentially the nearest-neighbor Pb/Bi separation are not explained.

They also made a series of calculations of the effect of the various crystallographic distortions. The orthorhombic and tetragonal phases, which result from the rotations of the oxygen octahedra, were found to have some splittings which were fairly large at the zone center but the effect on the Fermi surface for alloys with x in the range 0.2 < x < 1 was not important.

This would suggest that one cannot look to the octahedral rotations to give a substantial contribution to the random potential. Also, more importantly the crystallographic change from tetragonal to orthorhombic that occurs at the transition to the semiconducting phase should not appreciably affect the Fermi surface. Therefore we cannot look on this crystallographic change as the origin of the change to semiconducting behavior and the suppression of the superconducting T_c at $x > 0.3$.

The monoclinic phase which is found for x close to 1 has however a breathing mode distortion of the O lattice which causes neighboring Bi sites to be differentiated. This is a commensurate zone boundary distortion of the s.c. lattice of Bi atoms. If the $2p_z$-6s subband were an isolated s tightbinding band with only nearest-neighbor hopping, it would be perfectly nesting and we would expect a complete truncation of the Fermi surface by the monoclinic distortion. MATTHEISS and HAMANN [6] calculated the energy bands for the Ba Bi O_3 monoclinic structure reported by COX and SLEIGHT [5]. They indeed found a large effect on the Fermi surface. The splittings at the new zone boundary however were only $\simeq$ 1 eV which is small compared with the width of the upper $2p_z$-6s subband. The effect of such splittings is shown in Fig. 3. Since the energy splitting is rather small the question of how much of the Fermi survives is sensitive to exactly how good the nesting of the Fermi surface is. Thus they find a semimetal state for Ba Bi O_3 since the $2p_z$-6s subband differs somewhat in detail from the ideal s tight-binding form.However such details are not really the main point. Rather it is that the energy splitting due to the monoclinic distortion is small - much smaller than the $2p_z$-6s subbandwidth.

They also estimate the charge disproportionation between the inequivalent Bi sites and it is very small. In fact they report values of only ~ 0.01 electron transferred! Clearly their calculations do not support the strong coupling models but rather quite clearly predict a weak coupling situation. This very small value of the charge transfer is in agreement with the results of XPS measurements (WERTHEIM et al.[17]).

Therefore the band structure calculations offer no solution to the three main mysteries - why is T_c so high, why are the resistivities of the metallic alloys so high and lastly why does T_c collapse for $x \gtrsim 0.3$ and a semiconducting state ensue? The a priori electronic structure calculations rather predict very broad 6s-$2p_z$ bands and the general properties should be those of a free electron metal. Indeed MATTHEISS and HAMANN [6] can give no definite proposal as to the origin of the striking disagreement between the result of their band theory and experiment and suggest that the answers must lie in some underlying undiscovered structural property such as sample inhomogeneities,phase separation, ordered alloy phases,etc.

5. Conclusions

It is clear from the above discussion that there is at present no consensus on the theoretical model that should be used as a starting point to build a detailed theory. There are the two schools of thought; the first one ascribes the anomalous properties to strong electron-phonon coupling and the second one postulates that the high temperature superconductivity and its demise is due to an undetermined structural origin - possibly is due to the presence of a second or minority phase whose structure is as yet unknown. I will discuss the difficulties and advantages of each model in turn.

The strong coupling electron-phonon model is at present not well understood theoretically. The work of RICE and SNEDDON [16] on the narrow band limit needs to be extended to include the effects of the electron hopping term and to establish the limits of coupling constants, bandwidths, etc., to which the local instability model applies. Clearly if we are to succeed in explaining the properties of the alloys with the maximum T_c (i.e., $x \simeq 0.25$) then the very large resistivity rising with decreasing temperature means that one must postulate a strong modification of the metallic properties due to a static disordered displacement pattern even at $x \simeq 0.25$ and a key question then is to what extent this reacts back on the superconductivity. Clearly one cannot simply apply the usual theory of strong coupling superconductors to such a metallic phase. In addition this static disordered displacement pattern of the O atoms should be observable in X-ray and neutron scattering experiments.

The second point of view suggested in particular by METHFESSEL and METHFESSEL [18] ascribes the high temperature superconductivity to a minority phase. They were motivated partly by the absence of a specific heat discontinuity in the early measurements but the recent measurements of Batlogg and Ott, quoted above, have observed an anomaly of the expected size. Further BATLOGG et al [4,9] argue on the basis of their magnetic and tunneling experiments that the superconductivity is a bulk effect and not due to filaments etc. of a minority phase. One must also be aware that there have been no reports in the structural studies that there are obvious minority phases in the samples. Also it would appear that the majority phase, in this hypothesis, must be nonmetallic. For example, the absence of a linear term in the specific heat at low temperatures in the superconducting phase [9] rules out the presence of a majority normal metallic phase. This leads to a clear problem if one identifies the majority phase with the structure reported by COX and SLEIGHT [5] since the band structure predicts that this phase is a wide band metal. Again one comes back to the necessity of a reexamination of the structure. Such a reexamination is underway in Grenoble (MAREZIO [19]) and a significant diffuse scattering pattern has been found (BOUGEROL [20]) but not yet interpreted.

In conclusion, the properties of $Ba\,Pb_{1-x}\,Bi_x\,O_3$ pose some very intriguing puzzles and questions for further study. If it is a case that the superconductivity and metallic state reach an inherent strong coupling limit of the electron-phonon interaction then it would be very interesting theoretically. On the other hand, if the superconductivity is due to some as yet undetermined minority structure or other structural cause, then the conclusion is more optimistic in the sense that advances in materials preparation could lead to an even higher value of T_c.

6. Acknowledgements

I am especially grateful to B. Batlogg and H. R. Ott for numerous stimulating discussions and for permission to quote their data before publication and also to M. Marezio and L. F. Mattheiss for very useful conversations.

References

1 A.W. Sleight, J.L. Gillson and P.E. Bierstedt, Solid State Commun. 17, 27 (1975)

2 Y. Khan, K. Nahm, M. Rosenberg and H. Willner, Phys. Status Solidi, A39 , 79 (1977)

3 T.D. Thanh, A. Koma and S. Tanaka, Appl. Phys. 22, 205 (1980)

4 B. Batlogg, J.P. Remeika, R.C. Dynes, H. Barz, A.S. Cooper and J.P. Garno in Superconductivity in d- and f-band Metals ed. by W. Buckel and W. Weber (Kernforschungszentrum Karlsruhe, FRG, 1982) p. 401

5 D.E. Cox and A.W. Sleight in Proceedings of the Conference on Neutron Scattering, Gatlinberg Tenn. 1976 ed. by R.M. Moon (National Technical Information Service, Springfield, VA, 1976)

6 L.F. Mattheiss and D.R. Hamann, Phys. Rev. B26, 2686 (1982); ibid.B28, 4227 (1983)

7 D.E. Cox and A.W. Sleight, Solid State Commun. 19, 969 (1976)

8 D.E. Cox and A.W. Sleight, Acta Crystallogr. Sect. B35, 1 (1979)

9 B. Batlogg and H. R. Ott,private communication

10 For a review see A.A. Abrikosov, L.P. Gor'kov and I.E. Dzyaloshinskii, "Quantum Field Theoretical Methods in Statistical Mechanics (Pergamon Press, New York, 1965)

11 A. Alexandrov and J. Ranninger, Phys. Rev. B23, 1796 (1981)

12 I.O. Kulik and A.G. Pedan, Zh. Eksp. Teor. Fiz. 79, 1469 (1980) [Eng. Trans. JETP 52, 742 (1980)]; Physica, 107B, 665 (1981)

13 K. Kubo and S. Takada, to be published

14 B.K. Chakraverty and C. Schlenker, J. Phys. (Paris) Colloq. 37, C4-353 (1976); B.K. Chakraverty, J. Phys. (Paris) Lett. 40, 1-9 (1976); C. Schlenker, S. Ahmed, R. Buder and M. Gourmala, J. Phys. C 12, 3503 (1979)

15 S.K. Chan and V. Heine, J. Phys. F. 3, 795 (1973)

16 T.M. Rice and L. Sneddon, Phys. Rev. Lett. 47, 689 (1981)

17 G.K. Wertheim, J.P. Remeika and D.N.E. Buchanan, Phys. Rev. B26, 2120 (1982)

18 C. Methfessel and S. Methfessel in 'Superconductivity in d- and f-band Metals' ed. by W. Buckel and W. Weber (Kernforschungszentrum, Karlsruhe, FRG (1982) p. 393

19 M. Marezio, private communication

20 C. Bougerol, thesis, Lab. de Cristallographie CNRS, Grenoble, 1983

Superconductivity in Some Metal Hydrides

E. Matsushita and T. Matsubara

Department of Physics, Kyoto University, Kyoto 606, Japan

1. Introduction

It is known that some metal hydrides formed from particular metals by hydration exhibit certain characteristic properties in addition to those of their original metals. As indicated in the periodic table of Fig.1, there are several groups of metallic elements and their alloys which can make stable hydrides with variable hydrogen contents [1]. They are the following metal elements.

1. Transition metals.
2. Rare earth metals.
3. Actinide metals.

These metals are magnetic elements in a wide sense and furthermore many of them exhibit also superconductivity [2]. An interesting and important point is that hydration of these metals and alloys produces significant modifications in their electronic structures as well as their phonon spectra. Thus metal hydrides may be substances in which the interplay between magnetism and superconductivity can take place in various degrees.

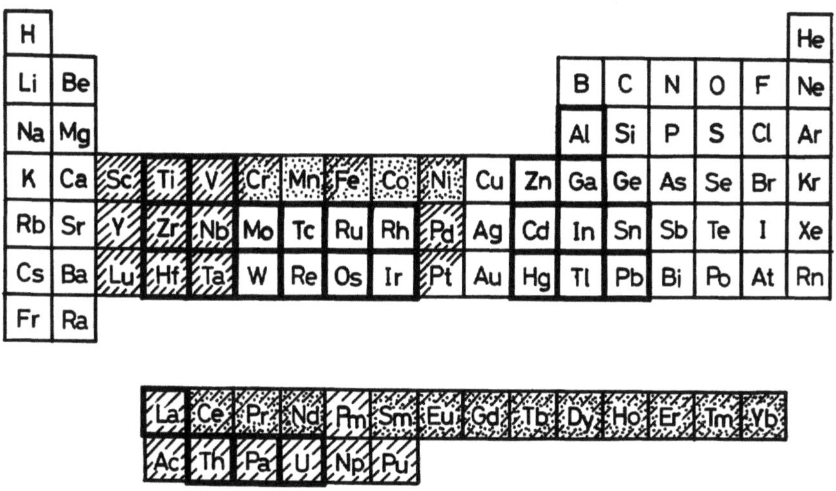

Fig.1 Periodic table covering transition, rare earth, and actinide metals: designates the exothermal formation of hydride: indicates that some hydrogen can be dissolved either in the pure element or in alloy thereof: superconducting metal: magnetic metal

Indeed, hydrogen atoms in metals act sometimes as electron donors and sometimes as electron acceptors depending on their environments, resulting in appreciable changes in the band structures of the host metals. At the same time, the lightness of nuclear mass of hydrogen produces the "optic phonons" which are well separated from the host acoustic phonons in energies and are strongly associated with apparently anharmonic proton vibrations. Thus depending on case by case, the changes in both electron and phonon states due to hydration give rise to the effects of enhancement or destruction of magnetic order or superconducting order.

2. Examples

Let us first mention a few examples which are closely concerned with the present problem.

(1) Hydrides with enhanced superconductivity
Most noticeable examples of the case in which hydration enhances the superconductivity are found in

$$Pd \rightarrow PdH_x \qquad (T_c^{max}=9K)$$

$$Pd_{1-x}M_x \rightarrow Pd_{1-x}M_xH_y \qquad (T_c^{max}=16.6K)$$

$$ThH_2 \rightarrow Th_4H_{15} \qquad (T_c=8K)$$

M = Noble metals.

The left-hand side shows non-superconductor, and the right-hand side superconductor. T_c^{max} means the value of the optimum superconducting transition temperature T_c for various compositions. It should be noted that superconducting $Pd_{1-x}M_xH_y$ consists of non-superconducting three elements Pd, M and H as in the case of the perovskite type oxide superconductors $BaPb_{1-x}Bi_xO_3$ which RICE has discussed in detail in this volume [3]. We will come back again to these alloys PdH_x in the later section.

(2) Hydrides with enhanced magnetic and destructed superconducting orders
Examples of enhanced magnetism and destroyed superconductivity are found in the following systems:

$$Th_7Fe_3 \rightarrow Th_7Fe_3H_{30} \text{ (Ferromagnet)}$$

$$Th_7Co_3 \rightarrow Th_7Co_3H_{30} \text{ (Strong paramagnet)}$$

$$Th_7Ni_3 \rightarrow Th_7Ni_3H_{30} \text{ (Medium paramagnet).}$$

The compounds at the left-hand side are superconducting at low temperature, while those of the right-hand side become not superconducting but magnetic. For instance , Th_7Fe_3 shows superconductivity below T_c=1.86K and no magnetism as shown in Fig.2 [4]. However,upon hydration to form $Th_7Fe_3H_{30}$, which is known as one of the most dense hydrogen media, it gains a strong ferromagnetism at the expense of losing superconductivity. It is suggested that in Th_7Fe_3 the 3d bands of Fe are filled up by electrons of rare earth metal so that magnetic moment is quenched. By doping hydrogen, however,a part of electrons of Th are transferred to anionic H atoms leaving some holes in Fe 3d bands which expose a ferromagnetic order. Other examples show almost similar behaviour except for the differences in magnetic orders. Three different magnetic outcomes from heavy hydration on the transition metal series Fe, Co, Ni clearly show that the phenomena depend crucially on the number of d electrons.

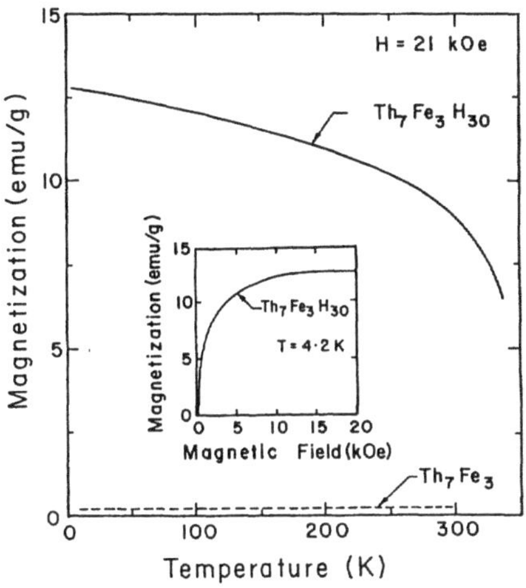

Fig.2 Magnetization curve for Th_7Fe_3 and $Th_7Fe_3H_{30}$ [4]

As these examples show, in some metal hydrides dramatic changes in superconductivity and magnetism can take place by hydration depending on the band structure of the host metals and on the concentration of hydrogen.

A conclusion which will be extracted from these examples is that an essential problem at hand is to study the competition between superconductivity and magnetism in the itinerant electron system under various conditions on electron and phonon states caused by hydration such as produced by

the changes in
- the density of states of electrons: $N(E)$
- the density of states of phonons: $F(\omega)$
- the strength of electron-phonon coupling: λ
- the degree of spin fluctuation: U.

3. Proposal of a general theoretical approach to the problem

Such a problem as the interplay between superconductivity and magnetism in the itinerant electron system was once a big topic many years ago on which many experimental and theoretical works were published. The relevant literature will be found in condensed form in some books; for example in [5] or [6].

However, attention should be paid to a fact that at present the study of the itinerant electron magnetism is in a new stage where the theories are well developed so that one can have a fairly general view of the field in a unified way. See for instance MORIYA's review [7].

The success in recent theories has been brought by using an ingenious technique of functional integral method which enables us to convert the difficult problem of interacting many-electron system into a single-electron problem in a certain random field. It will be natural, therefore, to utilize the same approach in the present problem: the competition between superconductivity and magnetism in the metal hydrides. If we follow this approach, as shown shortly, the problem will become precisely a similar problem to that of "magnetic superconductors" discussed in this symposium.

Start with a model Hamiltonian which contains the effective BCS interaction and Hubbard type electron correlation term [8]:

$$H = H_{BCS} + U \sum_i n_{i\uparrow} n_{i\downarrow} . \qquad (1)$$

The second electron correlation term on the right-hand side of (1) can be written as

$$U n_{i\uparrow} n_{i\downarrow} = \frac{1}{4} U n_i^2 - U(\vec{e}_i \cdot \vec{S}_i)^2 \qquad (2)$$

where

$$n_i = n_{i\uparrow} + n_{i\downarrow}$$

and $\vec{S}_i$ are respectively the electron number and electron spin operators at site i, and $\vec{e}_i$ may be arbitrary unit vector. By making use of the identity

$$\exp(a^2) = \int_{-\infty}^{\infty} dx \, \exp(-\pi x^2 - 2\sqrt{\pi} x a) \qquad (3)$$

the partition function of the system, $Z = \mathrm{Tr}[\exp(-\beta H)]$, may be put into a form (in Hubbard's notation)

$$\begin{aligned} Z &= \int \prod_i D\vec{v}_i(t) D\phi_i(t) \, \exp[-U^{-1} \sum_i \int_0^\beta \{\vec{v}_i^{\,2}(t) + \phi_i^2(t)\} dt] \\ &\quad \times \mathrm{Tr} \exp_+ [-\int_0^\beta \{H_{BCS}(t) + \sum_i i\phi_i(t) n_i(t) + \sum_i 2\vec{v}_i(t)\vec{S}_i(t)\} dt] . \end{aligned} \qquad (4)$$

If we content ourselves with the so-called static approximation where the imaginary time integration in (4) is omitted, Z reduces to

$$Z = \langle \mathrm{Tr} \exp[-\beta\{H_{BCS} + \sum_i (2\vec{v}_i \cdot \vec{S}_i + i\phi_i n_i)\}] \rangle \qquad (5)$$

where < > means to take average over the random electric field potential ϕ_i and random magnetic field $\vec{v}_i$ as indicated in (4). Thus the problem now becomes equivalent to that of BCS system under an influence of random magnetic field at each site governed by a Gaussian distribution law. If $\vec{v}_i$ is regarded as an exchange field exerted by a system outside of electrons (just as those from magnetic ions in magnetic superconductors), the thermodynamic properties of the electron system will be similar to those of a "fictitious magnetic superconductor" provided that suitable interpretation is made for the parameters associated with the "fictitious magnetic system". Along a line of such approach, a study of the problem on the competition between superconductivity and magnetism in the itinerant electron system is in progress. We hope to be able to report the detail of this trial in near future.

Here we only mention a result quickly derived from this approach. In the later discussion on PdH_x system, Pd is shown to play a special role from the viewpoint of superconductivity and magnetism. It is believed that Pd is characterized by the presence of large spin fluctuation (paramagnon effects) which is strong enough to suppress superconductivity in Pd but not strong enough to produce a magnetic order. When we start with the Hamiltonian

$$H = H_{BCS} + 2\sum_i \vec{v}_i \cdot \vec{S}_i \qquad (6)$$

which appears in the exponent of the right-hand side of (5), this characteristic property of Pd may be understood in the following way. Treating the second term in (6) as perturbation, we can apply ABRIKOSOV-GORKOV theory [9] of magnetic impurity in superconductor. Then the famous formula to give a decrease in superconducting transition temperature immediately follows as an universal function of the time-reversal symmetry breaking parameter a:

$$\ln(T_c/T_{co}) = \psi(\tfrac{1}{2}) - \psi(\tfrac{1}{2} + a) \qquad (7)$$

where T_{co} is the transition temperature without spin fluctuation effect (U=0), $\psi(x)$ the di-gamma function and a is now given by

$$a = (2\pi T_c \tau)^{-1}, \quad \frac{1}{\tau} = N(0)\langle \vec{v}_i^2 \rangle \quad . \qquad (8)$$

The average of $\vec{v}_i^{\,2}$ depends on U through the Gaussian distribution in (4) as

$$\langle \vec{v}_i^2 \rangle = \int v^2 \exp(-v^2/Uk_BT)dv \; / \; (\pi U k_B T)^{1/2}$$

and hence increases as U is increased. This means from Fig.3 that superconducting transition temperature T_c decreases with increasing U and eventually vanishes at a critical value U_S. On the other hand it is known that for sufficiently large value of U, say above U_F, the paramagnetic susceptibility diverges at T=0, thus a ferromagnetic order is produced. Let us estimate U_S and U_F in a crude approximation. HERTZ and KLENIN [10] have discussed the fluctuations in itinerant electron paramagnets by making use of the functional integral approach and with a simple model density of states

$$N_0(E) = (2\pi W^2)^{-\frac{1}{2}} \exp(-E^2/2W^2). \qquad (10)$$

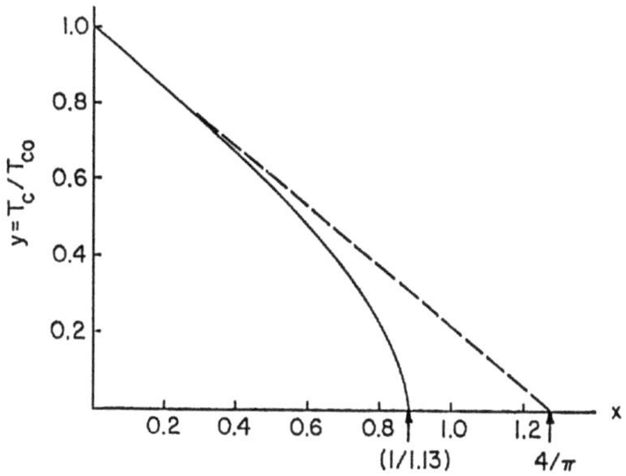

Fig.3 Relation between parameter a and T_c[9]. The curve is the solution of

$$\log y = \psi(\tfrac{1}{2}) - \psi(\tfrac{1}{2} + x/2\pi y)$$

They calculated the magnetic susceptibility as depicted in Fig.4 to show that the largest U for which the system remains paramagnetic at T=0 is given by

$$U_F = (2\pi W^2)^{\frac{1}{2}} = N(0)^{-1} \quad . \qquad (11)$$

Using the same model density of state as theirs, and performing the average of v_i^2 we have

$$\frac{1}{\tau} = 3N(0)Uk_BT = 3(U/U_F)k_BT \qquad (12)$$

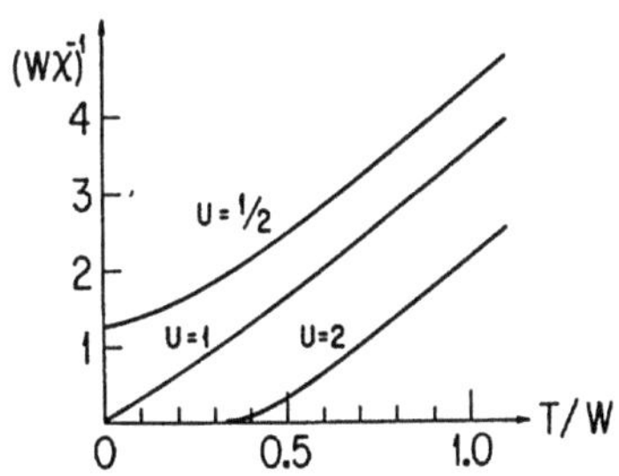

Fig.4 Inverse susceptibility as a function of temperature calculated in [10]

which yields

$$U_S/U_F = 2/3\pi . \qquad (13)$$

In this model, therefore, for $U_S<U<U_F$ the system is neither superconducting nor ferromagnetic as in the case of Pd (see Fig.5).

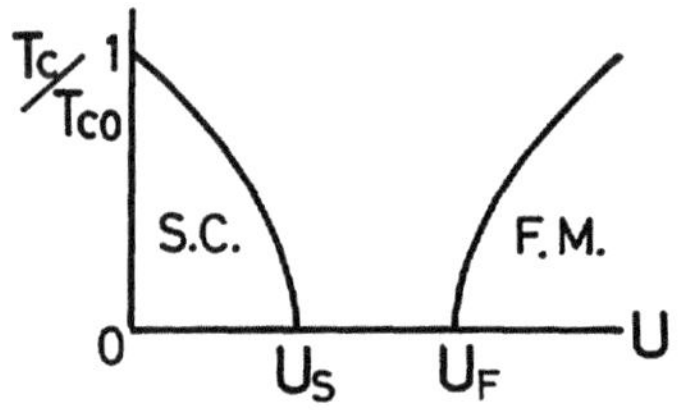

Fig.5 Schematic figure of Tc versus U

4. Superconductivity in PdH system

Hereafter we will focus our attention on the case of PdH_x, and give a short review of experiments and previous theories on the superconductivity in the PdH_x system, particularly for the stoichiometric case x=1. Before beginning with the detailed discussion on PdH_x, it will be instructive to present a table to compare this system with other "exotic" superconducting compounds which consist of non-superconducting elements.

Table 1. Comparison of various physical quantities collected from various sources for three types of exotic superconducting compounds consisting of non-superconducting elements

	$CeCu_2Si_2$	PdH	$Ba(Pb_{1-x}Bi_x)O_3$(x=0.3)	
T_c	0.6	9.62	13	[K]
$H_{c2}(0)$	28	0.98-2.5	7	[kOe]
$\frac{H_{c2}(0)}{-T_c(dH_{c2}/dT)_c}$	0.19	1.0	0.08	
$\Delta C/\gamma T_c$	1.4	1	1.43	
γ	1×10^3	1.7	0.15	[mJ/g-atom.K^2]
N(0)	0.7	0.24	0.03	[states/eV.spin]
$2\Delta/k_BT_c$	3.5	3.8	3.5	
ρ_0	3.5×10^{-6}	1×10^{-6}	3×10^{-3}	[Ωcm]
m*/m	220	1	0.5-0.8	

It is remarkable that $CeCu_2Si_2$ has large value of $H_{c2}(0)$ in spite of low T_c. It seems that this compound has very large mass enhancement factor m*/m as suggested from the enormous specific heat coefficient γ, reminding us of a heavy Fermion liquid. A theoretical account for of superconductivity in this compound is given by TACHIKI in this volume [11].

In contrast to this, PdH has very small value of $H_{c2}(0)$ in spite of rather high T_c. The experimental values of $H_{c2}(0)$ are different for different experimental methods and for different observers: Fig.6 is ob-

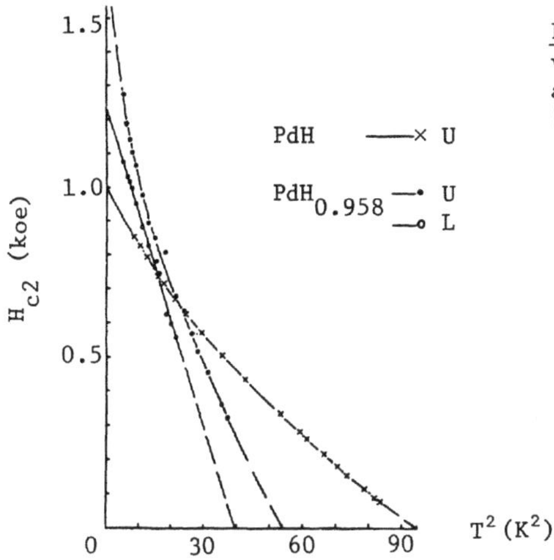

Fig.6 Experimental curve of H_{c2} versus T^2 derived by magnetization and susceptibility measurements [12]

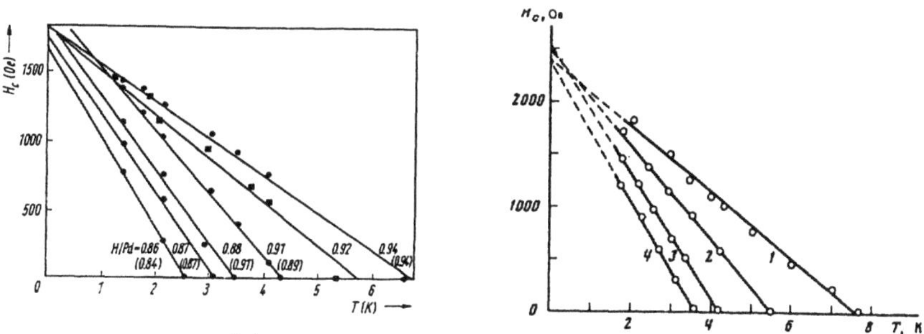

(a) (b)

Fig.7 H_{c2}-T curves for PdH_x with various hydrogen contents x obtained by the resistivity measurements in two different results (a)[13] and (b)[14]

tained by the magnetization and susceptibility measurements whereas Fig.7 (a) and (b) are by the resitivity measurements.

In any case, however, their values are very small compared with $H_{c2}(0)$ of other compounds which have similar values of T_c. This unusual smallness of $H_{c2}(0)$ can be most clearly seen by superposing the data of $H_{c2}(0)$ for PdH on the H_{c2}-T curve for other compounds as in Fig.8[15].

The experimental magnetization curves indicate that PdH_x are the type II superconductors (but very near to type I)[12]. On the other hand, if we estimate the value of T_c for PdH based on the BCS theory of type I superconductors using observed values of various physical quantities, it turns out that most of experimental measurements are rather consistent with each other. For instance the energy gap at T=0, $\Delta(0)$, may be related with $H_{c2}(0)$ by the equation

$$H_c^2(0) = 4\pi N(0)(m^*/m)\Delta^2(0) \qquad (14)$$

and using known values of N(0), m*/m and $H_{c2}(0)$ one can evaluate $\Delta(0)$ and then T_c through the BCS formula to obtain $T_c \sim 9$K which is near the observed

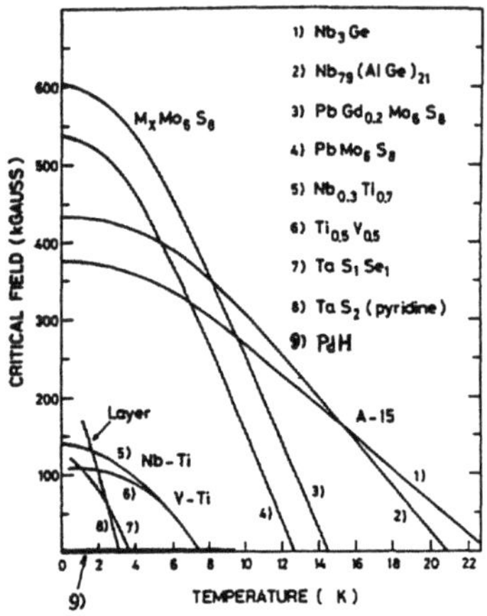

Fig.8 H_{c2} versus temperature curves. The data 1)-8) are taken from [15]. Superposing the data 9) for PdH on this figure reveals that the curve 9) is located at very low level value. This means that very small $H_{c2}(0)$ is originated from small N(0):γ for PdH in the Table 1 is much smaller than those for A-15 type compounds, for instance $\gamma \sim 13.0$-17.0 for V_3Si or 12.2-14.3 for Nb_3Sn in units of [mJ/g-atom.K^2][16].

$Ba(Pb_{1-x}Bi_x)O_3$ is similar to PdH but should have very large electron-phonon interaction in view of small N(0). Such a strong coupling limit is considered and discussed by RICE in this volume [3]. From the values of $\Delta C/\gamma T_c$ and $2\Delta/k_BT_c$ listed in the Table 1, which are near the BCS values $(\Delta C/\gamma T_c)(BCS)=1.43$ and $(2\Delta/k_BT_c)(BCS)=3.52$, it is inferred that the superconductivity of these materials can be accounted for by BCS theory. However a close examination of the experimental facts reveals that there is an extra mechanism characteristic only of the PdH_x system. That is, the principal phonon which mediates the electron-electron interaction in PdH_x is not an acoustic but optic phonon as discussed shortly.

Pd is known to become diamagnetic by alloying with noble metals as shown in Fig.9 or by doping H(D) atoms as shown in Fig.10. It should be noted

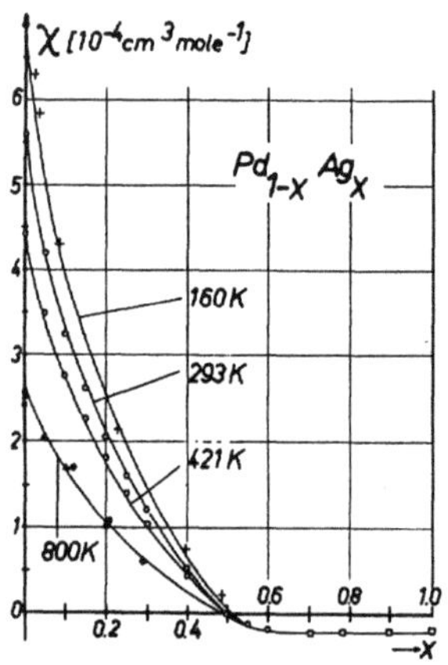

Fig.9 Magnetic susceptibility versus atomic composition x of $Pd_{1-x}Ag_x$ alloy [17]

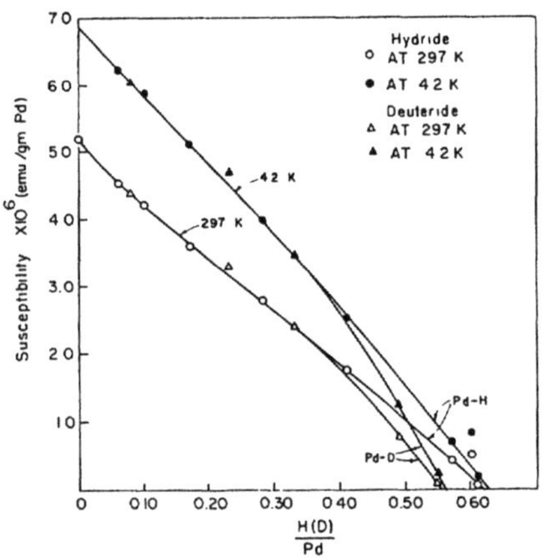

Fig.10 Magnetic susceptibility versus atomic ratio x=H(D)/Pd of $PdH_x(PdD_x)$[18]

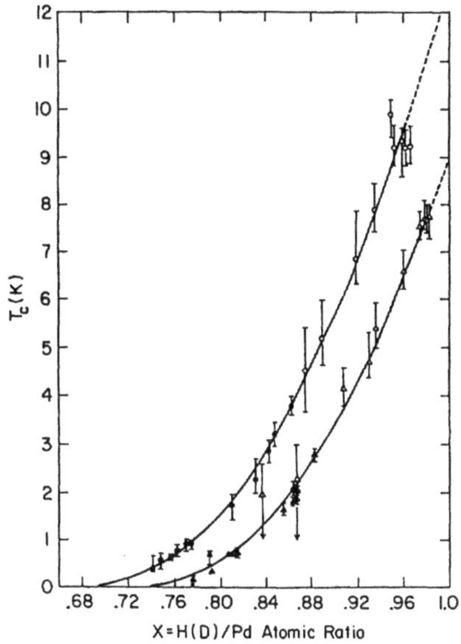

Fig.11 x-dependence of T_c for $PdH_x(PdD_x)$ system [19]

that superconductivity comes out only in the latter case. Figures 10 and 11 show the x-dependence of magnetic susceptibility and superconducting transition temperature for PdH_x and PdD_x.

These results suggest that by hydration (or deuteration), the large paramagnetism due to the spin fluctuation is suppressed to a diamagnetic level for $x>x_m(x_m(H)=0.62,\ x_m(D)=0.56)$, and superconductivity comes out for $x>x_s(x_s(H)=0.715, x_s(D)=0.668)$ in $PdH_x(D_x)$. The existence of a gap between x_m and x_s means that some kind of enhancement in electron-phonon coupling parameter λ (McMillan parameter defined later in eq.(16)) has to be created by additional hydrogen atoms in order to have superconductivity, since filling up of the d bands by hydrogen doping, associated with the appearance of the diamagnetism, must reduce considerably the value of N(0).

There exists a possibility for an enhancement of λ in PdH_x system because optic phonon mode is created in higher energy region when light atoms are added. Figure 12 shows the phonon dispersion curve of $PdD_{0.63}$ obtained

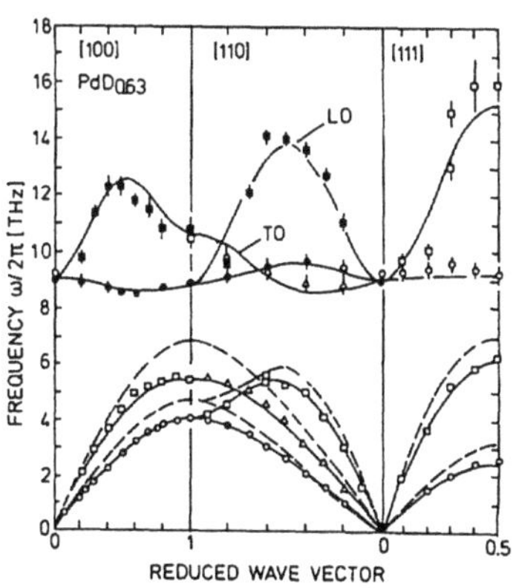

Fig.12 Phonon dispersion curve of $PdD_{0.63}$ [20]

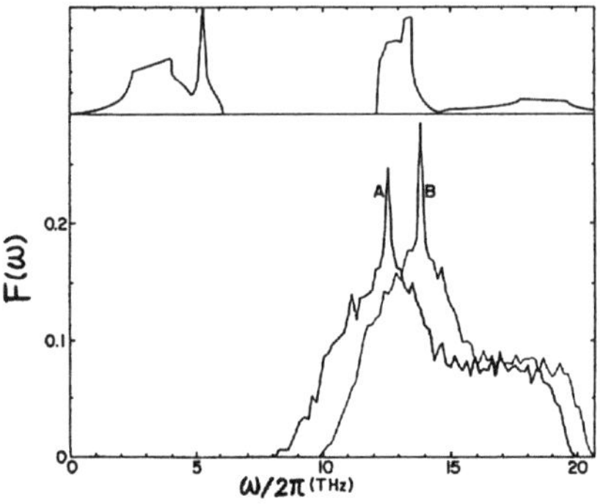

Fig. 13 Phonon spectrum density $F(\omega)$ for $PdD_{0.63}$ (curve A) and for $PdH_{0.63}$ (curve B) [21]

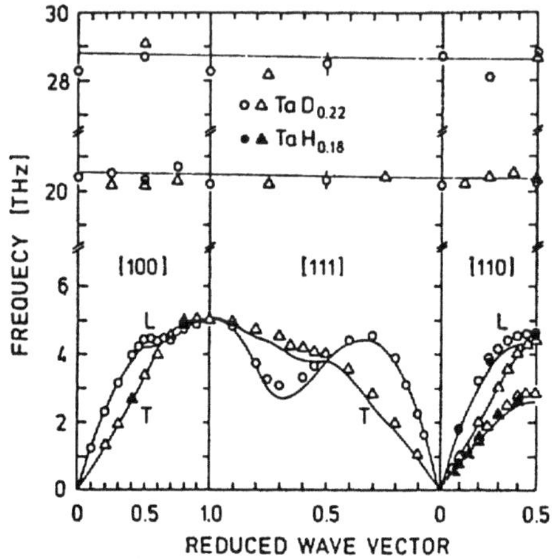

Fig.14 Phonon dispersion curves of $TaH_2(TaD_2)$[22]

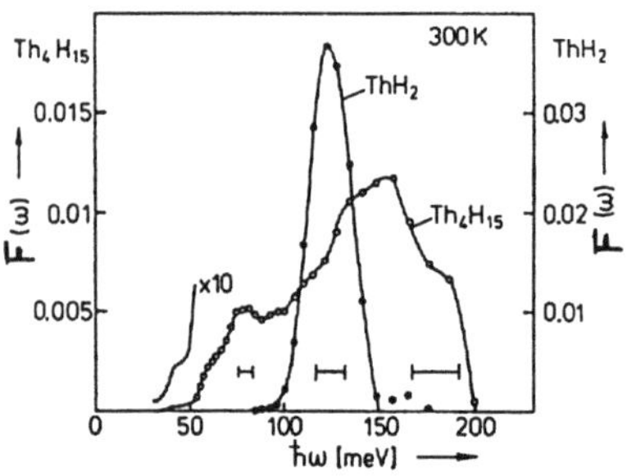

Fig. 15 Optic phonon spectrum density $F(\omega)$ for ThH_2 and Th_4H_{15}[23]

from inelastic neutron scattering experiments [20]. Fig. 13 is the phonon spectrum density $F(\omega)$ of $PdH_{0.63}$ determined from incoherent scattering cross section of pulsed neutron (curve B) which is compared with the spectrum density of $PdD_{0.63}$ (curve A) derived using the dispersion curves of Fig.12 [21]. Two separate branches of phonons are clearly seen in Fig.12: low frequency acoustic phonon mainly caused by Pd-atom vibration and high frequency optic phonon by H(D)-atom vibration because of the large difference in their masses.

Comparing with other metal hydrides, for instance TaH_2 (Fig. 14) or ThH_2 (Fig.15) which are both non-superconductors, we can note that the optic phonons in PdH_x or PdD_x have rather low frequencies and large dispersion. It is interesting to observe that the similar characteristic features also appear in the superconducting Th_4H_{15} in contrast to the non-superconducting ThH_2 as witnessed in Fig.15.

Thus it may be possible that the optic phonon is coupled to electrons to produce a new effect. The large anharmonic zero point vibrations of protons (or deuterons) will be very important in this case. As noticeable evidence of a role played by such optic phonon, we may mention the isotope effect in T_c which is evidently demonstrated in Fig.11 and now recognized by a term "inverse isotope effect" as one of the most conspicuous properties in $PdH_x(D_x)$ system. It has been established after many arguments that this isotope effect can be understood only when an unusual relation between the force constants associated with optic phonon for different isotopes

$$m_H\omega_H^2 \cong 1.2 m_D\omega_D^2 \tag{15}$$

is accepted, although this relation has been extracted from the inelastic neutron scattering experiments performed by SINHA's group [21].

We have tried to give a theoretical basis for the relation (15). Both atomic theory and electron theory are worked out to derive (15) by accounting for how the large anharmonic zero point vibrations give rise to mass-dependent effective potential for proton or deuteron motions. In our atomic theory [24], we base our argument on the self-consistent Einstein model developed by us where the effective force constant $\phi=m\omega^2$ is determined from a variation principle without recourse to harmonic expansion of potential. Assuming Morse type potential for interatomic potential with suitable parameters determined from various sources, we obtained a very satisfactory result $\phi_H/\phi_D \cong 1.2$.

To justify our atomic theory to account for (15) we worked out further an electron theory [25] to derive the effective adiabatic potential for proton motions, and qualitatively we were able to provide an electron theoretical ground for (15). Our electron theory is also extended to cover the cases $PdH_x(PdD_x)$ $x\neq1$ and $Pd_{1-y}Ag_yH_x$ to explain the isotope effect in T_c in these alloys [26].

When the force constants thus obtained are combined with the electron part $N(0)\langle I^2\rangle$ in the electron-phonon coupling constant

$$\lambda = \frac{N(0)\langle I^2\rangle}{m\overline{\omega}^2} \tag{16}$$

one can estimate the value of λ. The electron part $N(0)\langle I^2\rangle$ has been calculated in detail by PAPACONSTANTOPOULOS et al. [27], as given in the Table 2. Upon using Table 2, a rather large λ_H is deduced mostly because of small $m_H\overline{\omega}^2_H$, in spite of small $N_H(0)$. Thus it turns out that $\lambda_H=3\lambda_{Pd}$, which is also in agreement with the experimental results obtained from phonon resistivity measurement [28]. This means that the optic phonon contribution to superconductivity is more dominant than other phonon contributions in PdH_x (PdD_x) system.

Table 2 Theoretical results for electron part and phonon part of λ for Pd and PdH(PdD), where the relation (15) is used [27]

		$N(0)\langle I^2\rangle$ [eV/A^2]	$m\overline{\omega}^2$ [eV/A^2]	λ
	Pd	3.59	7.55	0.47
PdH(D)	Pd	0.865	4.951	0.175
	H	0.392	1.062	0.369
	D	0.392	0.871	0.450
		$\lambda(PdH)=\lambda_H+\lambda_{Pd}=0.544$	$\lambda(PdD)=\lambda_D+\lambda_{Pd}=0.625$	

5. Simple theory of superconductivity in PdH

The essential physics to account for the superconductivity in the PdH_x system seems to be given by the above argument. In this section, a simplified theory is presented in order to understand this phenomena in a more simple and intuitive way. For this purpose it is useful to adopt the site representation formalism developed by APPEL and KOHN [29]. Within the framework of BCS theory, Appel and Kohn gave a homogeneous integral equation in site representation for the vertex part Γ (AK equation), from which critical temperature T_c is calculated. It is expressed in a schematic diagram form as shown in Fig.16, where I denotes the irreducible effective electron-electron interaction vertex, G electron Green's function and ω the Matsubara frequency.

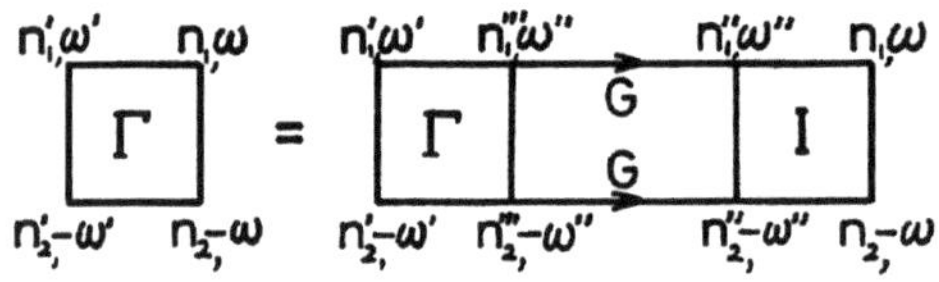

Fig. 16 Diagram representation of AK equation

The site representation is referred to as an orthonormal set of the Wannier functions $W_I(r-n_I)$ at Pd sites n_I on a FCC lattice and $W_{II}(r-n_{II})$ at H sites n_{II} on another FCC lattice. To simplify the argument, we consider non-degenerate orbitals for Pd sites. After a lengthy manipulation, the final form for AK equation is expressed (with $n=n_1-n_2$, etc.) as

$$\Gamma(n,\omega)=\beta^{-1}\sum_{\substack{n'\\ \omega'}}\Gamma(n',\omega')K(n',\omega';n,\omega) \quad \text{where} \tag{17}$$

$$K(n',\omega';n,\omega) = -\sum_{n''} F(n'',\omega')I(n'-n'',\omega';n,\omega). \tag{18}$$

The function

$$F(n,\omega) = \sum_{m_2} G(m_2+n,\omega)G(m_2,-\omega) \tag{19}$$

describes the propagation of an electron pair. The contracted interaction vertex

$$I(n'\omega';n\omega) = \sum_{n_2''} I(n'+n_2'',n_2''\omega';n,0;\omega) \tag{20}$$

is divided into two parts: phonon part and coulomb part as

$$I(n'\omega';n\omega) = I_{ph}(n'\omega';n\omega) + I_c(n'\omega';n\omega). \tag{21}$$

The phonon part is represented in Fig.17 which describes a process where electron 1 transfers from n_1' to n_1, electron 2 from n_2' to n_2, while a lattice

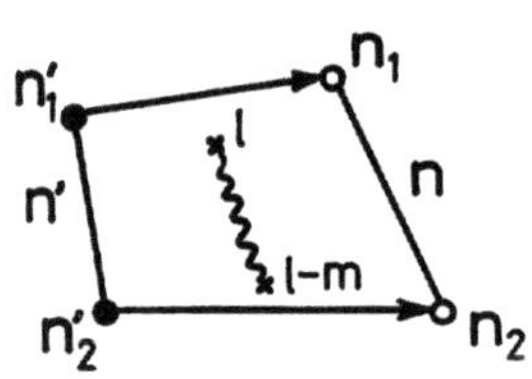

Fig.17 The phonon part of interaction vertex

distortion propagates between 1 and 1-m. $I_{ph}(n'\omega';n\omega)$ is obtained by sum of these diagrams over 1, m, n_1-n_1' and n_2-n_2'.

The coulomb part may be assumed to be of short range. With the contact model it is put as

$$I_c(n',n;\omega) = \delta_{n',0}\delta_{n,0}U. \tag{22}$$

As proved in [29], most dominant contributions are certainly such that an electron pair composed of electrons sitting on the same atom is transferred together to another atom (that is n'=n=0), and retaining such most dominant contribution leads to a simplified equation which is one-dimensional in energy space and easily managed. A merit of site representation formalism in the present problem is that the phonon part of interaction vertex takes on a form such that the roles played by acoustic and optic phonons may be separated. To be more precise, write down the expression for I_{ph} explicitly:

$$I_{ph}(n_1',n_2';n_1,n_2,\omega) = \sum_{11'} L_{\alpha\beta}^{11'}(m,\omega)(n_1'|\nabla_\alpha V(r-1)|n_1)(n_2'|\nabla_\beta V(r-1')|n_2) \tag{23}$$

where $L_{\alpha\beta}^{11'}(m,\omega)$ is a phonon Green's function in site space with m=1-1'

$$L_{\alpha\beta}^{11'}(m,\omega)=N^{-1}\sum_{qs}\frac{e_{\alpha s}^*(q,1)e_{\beta s}(q,1')}{[M(1)M(1')]^{\frac{1}{2}}[\omega^2-\omega_s^2(q)]}\exp[-iq(1-1')], \tag{24}$$

s being the index for phonon branch, M(1) the atomic mass at site 1 and

$$(n'|\nabla_\alpha V(r-1)|n) = \int W^*(r-n')\nabla_\alpha V(r-1)W(r-n)dr. \tag{25}$$

Due to the large mass difference between Pd and H, the eigenvectors of acoustic phonon $\vec{e}_a(q,1)$ have large amplitudes on Pd site whereas those of optic phonon $\vec{e}_o(q,1)$ at H sites. This is reflected to the site- and mode-dependence of phonon Green's function and to the orbital dependence through (23) and (25).

Roughly speaking, when optic phonon propagation is playing primary role in maintaining the electron pair correlation, most strong effect of the lattice distortions will be received by an electron pair which is located at H site. This is the main origin of the x-dependence of I_{ph} and hence of T_c in PdH_x. Another important point to be noted is a fact that cutoff frequency below which the effective interelectron interaction becomes attractive is different for optic phonon and acoustic phonon. Let the cutoff frequencies of acoustic and optic phonons be respectively ω_a and ω_o with

$$\omega_a < \omega_o . \tag{26}$$

Then it is guessed from (24) that optic phonon will give rise to attractive interaction for wider range of electron pair energy.

On the basis of the above considerations, let us give an expression for T_c and discuss why the optic phonon is important in the present case and why T_c is rather strongly dependent on hydrogen concentration. Upon adopting the contact approximation on (17) and (21), the fundamental equation is now written as

$$\Gamma(\omega) = -\beta^{-1}\sum_{\omega'}\Gamma(\omega')F(\omega')[I_{ph}(\omega',\omega) + U]. \tag{27}$$

As in the BCS theory of weak coupling superconductor, we take the phonon part of interaction vertex constants such as

$$I_{ph}(\omega',\omega) = \begin{cases} -(I_o + I_a) & 0<|\omega|,|\omega'|\leq\omega_a \\ -I_o & \omega_a<|\omega|,|\omega'|\leq\omega_o \\ 0 & \text{otherwise.} \end{cases} \tag{28}$$

Then the solution of eq.(27) should have a form

$$\Gamma(\omega) = \begin{cases} \Gamma_c + \Gamma_o + \Gamma_a & 0 < |\omega| \leq \omega_a \\ \Gamma_c + \Gamma_o & \omega_a < |\omega| \leq \omega_o \\ \Gamma_c & \omega_o < |\omega| \leq \omega_F \\ 0 & \text{otherwise} \end{cases} \tag{29}$$

where Γ_a, Γ_o and Γ_c are contributed by acoustic phonon, optic phonon and coulomb force respectively, and the cutoff frequency for electron-electron coulomb interaction is denoted by ω_F. Following AK, it is convenient to define for optic phonon

$$\beta^{-1} \sum_{|\omega|\leq\omega_o} F(\omega) = N(0)\ln\alpha\beta\omega_o \equiv f_o \tag{30}$$

with

$$\alpha = 2e^{\gamma}/\pi = 1.13 \tag{31}$$

(γ=Euler's constant), and similarly for acoustic phonon f_a and for coulomb part f_F which are obtained by replacing ω_o in f_o by ω_a and ω_F respectively. Then the equation (27) is transcribed into coupled three linear equations for three quantities Γ_s (s=a,o,c):

$$\begin{aligned} \Gamma_c &= -U(f_a\Gamma_a + f_o\Gamma_o + f_F\Gamma_c) \\ \Gamma_o &= I_o(f_a\Gamma_a + f_o\Gamma_o + f_o\Gamma_c) \\ \Gamma_a &= I_a f_a(\Gamma_a + \Gamma_o + \Gamma_c) \ . \end{aligned} \tag{32}$$

It is evident that the transition temperature T_c should be determined from the solubility condition for the homogeneous linear equations (32) and it needs only short algebra to have a condition

$$1 = f_a[I_a + \frac{I_o - U^*}{1-(I_o-U^*)N(0)\ln(\omega_o/\omega_a)}] \tag{33}$$

with

$$U^* = \frac{U}{1 + N(0)U\ln(\omega_F/\omega_o)} \ . \tag{34}$$

Here, to eliminate f_o and f_F in terms of f_a, use has been made of the relations

$$f_F = f_o + N(0)\ln(\omega_F/\omega_o) \quad \text{and}$$

$$f_o = f_a + N(0)\ln(\omega_o/\omega_a).$$

Thus, by solving eq.(33) for T_c, the transition temperature is given as

$$k_BT_c = 1.13\omega_a \exp[-1/N(0)I_{eff}] \tag{35}$$

where

$$I_{eff} = I_a + \frac{I_o - U^*}{1 - (I_o - U^*)N(0)\ln(\omega_o/\omega_a)} \tag{36}$$

which is the same form with the generalized BCS theory given by RIETSCHEL [30].

We observe in (35) and (36) that the optic phonon contribution to T_c is indeed enhanced by a factor $\ln(\omega_o/\omega_a)$, and I_o can also include x-dependence. Therefore for sufficiently large x, I_{eff} becomes positive primarily because of the presence of optic phonon, and then T_c may increase with increasing x rapidly as in Fig. 11. In this way the role played by optic phonon in the present case is understood. Similar situation may occur in the electron-electron attraction induced by the exchange of some high energy bosons such as plasmons [31] or excitons [32].

References

1. F. E. Wagner and G. Wortmann: In Hydrogen in Metals I, Basic Properties, in Topics in Applied Physics, Vol. 28, ed. by G. Alefeld and J. Völkl (Springer-Verlag, Berlin, Heidelberg, New York 1978) p.131
2. G. Gladstone, M. A. Jensen and J. R. Schrieffer: In Superconductivity, Vol.2, ed. by R. D. Parks (Marcel Dekker, Inc., New york 1969) p.665
3. T. M. Rice: in this volume
4. S. K. Malik, W. E. Wallace and T. Takeshita: Solid State Commun. 28, 359 (1978)
5. R. D. Parks (ed.): Superconductivity, Vol. 2 (Marcel Dekker, Inc., New york 1969)
6. G. T. Rado and H. Suhl (ed.): Magnetism, Vol.V (Academic Press, New York and London 1973)
7. T. Moriya: J. Magn. Magn. Mat. 31-34, 10 (1983): Proceedings of the International Conference on Magnetism, Sep. 6-10 1982, Kyoto, Japan
8. J. Hubbard: In Electron Correlation and Magnetism in Narrow-Band Systems, in Springer Series in Solid-State Sciences 29, ed. by T. Moriya (Springer-Verlag, Berlin, Heidelberg, New York 1981)
9. A. A. Abrikosov and L. P. Gor'kov: Sov. Phys. JETP 9, 220 (1959): 12, 1243 (1961)
10. J. A. Hertz and M. A. Klenin: Phys. Rev. B10, 1084 (1974)
11. M. Tachiki: in this volume
12. D. S. McLachlan and T. B. Doyle: In Proceedings of the 14th International Conference on Low Temperature Physics, Aug. 14-20 1975, Otaniemi, Finland. Vol. 2, ed. by M. Krusius and M. Vuorio, p.44
13. T. Skośkiewicz: Phys. Stat. Sol. (b) 59, 329 (1973)
14. N. E. Alakseevskiǐ, Yu. A. Samarskii, H. Wolf, V. I. Tsebro and V. M. Zakosarenko: JETP Lett. 19, 350 (1974)
15. Ø. Fischer: In Proceedings of the 14th International Conference on Low Temperature Physics, Aug. 14-20 1975, Otaniemi, Finland. Vol. 5, ed. by M. Krusius and M. Vuorio, p.172
16. A. Junod, T. Jarlborg, and J. Muller: Phys. Rev. B27, 1568 (1983)
17. W.Sänger and J.Voitländer: Z. Physik B44, 283 (1981)
18. R. J. Miller and C. B. Satterthwaite: Phys. Rev. Lett. 34, 144 (1975)
19. R. W. Standley, M. Steinback, and C. B. Satterthwaite: Solid State Commun. 31, 801 (1979)

20. J. M. Rowe, J. J. Rush, H. G. Smith, M. Mostoller, and H. E. Flotow: Phys. Rev. Lett. 33, 1297 (1974)
21. A. Rahman, K. Sköld, G. Pelizarri, S. K. Sinha and H. E. Flotow: Phys. Rev. B14, 3630 (1976)
22. A. Magerl, N. Stump, W. D. Teuchert, V. Wagner and G. Alefeld: J. Phys. C: Solid State Phys. 10, 2783 (1977)
23. M. Dietrich, W. Reichardt and H. Rietschel: Solid State Commun. 21, 603 (1977)
24. E. Matsushita and T. Matsubara: Prog. Theor. Phys. 63, 357 (1980)
25. E. Matsushita : Solid State Commun. 38, 419 (1981)
26. E. Matsushita and T. Matsubara: J. Magn. Magn. Mat. 31-34, 511 (1983): Proceedings of the International Conference on Magnetism, Sep. 6-10 1982, Kyoto, Japan
27. D. A. Papaconstantopoulos, B. M. Klein, E. N. Economou and L. L. Boyer: Phys. Rev. B17, 141 (1978)
28. D. S. MacLachlan, R. Mailfert, B. Souffaché and J. P. Burger: In Proceedings of the 14th International Conference on Low Temperature Physics, Aug. 14-20 1975, Otaniemi, Finland. Vol. 2, ed. by M. Krusius nad M. Vuorio, P.40 See also D. S. MacLachlan, R. Mailfert, J. P. Burger and B. Souffaché: Solid State Commun. 17, 281 (1975)
29. J. Appel and W. Kohn: Phys. Rev. B4, 2162 (1971)
30. H. Rietschel: Z. Physik B22, 133 (1975)
31. H. Rietschel and L. J. Sham: Phys. Rev. B28, 5100 (1983)
32. W. A. Little: Phys. Rev. 134, A1416 (1964)
D. Allender, J. Bray and J. Bardeen: Phys. Rev. B7, 1020 (1973)

Appendix

A New Research Project in Japan on Superconducting Materials

Sadao Nakajima

Institute for Solid State Physics, University of Tokyo, Minato-ku
Tokyo 106, Japan

1. Introduction

Presently I am organizing a group of about forty senior Japanese scientists in universities and some other institutions to work on new superconducting materials towards higher transition temperatures. The three-year research project, which we believe to be the first step of a longer term research activity, will start from April 1984 with financial support of about \$2.4 x 10^6 from the Ministry of Education, Science, and Culture, Japan.

In this paper, I will briefly describe the organization and underlying physics of our three-year project since I hope it will be interesting even for foreign scientists working on superconductivity and we welcome any comments and advice which may help us to improve our project. I am grateful to Professor T. Matsubara and the Organizing Committee of the Symposium for allowing me to present this unusual paper which reports on what we want to do in future rather than what we have done already.

2. Organization

For university scientists it is a rather expensive and time-consuming task to develop new materials which require high purity and special facilities. Moreover, in order to do it in a systematic way, we need a collective approach without relying too much upon such individual geniuses as Matthias. Our three-year project should make a breakthrough in establishing collaboration among people in different fields, for searching new materials, understanding their solid state properties, and aiming at higher transition temperatures. Thus it will be carried out by the following five groups in close collaboration with each other.

Group 1: Theory

The theory group is expected to collaborate with each of the four experimental groups and play the role of glue to keep unity of our project as I will discuss in more detail later on.

Group 2 : Intermetallic Compounds and Magnetic Superconductors

This group will try to develop further the investigations on high T_c A15 compounds including Nb_3Si, on amorphous MoSi, MoC, and fcc MoN, and also on magnetic superconductors including heavy Fermion systems in ternary compounds.

Group 3 : Inorganic Superconductors

One of interesting materials to be studied in this group is $BaPb_{1-x}Bi_xO_3$, which might be the case of very strong electron-phonon coupling as was discussed by Professor Maurice Rice at this Symposium. The subjects investigated by this group include the interplay between superconductivity and CDW(charge density wave) in $NbSe_3$, the synthesis of metastable superconductors by very high pressure and laser pulse, and the search for electron conduction along dislocations in polar semiconductors.

Group 4 : Organic Materials and Salts

In this group, syntheses of Pt complex salts without the disorder in ionic positions and of $(TMTSF)_2X$, $(BEDT\text{-}TTF)_2X$, and new family of organic conductors are going on in conjunction with their characterization of structure, conductivity, magnetism, etc. The relation between structural fluctuation and superconductivity (or SDW, spin density wave) will also be studied by X-ray analysis.

Group 5 : Development of New Methods in Synthesis

This group will deal with single crystals of high melting points (e.g. Chevrel compounds, Nb compounds), with materials of extraordinary cluster structure, with ultra-fine particles, with ultra-thin films for interface superconductivity, etc.

3. Why Higher T_c?

I think I need to add a few remarks from theorist's point of view on what we precisely mean by superconductivity at "higher" temperatures, since I fear the word can be taken as sensationalism.

As Ginzburg has stated most explicitly, the problem of high temperature superconductivity, apart from its technological implications, still remains a challenge to solid state scientists. We don't know theoretically how high and in what materials we can raise the transition temperature. Neither do we have the well-established empirical rules to control the transition temperature effectively. Thus the problem of high T_c at the present stage is to investigate the mechanism which controls T_c in each type of materials, rather than trying to raise T_c itself. It should be regarded as a great achievement in solid state science even if we find eventually that T_c cannot be raised much higher than 23K of Nb_3Ge, the highest value presently known.

Then why do we need to say "higher" T_c? Because it will provide our long term research activity with integrity and continuity which may be easily lost when scientists in diverse fields of speciality have to work together with their own methods on so different materials. A good example is the proposal by LITTLE[2] which has obviously stimulated the investigations on organic conductors through the cooperation between physicists and chemists, given rise to the new research field of synthetic metals, and established a new state of matter CDW, although his original idea of high T_c organic superconductors has not been realized as yet.

As is pointed out also by GINZBURG[1], the situation is different as regards the problem of high critical magnetic field. It has already reached to the level of technological development and improvement as can be seen from the large scale application of superconducting magnets in high energy accelerators, nuclear fusion projects, or the Japan National Railway project of high speed train levitation. From the technological point of view, therefore, the "higher T_c" is a relative concept, which depends on the method of refrigeration. We are in possession of fairly well-developed superconductor technology, provided that we can be satisfied with the liquid helium refrigeration on economical as well as technological grounds. In other words, we don't need to aim at raising T_c directly up to the liquid nitrogen range; superconductivity at liquid neon or even at liquid hydrogen temperatures can be quite useful and valuable.

4. A New Theoretical Aspect

From the theoretical point of view, there are two fundamental questions concerning high temperature superconductivity.

1. How high can we raise T_c by the phonon-mediated attractive force between electrons?

2. Is there any really effective mechanism of attraction other than the phonon-mediated force?

As for the first question, we have a number of theoretical contributions including the well-known work of McMILLAN[3] and also that of ALLEN and DYNES[4]. Thanks to the progress in electronic computation, not only the numerical solutions of Eliashberg equation, but also the band theoretical part of the calculation including the electron-phonon coupling matrix elements are now very realistic and reliable. A test will be the experimental confirmation of the prediction [5] that MoN would have T_c around 30K, although it seems rather difficult experimentally to produce stoichiometric, cubic MoN. If confirmed, it would be the first "high T_c" (in the liquid neon range) superconductor produced intentionally.

When we discuss the possibility of "high" T_c superconductors, however, we should remember one very important fact which has been revealed by the development of novel superconducting materials in 1970's. Thus, in addition to the Cooper pairing we are interested in, we should also take into account other types of ordering, which electrons might energetically prefer, and also the possibility of Anderson localization [6] caused by the impurity scattering. They usually compete with superconductivity, but may coexist in some cases.

Both examples are found in magnetic superconductors as we have heard at this Symposium. Organic superconductors, which were discussed by Dr. Fenton, are also interesting [7]. In the case of $(TMTSF)_2X$'s, the Cooper pairing is stable only under higher pressure, whereas SDW is more stable under lower pressure where the condition of the Fermi surface nesting must be satisfied better [8]. The problem is rather subtle in the case of X = ClO_4, in which we have an order-disorder transition of ClO_4 orientation. The Cooper pairing is stable even under ambient pressure when ClO_4's are orientationally ordered. If the sample is quenched so that ClO_4's remain disordered at lower temperatures, however, the SDW phase is stabilized under lower pressure.

The last example is instructive. The Eliashberg equation to determine T_c contains the averages over the Fermi surface and therefore is not sensitive to the band theoretical details. This does not apply in general to other types of order, e.g. SDW and CDW, which are to be compared with the Cooper pairing. In other words, when we think of the competition between superconductivity and other types of order, we should be more careful about the details of our system. Our problem of higher T_c is thus one of few fundamental problems remaining in solid state science to be solved by long term basic research.

Finally, as regards the new mechanism of attraction between electrons, the microfabrication techniques developed in semiconductor technology will help us in simulating Ginzburg's interface superconductor or Little's superconducting chain. The latter might be simulated also by the electron conduction along dislocations in polar semiconductors, as was pointed out by H. Fukuyama.

Acknowledgment

I wish to thank Professor H. Fukuyama, Professor Y. Muto, Professor S. Tanaka, Professor I. Tsujikawa, and Professor T. Yasukochi for their cooperation in preparing our proposal to the government.

References

1. V.L. Ginzburg: "High Temperature Superconductivity", Ed. V.L. Ginzburg and D.A. Kirzhnits. Transl. A. K. Agyel and J.L. Birman (consultants Bureau, New York and London, 1982) p.1.
2. W.A. Little: Phys. Rev. A134, 1416 (1964).
3. W.L. McMillan: Phys. Rev. 167, 331 (1968).
4. P.B. Allen and R.C. Dynes: Phys. Rev. B12, 905 (1975).
5. W.E. Pickett, B.M. Klein, and D.A. Papaconstantspoulas: Physica B + C 107, 667 (1981).
6. S. Maekawa and H. Fukuyama: J. Phys. Soc. Japan 51, 1380 (1982). H. Takagi and Y. Kuroda: Solid State Commun. 41, 643 (1982).
7. K. Bechgaard: Mol. Cryst. Liq. Cryst. 79, 1 (1982). D. Jerome: ibid. 79, 155 (1982). See E.W. Fenton's paper in this Proceedings.
8. See K. Yamaji's paper in this Procedings.

Index of Contributors

Springer Series in Solid-State Sciences

Editors: M. Cardona P. Fulde H.-J. Queisser

MIX
Papier aus verantwortungsvollen Quellen
Paper from responsible sources
FSC® C105338

If you have any concerns about our products,
you can contact us on
ProductSafety@springernature.com

In case Publisher is established outside the EU,
the EU authorized representative is:
Springer Nature Customer Service Center GmbH
Europaplatz 3, 69115 Heidelberg, Germany

Printed by Libri Plureos GmbH
in Hamburg, Germany